Science for Sustainable Societies

Scope of the Series

This series aims to provide timely coverage of results of research conducted in accordance with the principles of sustainability science to address impediments to achieving sustainable societies – that is, societies that are low carbon emitters, that live in harmony with nature, and that promote the recycling and re-use of natural resources. Books in the series also address innovative means of advancing sustainability science itself in the development of both research and education models.

The overall goal of the series is to contribute to the development of sustainability science and to its promotion at research institutions worldwide, with a view to furthering knowledge and overcoming the limitations of traditional discipline-based research to address complex problems that afflict humanity and now seem intractable.

Books published in this series will be solicited from scholars working across academic disciplines to address challenges to sustainable development in all areas of human endeavors.

This is an official book series of the Integrated Research System for Sustainability Science (IR3S) of the University of Tokyo and the Institute for Global Environmental Strategies (IGES).

More information about this series at http://www.springer.com/series/11884

Alexandros Gasparatos • Abubakari Ahmed •
Merle Naidoo • Alice Karanja •
Kensuke Fukushi • Osamu Saito •
Kazuhiko Takeuchi
Editors

Sustainability Challenges in Sub-Saharan Africa I

Continental Perspectives and Insights from Western and Central Africa

Springer

Editors
Alexandros Gasparatos
Institute for Future Initiatives (IFI)
The University of Tokyo
Tokyo, Japan

Abubakari Ahmed
Department of Planning
University for Development Studies
Wa, Ghana

Merle Naidoo
Graduate Programme in Sustainability
Science – Global Leadership
Initiative (GPSS-GLI)
The University of Tokyo
Tokyo, Japan

Alice Karanja
Institute for Future Initiatives (IFI)
The University of Tokyo
Tokyo, Japan

Kensuke Fukushi
Institute for Future Initiatives (IFI)
The University of Tokyo
Tokyo, Japan

Institute for the Advanced Study
of Sustainability (UNU-IAS)
United Nations University
Tokyo, Japan

Osamu Saito
Institute for Global Environmental
Studies (IGES)
Hayama, Japan

Institute for Future Initiatives (IFI)
The University of Tokyo
Tokyo, Japan

Kazuhiko Takeuchi
Institute for Global Environmental
Studies (IGES)
Hayama, Japan

Institute for Future Initiatives (IFI)
The University of Tokyo
Tokyo, Japan

ISSN 2197-7348 ISSN 2197-7356 (electronic)
Science for Sustainable Societies
ISBN 978-981-15-4457-6 ISBN 978-981-15-4458-3 (eBook)
https://doi.org/10.1007/978-981-15-4458-3

This Springer imprint is published by the registered company Springer Nature Singapore Pte Ltd.
The registered company address is: 152 Beach Road, #21-01/04 Gateway East, Singapore 189721, Singapore

Acknowledgements

We wish to thank our Springer editorial team for the guidance, support, and patience throughout the development of the volume. We wish to thank particularly Sushil Kumar Sharma and Mei Hann Lee, for their endless patience throughout the development of these two volumes. We also thank Marcin Jarzebski for designing many of the figures included in these two volumes.

The two volumes were conceptualized during the development and follow-up to the Sixth International Conference on Sustainability Science (ICSS 2016) in Stellenbosch, South Africa (March 2016). We sincerely thank the conference partners and participants for their suggestions and inspiration that kick-started the process of these books.

The two volumes would not have been possible without the generous support of the Japan Science and Technology Agency (JST) for project FICESSA, Japan International Cooperation Agency (JICA) for project USiA (Urban Sustainability in Africa based on inter-linkages of SDGs), and the Japan Society for the Promotion of Science (JSPS) for Grant-in-Aid for Young Scientists A (Project 17H05037), and the core-to-core project "Establishment and Advancement of a Global Meta-Network on Sustainability Science" (Project 23001).

Merle Naidoo, Alice Karanja, and Abubakari Ahmed received Monbukagakusho PhD scholarships offered by the Japanese Ministry of Education, Culture, Sports, Science, and Technology (MEXT) through the Graduate Program in Sustainability Science—Global Leadership Initiative (GPSS-GLI).

Contents

About the Authors

Abubakari Ahmed is a Lecturer at the University for Development Studies.

Julie C. Aleman is a Visiting Scholar at the University of Montreal and was a Postdoctoral Associate at the University of Liège during the time the chapter was written.

Marta Antonelli is a Senior Research Associate at the Euro-Mediterranean Center on Climate Change.

Gideon Baffoe is a Research Fellow at the University of Glasgow.

Boubacar Siddighi Balde is a Project Researcher at the University of Tokyo.

Bothwell Batidzirai is a Senior Research Officer at the University of Cape Town.

Yaw Agyeman Boafo is a Research Fellow at the University of Ghana.

Stefania Bracco is a Consultant for the UN Food and Agriculture Organization (FAO).

Linda Chinangwa is a Postdoctoral Fellow at the United Nations University.

Rodolfo Dam Lam is a PhD Candidate at the University of Tokyo.

Sileshi Degefa is a PhD Candidate at the United Nations University.

Eric Brako Dompreh is a PhD Candidate at the University of Tokyo.

François Manga Essouma is a PhD candidate at the University of Yaoundé I.

Adeline Fayolle is a Lecturer in Forest and Biodiversity Sciences at the University of Liège.

Kensuke Fukushi is a Professor at the University of Tokyo.

Alexandros Gasparatos is an Associate Professor at the University of Tokyo and a Visiting Associate Professor at the United Nations University.

Coert J. Geldenhuys is an Extraordinary Professor in Forest Science at Stellenbosch University.

Christopher Gordon is the CDKN Country Engagement Lead at the University of Ghana.

Miyuki Iiyama is Director of the Research Strategy Office at the Japan International Research Center for Agricultural Sciences (JIRCAS).

Marcin Pawel Jarzebski is a Project Researcher at the University of Tokyo.

Godfred Seidu Jasaw is a Senior Lecturer and the Director of the Kazuhiko Takeuchi Centre for Sustainability and Resilience, both at the University for Development Studies.

Francis X. Johnson is a Senior Research Fellow at the Stockholm Environment Institute (SEI).

Denabo Juju is an Assistant Professor at Hawassa University.

Alice Karanja is a Project Researcher at the University of Tokyo.

Hajime Kobayashi is a Professor Emeritus of Tottori University.

Gordana Kranjac-Berisavljevic is a Professor of Agricultural Engineering at the University for Development Studies.

Julia Lopes is an MSc student at the United Nations University.

William Armand Mala is an Associate Professor in Agroforestry at the University of Yaoundé I.

Adelina Mensah is a Senior Research Fellow at the University of Ghana.

Linus Mofor is a Senior Environmental Affairs Officer at the United Nations Economic Commission for Africa (UNECA).

Geetha Mohan is a Research Fellow at the United Nations University.

Merle Naidoo is a Visiting Researcher at the University of Tokyo and a Research Fellow at the University of the Free State.

Caroline A. Ochieng is currently on leave from the Stockholm Environment Institute (SEI), at the Energy Sector Management Assistance Program (ESMAP) of the World Bank.

Olle Olsson is a Research Fellow at the Stockholm Environment Institute (SEI).

Ravi Prabhu is the Deputy Director General (Research) at the World Agroforestry Centre (ICRAF).

Osamu Saito is a Principal Policy Researcher at the Institute for Global Environmental Strategies (IGES).

Albert Novas Somanje is an MSc student at the United Nations University.

Kazuhiko Takeuchi is a Project Professor at the University of Tokyo and the President of the Institute for Global Environmental Strategies (IGES).

Margherita Emma Turvani is a Professor of Economics at Venice International University.

Esteban Vclez is an MSc student at the United Nations University.

Antonio Vicario is an ICT & GIS project manager, data analyst, and land planning specialist at Venice International University.

Gerald A. B. Yiran is a Senior Lecturer at the University of Ghana.

Part I
Continental Perspectives

Chapter 1
Sustainability Challenges in Sub-Saharan Africa in the Context of the Sustainable Development Goals (SDGs)

Denabo Juju, Gideon Baffoe, Rodolfo Dam Lam, Alice Karanja, Merle Naidoo, Abubakari Ahmed, Marcin Pawel Jarzebski, Osamu Saito, Kensuke Fukushi, Kazuhiko Takeuchi, and Alexandros Gasparatos

1.1 Introduction

Sub-Saharan Africa (SSA) comprises 46 countries that are located south of the Sahara desert. SSA is the poorest and least developed region of the world across a series of socio-economic metrics (UNDP 2018a). For example, it is characterized by

D. Juju
Hawassa University, Hawassa, Ethiopia

G. Baffoe
University of Glasgow, Glasgow, UK

R. Dam Lam · M. P. Jarzebski
University of Tokyo, Tokyo, Japan

A. Karanja · A. Gasparatos (✉)
Institute for Future Initiatives (IFI), The University of Tokyo, Tokyo, Japan
e-mail: gasparatos@ifi.u-tokyo.ac.jp

M. Naidoo
Graduate Programme in Sustainability Science – Global Leadership Initiative (GPSS-GLI), The University of Tokyo, Tokyo, Japan

A. Ahmed
Department of Planning, University for Development Studies, Wa, Ghana

O. Saito · K. Takeuchi
Institute for Global Environmental Studies (IGES), Hayama, Japan

Institute for Future Initiatives (IFI), The University of Tokyo, Tokyo, Japan
e-mail: saito@unu.edu; atake@mail.ecc.u-tokyo.ac.jp

K. Fukushi
Institute for Future Initiatives (IFI), The University of Tokyo, Tokyo, Japan

Institute for the Advanced Study of Sustainability (UNU-IAS), United Nations University, Tokyo, Japan
e-mail: fukushi@ifi.u-tokyo.ac.jp

A. Gasparatos et al. (eds.), *Sustainability Challenges in Sub-Saharan Africa I*,
Science for Sustainable Societies, https://doi.org/10.1007/978-981-15-4458-3_1

high and endemic poverty (World Bank 2016), high food insecurity (FAO and UNECA 2018), low educational attainment (World Bank 2018a) and low access to basic infrastructure and social services (WHO 2018b), among others. Indeed many international organizations, national governments, civil society organizations and academics agree that SSA is lagging so far behind other regions of the world and that multiple development priorities must be tackled to ensure sustainable development across the continent (Table 1.1). However, as outlined below numerous and highly interlinked challenges/barriers and rapid demographic, socio-economic and environmental change complicate the attainment of these multiple sustainability objectives (UNDP 2018a).

SSA currently undergoes major demographic transitions underpinned by very high fertility and urbanization rates (UN-DESA 2018a). With an estimated annual population growth rate of 2.7% (one of the highest in the world), SSA already has a total population of over one billion people (World Bank 2018a). It has been projected that more than 30 SSA countries could double their population by 2050 (Mutunga et al. 2012). Currently, a very large fraction of the population comprises of dependent youth and children, making SSA the region with the youngest population globally (Canning et al. 2015). In fact, SSA's growing youth population (0–24 years) accounts for about 60% in the total population and 54.3% of the labour force (ILO 2017; UNECA 2016a). Similarly, urbanization rates are currently some of the highest in the world (UN-DESA 2018a). Rural to urban migration has been accelerating in many SSA countries and has contributed to the unprecedented expansion of urban slums and informal settlements, which now host most of the poor urban dwellers across the continent (Cobbinah et al. 2015; UN-DESA 2018a; UN 2018). Many of the rapidly expanding urban areas, and especially slums, experience high levels of pollution and a lack of basic services related to sanitation, waste management, education and health, among others (UN 2018; UN-HABITAT 2016). With the urban population projected to reach about 1.3 billion people by 2050 in SSA, there are real concerns that living conditions will further deteriorate in most urban centres, especially those that have already surpassed their capacities (Cobbinah and Erdiaw-Kwasie 2016; UN-HABITAT 2016).

At the same time, SSA has been experiencing remarkable economic growth over the past decades, which has lifted many of its residents out of poverty (AfDB 2019a). However, despite some SSA economies still being among the fastest growing globally, the economic growth has decelerated in many parts of the continent (Bluhm et al. 2018; AfDB 2019a). Agriculture remains the main source of livelihoods for about two-thirds of the population, contributing approximately one-third of the Gross Domestic Product (GDP) and occupying about 60% of the population (AfDB 2018). SSA is also considerably rich in natural resources accounting in 2010, for example, about 12% of the global mineral operations, including 7% of natural gas, 11% of oil, 19% of gold, 54% of diamonds, 62% of cobalt and 74% of platinum (AfDB 2016). The extraction of these non-renewable natural resources accounts for approximately 10% of the total annual economic output and 50% of the exports, with, however, these numbers varying extensively by country (Lundgren et al. 2013; Ericsson and Löf 2019). That said, despite the significant economic expansion, few countries have managed to diversify their national economies beyond the primary economy sectors (Alobo Loison 2015).

Table 1.1 Priority sustainable development areas, relevant SDGs and key challenges and constraints in Sub-Saharan Africa

Priority areas	Relevant SDGs	Key challenges and constraints
Maintain peace and security	16	Low political commitment, democratic deficits, conflicts, corruption
Ensure good governance and build strong institutions	17	Low political commitment, weak institutions, corruption, limited capacity, poor public service provision
Attract sustainable financing	All	Poor capacity to mobilize domestic resources, high dependency on foreign funding, low capacity to use funding effectively
Develop human capacity	All	Financial constraints, lack of expertise and infrastructure
Foster technology transfer	17	Weak partnerships, lack of expertise and infrastructure
Combat climate change	13	Lack of reliable data, poor financial and technical capacity
Mitigate environmental problems	14,15	Lack of reliable data, poor financial and technical capacity
Catalyse/sustain economic growth	1,8,10	Low economic diversification from primary sectors, low value addition
Improve quality of education	4	Financial constraints, lack of expertise and infrastructure
Generate employment opportunities	8,10	Low economic diversification from primary sectors, lack of expertise and capacity (especially for youth)
Ensure food security	2,3	Poor investment in agriculture, population growth, restrictive policies, climate change
Promote modern energy options	3,7,15	High dependency on traditional bioenergy, restrictive policies, poor financial and technical capacity
Reduce gender inequality	5	Low political commitment, lack of education, restrictive institutions
Alleviate poverty	1,10	Poor economic performance and employment generation, population growth, vulnerability of livelihoods
Improve health and nutrition	3	Financial constraints, lack of expertise and infrastructure, lack of education
Deliver infrastructure	3,6,9,11	Financial constraints, lack of expertise and infrastructure, population growth, climate change
Boost agricultural productivity	1,2	Low investment, land degradation, poor institutional support, climate change
Develop sustainable cities	3,11,12	Rapid urbanization, financial constraints, lack of expertise and infrastructure

Source: Adapted from Boafo et al. (2019)

Environmental change is a reality in many parts of the continent and is exacerbated to a large extent by the demographic and socio-economic changes outlined above (IPBES 2018). On the one hand, unsustainable agricultural practices and resource extraction (e.g. mining, fuelwood/charcoal production and use) drive land use change and land degradation, resource overexploitation, pollution and biodiversity loss across SSA (IPBES 2018). Climate change has an increasingly serious effect on many rural and urban areas, with recurring climatic hazards compromising rural livelihoods and urban activity (Henderson et al. 2017; IPCC 2014). For example, many studies have outlined how climate change is expected to affect subsistence and commercial agriculture and food security across the continent (Müller et al. 2011; Challinor et al. 2014). Furthermore, with most major cities located along the coast, there are fears of increased flooding from storms, coastal erosion, loss of coastal aquifers, as well as other threats associated with future climate change such as sea level rise, extreme heat waves, population displacement, loss of livelihoods and destruction of property (Parnell and Walawege 2011; Pauleit et al. 2015).

The low level of development and rapid ongoing change combine with existing constraints and failures, collectively put further barriers in meeting the major sustainable development priorities (Table 1.1). For example, the efforts to sustain economic growth and reduce poverty have been largely derailed due to low diversification from primary economic sectors, economic recessions in other parts of the world, declining commodity prices, low value addition, poor infrastructure and funding constraints, among others (Lange and Klasen 2017). Ensuring the fair distribution of economic gains is far from reality in many countries mainly due to institutional failures such as corruption and lack of good governance (Arndt et al. 2018), undermining the potential economic benefits from its rich natural resource endowment (Lundgren et al. 2013). Furthermore, the constraints (and ineffective use) of financial resources, and the lack of expertise and capacity have contributed to the ineffective delivery of infrastructure and social services and the lack of adaptation to environmental change. This in turn adversely affects the realization of other sustainable development priorities related to health, education and poverty reduction (Ndikumana and Pickbourn 2017; Dreibelbis et al. 2013; Omisore 2018).

The recent adoption of the Sustainable Development Goals (SDGs) provides a golden opportunity to tackle these interlinked challenges in SSA, and help achieve sustainable development and the Agenda 2063 of the African Union (AfDB 2015a; UNDP 2018a). The SDGs represent the global development agenda following the Millennium Development Goals (MDGs) (Omisore 2018). The SDGs aim at addressing complex global developmental challenges and encompass a new vision of sustainable development for the world as a whole, while building on the successes and compensating for the shortcomings of the MDGs (Mainali et al. 2018; UNDP and World Bank 2016). The SDGs contain 17 goals and 169 specific targets that span different economic, social and environmental domains. The SDGs can be broadly divided across five major categories, namely (a) unfinished business of the MDGs: people (SDG 1–5); (b) new areas: prosperity (SDG 6–11); (c) green agenda: planet (SDG 12–15); (d) governance: peace (SDG16); and (e) means of implementation: partnership (SDG17) (Fig. 1.1). However, the SDGs and underlying targets are highly interconnected, as often the key to achieving one goal involves tackling issues associated with another (ICSU 2017).

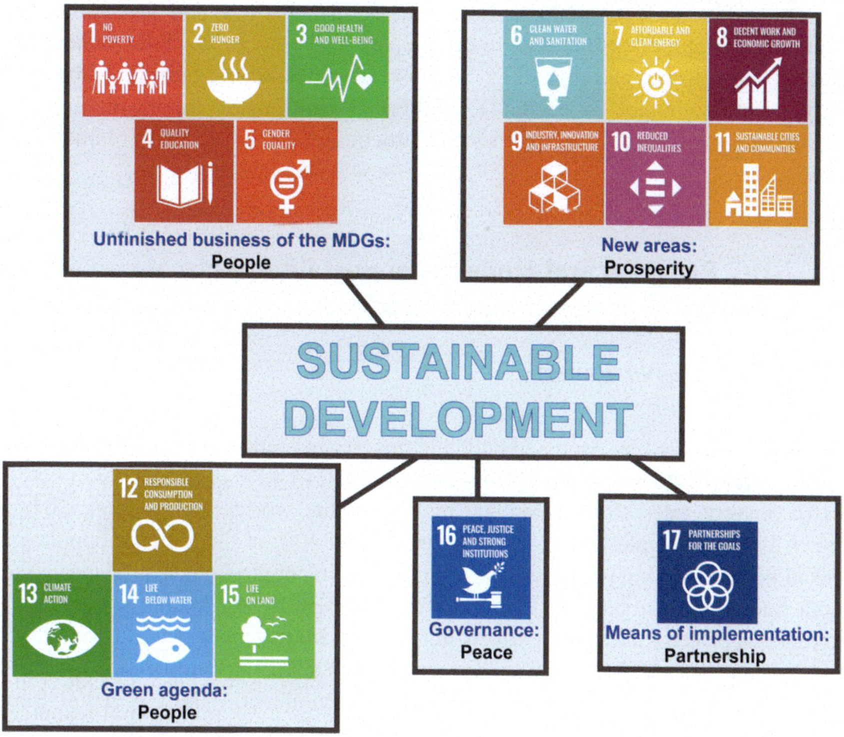

Fig. 1.1 Categories of sustainable development goals

However, for many SSA countries some of the SDGs (and underlying targets) are overambitious, and hence unlikely to be achieved due to the very low starting points (Lange and Klasen 2017; *SDG Centre for Africa* 2019). For example, the likelihood of countries such as Ethiopia meeting SDG targets related to access to clean water and sanitation is remote mainly due to its very low starting base, i.e. only 7.1% of the national population has access to basic sanitation (WHO and UNICEF 2018). Likewise, the target for the under 5-year-old child mortality appears unrealistic for many SSA countries (Lange and Klasen 2017). This seems to imply a similar trajectory with some of the MDGs, considering that many SSA countries failed by a wide margin in meeting them (UN 2015).

This two-volume edited series explores some of the main sustainability challenges facing SSA countries in the context of the SDGs. In particular each individual chapter tackles different sustainability challenges across the full spectrum of the SDGs. The thematic, methodological and geographical focus varies substantially between chapters in an effort to appreciate the full breadth of the current sustainability challenges across SSA, as well as of the different methods that can be used to understand them and solutions to tackle them.

The aim of this first chapter is to identify and critically discuss some of the current trends and challenges for meeting the different SDGs in SSA. Considering the many different countries, local contexts and sustainability challenges in the region, this introduction chapter does not seek to be comprehensive but to highlight some of the most important trends and prevailing circumstances. Section 1.2 introduces the main patterns and challenges for each SDG. Section 1.3 introduces briefly the different chapters and the overall aims, themes and structure of the two edited volumes.

1.2 SDG Progress and Underlying Sustainability Challenges in Sub-Saharan Africa

1.2.1 SDG 1: No Poverty

SSA is characterized by deep and endemic poverty in many areas. Even though the overall poverty rate has dropped from 57% in 1990 to 43% in 2012 (Hakura et al. 2016), poverty rates still remain the highest in the world (World Bank 2018a) (Fig. 1.2). For example, in 24 countries, more than 40% of the national population lives in poverty, with poverty rates as high as 70% in countries such as Madagascar, South Sudan and Zimbabwe (*SDG Centre for Africa* 2019). Actually, despite the pronounced poverty alleviation efforts through the MDGs, between 2002 and 2012 poverty reduced in SSA only by an average of 1.5% annually, compared to the 2.7% average reduction in all other developing regions combined (UNECA 2017). Similarly, despite the increasing per capita income in many parts of the continent, absolute income levels are still quite low in many countries (less than USD 2000 per year) and vary substantially between sub-regions (Fig. 1.3). Figures 1.2 and 1.3 highlight some of the high-income and poverty variation in some of the major economies across the different sub-regions.

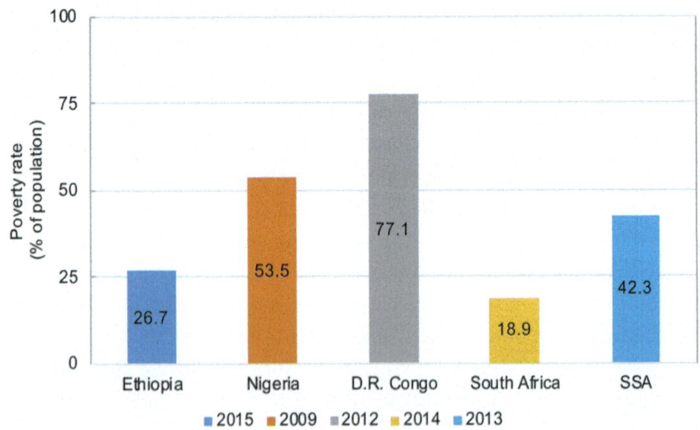

Fig. 1.2 Poverty headcount ratio for SSA and selected countries (in % of the population living below USD 1.90 per day, 2011 PPP). Source: (World Bank 2018a)

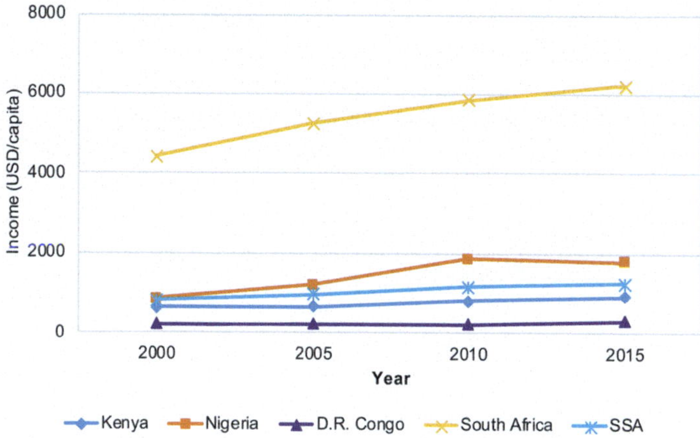

Fig. 1.3 Adjusted net national income per capita for SSA and selected countries (constant 2010 USD/capita). Source: (World Bank 2018a)

Some of the key challenges perpetuating the high incidence of poverty include, among others, the rapid population growth, environmental change, income inequalities, fragility to shocks, inadequate economic diversification from primary economic sectors and lack of (and ineffective use) of resources (AfDB 2018; Sembene 2015; UNECA 2017; Hakura et al. 2016; Alobo Loison 2015) (Table 1.1). Many national, regional and international actors have sought for decades to reduce poverty with large-scale policies, programs and plans or small tailored interventions (World Bank 2019a; Sembene 2015). However such efforts often underperform or even fail due to the lack of evidence, capacity, political commitment and/or funding (World Bank 2017). Furthermore, there is a critical lack of updated, reliable and comparable data across geographic scales to understand better the current poverty patterns and its determinants (*SDG Centre for Africa* 2019).

The above are compounded by the fact that SSA still has a predominately rural population (Amoah and Phillips 2018). Rural settlements are scattered and lack proper infrastructure, hence it is not easy to deliver basic services vital for poverty reduction such as education, health and access to clean drinking water and energy, or to generate jobs and income opportunities, creating thus a vicious cycle (see Sects. 1.2.6, 1.2.7, 1.2.9 and 1.2.11). Furthermore, as rain-fed subsistence agriculture and pastoralism are the primary livelihoods across much of the continent, their increasing vulnerability to climate change, land degradation and social conflicts poses further barriers to rural poverty alleviation (IPBES 2018) (Sects. 1.2.2, 1.2.8, 1.2.13 and 1.2.15). For example, poverty for as many as 52% of the pastoralist and 42% of the agro-pastoralist communities in the region might be due to drought-related crop failures and massive death of livestock (Middleton and Sternberg 2013). Indeed, communities in semi-arid areas lag for most of the development indicators and hence are highly vulnerable to extreme poverty (Middleton and Sternberg 2013).

Despite the growing urban population (Sect. 1.1), the dynamics and extent of urban poverty have not received due attention in research and policy (Lucci et al.

2018). Overall, due to rapid urbanization, high unemployment (over 7.3%), income inequality and resource limitations poverty persists and thrives in urban SSA (Bluhm et al. 2018). However, there is lack of accurate information and reliable data about poverty in urban slums, which could compromise the efforts to develop efficient and inclusive policies and interventions to reduce urban poverty (Lucci et al. 2018).

1.2.2 SDG 2: Zero Hunger

SSA registers some of the highest levels of under-nutrition, malnutrition and food insecurity globally. Despite substantial progress reducing under-nutrition (decrease from 30% to 22.9% between 2000 and 2016) the levels of child stunting and food insecurity are both still highly variable between countries and unacceptably high compared to other parts of the world (UNECA 2017). For example, more than a quarter of the population is severely food insecure in each sub-region compared to the global average of about 10%: Eastern Africa (32.2%), Central Africa (32.8%), Southern Africa (24.8%) and Western Africa (25.8%) (*SDG Centre for Africa* 2019) (Fig. 1.4). The number of stunted children under 5 years of age is quite high, standing at 57.3 million in 2014 (up from 55.8 million in 2010 and 50.1 million in 2000), but the overall stunting rates were 8.3% (in 2014), which were slightly above the global average of 7.5% (UNECA 2017) (Fig. 1.5). At the same time, public investment in agriculture has declined gradually and more pronounced than the global average, while the flow of overseas development assistance (ODA) to agriculture declined from a peak of approximately 25% in the early 1980s to approximately 6% in 2017 (UNECA 2017).

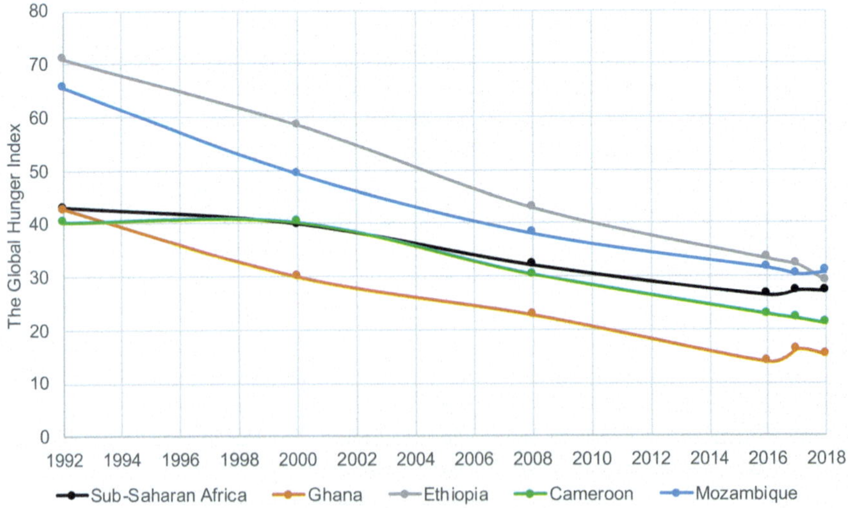

Fig. 1.4 Global Hunger Index for SSA and selected countries

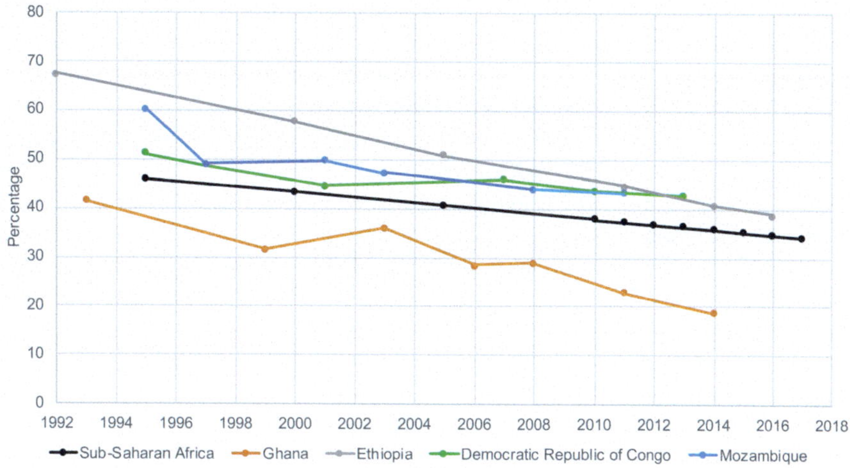

Fig. 1.5 Child stunting rates for SSA and selected countries (in %)

Owing to population growth, food demand in SSA is expected to increase by 60% in the next decades (Fan and Brzeska 2016). For example, maize demand in the region is expected to triple by 2050 (Ekpa et al. 2018). In general, the high poverty, population growth, severe land degradation, climate change, trade imbalances, weak institutions, low investment in agriculture and poor infrastructure are some of the main interrelated challenges for ensuring food security in SSA as discussed below (Webber et al. 2014; UNECA 2017; Onyutha 2018a, b).

When it comes to the agricultural sector, some of the many interrelated challenges for enabling the production of sufficient amounts of food include the inadequate access to productive inputs, constraining legal and land policy frameworks and environmental change (Fan et al. 2013). For example, increasing food production, while avoiding environmental degradation, remains a formidable challenges for many SSA countries (Fan and Brzeska 2016; IPBES 2018) (see also Sect. 1.2.15), not only for meeting the growing food demand in absolute terms but also for coping with the evolving consumer preferences (Ortega and Tschirley 2017). Environmental change is increasingly exerting a huge pressure to the agricultural sector across the continent, affecting yields and the availability of critical inputs such as irrigation water (Xie et al. 2014). This was particularly evident during the recent 2015–2016 El Niño event which induced the worst drought in the last 30 years, leading to massive crop failures and livestock deaths across the continent (Rembold et al. 2016; FAO 2016). This forced many countries, especially in southern Africa, to divert resources from development projects to emergency food aid for the drought victims (FAO 2016).

Other challenges stem from systemic issues, such as the fact that agricultural production is dominated by smallholder subsistence farmers that often lack the financial capacity, knowledge and skills to invest in the agricultural sector and unleash its vast potential (Kuivanen et al. 2016). Furthermore, the inadequate access to credit, low use of agricultural inputs, lack of market information and dominance of antiquated technologies are some more factors impeding the production capability of smallholder farmers both for food crops and industrial crops (Fan and Brzeska 2016; Kuivanen et al. 2016). Increasing agricultural productivity would require massive investments in irrigation and other inputs, but most countries in the region, have limited capacity to do so (Xie et al. 2014).

Other challenges for ending hunger and food insecurity in SSA include unexpected and severe food price fluctuations, conflicts, political instability, post-harvest losses and food waste. For instance, the price of staple food items increased four times between 2008 and 2010 in Eastern Africa (UNECA 2015b). Furthermore, a substantial amount of food is lost mainly during harvesting (and secondarily during consumption), due to the predominant use of traditional technologies to harvest, prepare and store food (Parfitt et al. 2010) (Sect. 1.2.12). Even though the exact amount and quality of the wasted food are not known in most SSA countries due to the lack of reliable information and data (Sheahan and Barrett 2017), this likely discourages the potential investments in technologies that could minimize potential food losses (*Deloitte* 2015).

Finally, there is a lack of political will and a (sometimes) intentional use of hunger as an instrument of punishment in some parts of the region. For example, some dictatorships and rebel groups use hunger and mass starvation as a weapon during conflicts to quell potential political dissent. This strongly implies that ensuring peace and security would be a necessary pre-condition for achieving food security in many parts of the region, but this could prove to be very challenging considering the very diverse national, political and tribal tensions (Sects. 1.2.16 and 1.2.17).

1.2.3 SDG 3: Good Health and Well-being

Despite some promising public health improvements during the MDGs process (UN 2015), SSA still has the highest rate of child mortality for infants and under 5 year olds in the world (WHO 2018a). Only a handful of countries rank below the global average for these indicators (*SDG Centre for Africa* 2019; UNECA 2017). This low performance is partly because of the very low starting base for most SSA countries (Ndikumana and Pickbourn 2017; Lange and Klasen 2017) (Fig. 1.6), but also of the low proportion of births attended by skilled personnel (UNECA 2017). Many countries have had good progress in eradicating preventable childhood illnesses such as polio and measles, through increased immunization coverage (Nakakana et al. 2015), and reduced the incidence of major epidemics such as HIV/AIDS (from 3.87 new infections per 1000 uninfected people in 2000, to 1.48 in 2015) (UNECA 2017; WHO 2018a). However, in absolute terms, most HIV/AIDS patients and more than 90% of the estimated 300–500 million malaria globally, occur in SSA (WHO 2017). At the same time, despite the lack of comprehensive statistics, the prevalence of previously uncommon non-communicable diseases such as hypertension and diabetes is rapidly rising among the young

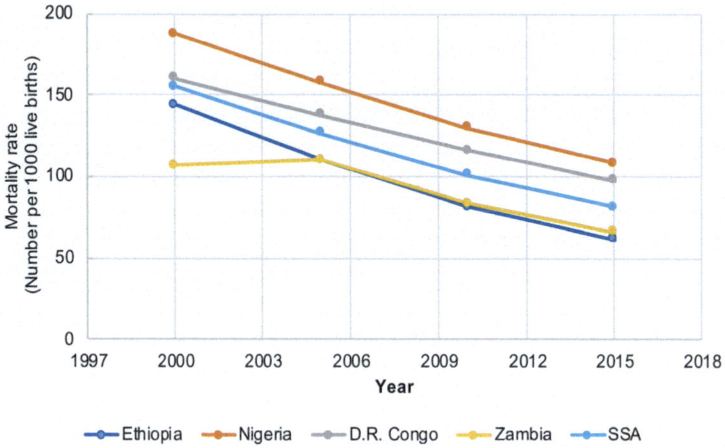

Fig. 1.6 Mortality rates for children under 5 years old in SSA and selected countries (in number per 1000 live births). Source: (World Bank 2018a)

population, mainly due to changes in lifestyle (e.g. unhealthy eating habits, sedentary lifestyles) (WHO 2018a). Interestingly SSA has by far the largest fraction of road accident fatalities globally (UNECA 2017).

Population growth, low health literacy,[1] rapid urbanization and the lack of comprehensive national civil registration systems and vital statistics, are some of the many challenges for expanding healthcare systems, improving public health, and essentially meeting SDG3 across the continent (de Vries et al. 2014; Amoah and Phillips 2018). For instance, only one-third of the Ghanaian population has adequate health literacy, which constraints the performance of a national health system that is considered as one of the most promising in the region (Amoah and Phillips 2018). Similarly, traditional beliefs and negative perceptions (e.g. newborn babies not regarded as worthy family members in rural Ethiopia), are major impediments for expanding health care services to newborns (Onarheim et al., 2017).

Inadequate finance, dysfunctional logistics and poor infrastructure (e.g. inefficient telecommunication networks) have rendered the delivery and monitoring of healthcare services costly in most SAA countries (Brinkel et al. 2014). For example, in Namibia the ongoing efforts to deal with pressing health issues such as HIV/AIDS are constrained by the lack of skilled health professionals, inadequate health infrastructure and financial resources to reach out to marginalized communities (King et al., 2017). It is expected that the increasing prevalence of

[1]Health literacy refers to an individual's knowledge and competence to understand and use modern health care services (Sørensen et al., 2012).

non-communicable diseases will put an even higher burden on the already underfunded national health systems (WHO 2018a).[2]

Poor service delivery from the public sector is another major challenge, contributing to the ineffective utilization of the scarce financial resources (Ssozi and Amlani 2015). This is possibly due to the lack of competent health workers, inadequate incentives for qualified health professionals and broader institutional constraints (Lange and Klasen 2017). Women often do not receive adequate maternal care due to the barriers posed by long distances to health centres, long waiting times, inadequate health care facilities and incompetent health professionals (Mohale et al. 2017). Similarly, major obstacles for effectively dealing with epidemics such as HIV/AIDS, and more recently Ebola, include the limited resources for the extended duration therapies, the high cost (or even absence) of medicine, the shortage of medical facilities and personnel, as well as the poor health literacy, social isolation and stigma (WHO 2018a).

1.2.4 SDG 4: Quality Education

Even though no SSA country has achieved universal enrolment in primary education, more than half of the countries have achieved enrolment rates of greater than 90% (*SDG Centre for Africa* 2019).[3] In fact, since 2000 most SSA countries have achieved impressive progress towards universal access to primary education, but progress for other Education for All (EFA) goals is slow (Africa 2030 report, 2017). Out of the 18 countries globally that are far from achieving the quantifiable EFA goals, 10 are located in West and Central Africa, and are characterized by very low literacy rates mostly for male students (UNESCO 2015). More than half of the children globally that are out-of-school live in SSA, with most of them being girls and an estimated 50% located in conflict-affected regions (UN 2015, 2019). It is worth mentioning that despite this impressive progress in enrolment, an estimated 88% of the children (202 million) in primary and lower secondary school age in SSA are not proficient in reading, and 84% (193 million) are not proficient in mathematics (the highest such levels in the world) (*SDG Centre for Africa* 2019).

A series of challenges pose obstacles for a more significant progress in meeting SDG4. For example, disparities in education along the lines of gender, wealth and location (e.g. urban vs. rural areas, developed vs. underdeveloped areas) are an unfortunate reality, and vary widely between countries and years. For instance, in Nigeria, enrolment in basic education dropped by 8.24% in 2015, but increased by

[2]For example, the prevalence of hypertension is likely to increase in SSA by 89% between 2000 and 2025, which is much higher than the 24% projected increase for the developed world in the same time period (WHO 2018a).

[3]There are large disparities between countries. For example, there is low enrolment in countries such as Eritrea, Djibouti, Burkina Faso, Mali and Niger (less than 70%), and very high in countries such as Malawi, Ghana and Sierra Leone (close to 100%) (*SDG Centre for Africa* 2019).

6.74% in 2016, due to the conflicts in the North East and the inability of some states to submit data on private schools (Federal Ministry of Education 2016). In general, educational challenges mostly affect the poorer strata of society, and especially girls (Save the Children International 2013), e.g. for every 100 boys of primary school age out of school in 2017, 121 girls were denied the right to education (the second highest such fraction globally) (UN 2019).

Other major challenges include stagnant learning outcomes, lack of effective teacher training, low school readiness and lack of infrastructure. In most SSA countries, only a relatively low percentage of teachers in pre-primary, primary and secondary education have adequate training (48%, 64% and 50% respectively) (UN 2019). The majority of schools in the region, and especially in rural areas, do not have access to electricity, potable water or the internet further taking a toll on school attendance (UN 2019). For example, at the upper secondary level, 57% of schools have electricity, but only 25–50% have access to drinking water, hand washing facilities, computers and the Internet (UN 2019).

1.2.5 SDG 5: Gender Equality

Women inequality is pervasive across SSA. As discussed above, despite the significant progress in bridging gender disparity in primary education (almost parity between male and female students), there is a markedly lower registration of female students in secondary and tertiary education (UNDP 2016). On a more positive note, women are well represented in the labour market. In particular, SSA has one of the highest representation of women in the labour force, accounting for as much as 61% of the total labour force (World Bank 2014). However, these jobs are usually lowly paid, insecure and rarely reaching the managerial level (UNDP 2016; UN 2019). On the contrary, many SSA countries have some of the highest proportions of female members of parliament, with the average across SSA countries (24.2% of members of parliament are female) exceeding the global average (23.6%)[4] (*SDG Centre for Africa* 2019).

Gender inequality is a cross-cutting issue in SSA that, if tackled effectively, can have positive lasting consequences for many of the gender-based aspects of other SDGs such as education, health, poverty and food security among others (Nhamo et al. 2018). Fully incorporating women into the sustainable development agenda, increases also the probability of leaving no one person behind, including children, elders and disabled, considering that women play a major role as care-takers in the absence of strong social services in many parts of the continent (Rosche 2016).

---

[4]Rwanda has by far the greatest female representation in the parliament (61.3%), with Senegal and Mozambique having more than 40%. Still, there are several countries where women hold fewer than 10% of parliamentary seats (e.g. Congo, Central African Republic, Swaziland, Nigeria).

Although it is not easy to pin down all of the underlying challenges limiting the capability of women in SSA, it is safe to say that the lack of access to legal rights, land tenure, sexual and reproductive health services, as well as constraints related to freedom of mobility (World Bank 2014; UNDP 2016). Women and girls also generally have limited control over resources and are disproportionally involved in unpaid care work, making it not only difficult to obtain decent livelihoods and self-care but also restrain their active participation in the labour market (UNDP 2016).

Pervasive gender discrimination and legal injustices are still the norm in many countries across the region in various aspects related to financial inclusion and inheritance rights. For example, in Kenya, married men who hold *M-Pesa* and *M-Shwari* mobile banking accounts with Safaricom cannot legally name their wives as beneficiaries, meaning that widows cannot access or automatically claim their deceased husbands' bank accounts (World Bank 2014). Even though many states and regional bodies have put in place a wide array of legal norms, precedents and legislation promoting gender equality, there are major challenges in ensuring that these standards are advocated, accepted and integrated into national laws and regulations, and then fully implemented and enforced (UNDP 2016). This gap between legal rights and expectations, with prevailing practices and behaviours embedded in social and cultural norms pose perhaps the greatest challenge for accelerating gender equality across SSA (UNDP 2016; World Bank 2014).

1.2.6 SDG 6: Clean Water and Sanitation

Safe water and sanitation are vital for human health, social dignity, healthy ecosystems and productive livelihoods, but their lack is very prevalent in rural and informal urban settlements across the continent. Even though the proportion of the SSA population that has access to at least one safe source of water has increased from 17.9% in 2000 to 23.7% in 2015, this is much lower than the global average (71.2% in 2015) (UNECA 2018). Despite the lack of comparable and consistent data between countries, it is believed that more than 70% of the regional population does not have access to basic sanitation facilities, with the rates being higher in some countries such as Ethiopia and DRC (Fig. 1.7). Furthermore, less than half of the population has access to at least basic sanitation infrastructure in 39 countries (as low as 7.1% in Ethiopia), with great divides between urban and rural population (UNECA 2018).

Population growth, inadequate financing, rapid urbanization, growing water scarcity, uneven distribution of water resources, extreme climate change, water pollution and improper utilization are some of the main challenges that potentially undermine the achievement of water and sanitation SGD targets in SSA (Mugagga and Nabaasa 2016; Dos Santos and Gupta 2017; UNECA 2018). At the same time, the inadequate access to clean water and sanitation could compromise the achievement of multiple other SDGs related to health, education and poverty reduction, especially for women and girls. For example, women and girls are disproportionately

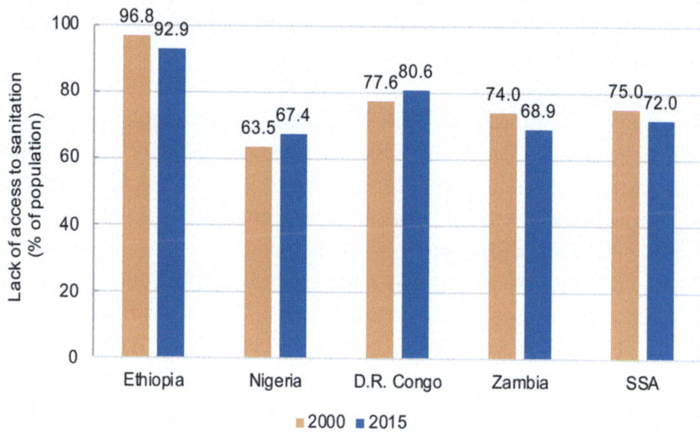

Fig. 1.7 Population without access to basic sanitation in SSA and selected countries (in % of the population). Source: (WHO and UNICEF 2018)

impacted by the rampant scarcity of clean water and adequate sanitation, as they spend disproportionate amount of time fetching water (Demie et al. 2016).

Technical inefficiencies (e.g. poor infrastructure, incompetent staff), corruption, financial constraints and institutional failures are some further challenges that impede the expansion of sanitation and drinking water delivery systems, especially in rural areas (Marson and Savin 2015; Ndikumana and Pickbourn 2017; UNECA 2018). For example, a study in six West African countries (i.e. Nigeria, Niger, Chad, Togo, Sierra Leone, Mauritania) identified poor infrastructure, bureaucratic inefficiencies and low participation of the private sector in water supply systems as the primary challenges for achieving the water and sanitation targets (Alagidede and Alagidede 2016). To overcome the major funding constraints, some SSA countries try to use the revenue generated from water utilities (mainly in urban areas) to finance the expansion of water and sanitation services to new areas. Yet ensuring a fair and efficient utilization of such funds remains a significant challenge in most countries (Alagidede and Alagidede 2016). Contrary to urban areas, the privatization of water services in rural areas appears infeasible due to low profit, which possibly further constraints the delivery of adequate sanitation and drinking water services in rural areas (Bayliss, 2003; Budds and McGranahan 2003).

With a large and expanding fraction of the SSA urban population living in slums (Sect. 1.1), ensuring the adequate access to clean water and sanitation in such urban informal settlements becomes a priority (Mboumboue and Njomo 2016). However, it is often difficult to find appropriate locations in slums to install water and sanitation facilities and ensure their proper management. Furthermore, most slum dwellers cannot afford to pay utility bills and adequately maintain the water and sanitation facilities (Cobbinah et al. 2015; Tsinda et al. 2013). There are valid concerns that due to the rapid slum growth and expansion, most of the SSA governments will not be able to provide these services for free or subsidize them, unless supported by external donors, which could prove to be difficult considering the already large sums allocated and the leveling of disbursements (UNECA 2018; UN 2019) (Sect. 1.2.17).

It is worth mentioning that often a major challenge for effectively delivering clean water and sanitation solutions in SSA is the actual approach for designing and implementing these solutions (GIZ 2019). Various scholars have pointed that many of the current sanitation solutions deployed across SSA might not reflect necessarily the needs and values of local communities. On the contrary there are biases towards solutions and sanitation 'ideals' that were developed in the global North and are far removed from local realities and sensibilities (Chap. 4 Vol. 2). There have been many examples of such systems being underutilized upon completion, suggesting the loss of valuable resources (WHO 2018b; AfDB 2015b). In this respect, a major challenge would be to include meaningfully local communities in the design, delivery and operation of appropriate systems that meet local needs, as a means of ensuring their sustained use (WHO 2018b).

Finally, data and knowledge gaps are another major challenge both for understanding current baseline conditions and the progress needed to meet SDG targets (UNECA 2018), as well as designing and implementing appropriate interventions (WHO 2018b). For example, data about the proportion of the population using safely managed drinking water services are available only for a handful of SSA countries, with many similar data gaps for other crucial indicators (UNECA 2018).

1.2.7 SDG 7: Affordable and Clean Energy

In many SSA countries, energy systems are antiquated, not dependable and cannot meet the demand of the growing and increasingly urbanized population (IEA 2019). Despite the substantial progress in electrification over the past decades (Fig. 1.8), only 43% of the population had access to electricity in 2016 (ESMAP 2019).[5] This is by far the lowest electrification rate among regions globally (UN 2019). The absolute number of electricity access-deficit people peaked in 2015 at 595.3 million people, and began to decrease for the first time in 2016 (by 28.5 million people) (ESMAP 2019). Rural areas have much lower electrification rates (17% in 2014) compared to urban areas (70%) (UNECA 2018). Similarly, only a small fraction of the population has access to clean cooking options, relying instead on traditional biomass fuels and inefficient stoves that are damaging to human health and the environment (UN 2019; UNECA 2018). For example, only 26% of the SSA population has access to clean cooking technologies and fuels such as *liquefied petroleum gas* (LPG), biogas, solar and/or improved biomass cookstoves (ESMAP 2019). It is worth mentioning that despite the large share of renewable energy in total final energy consumption

[5]Electrification rates vary significantly between sub-regions, ranging from 20% in central Africa, to 37% in eastern Africa, 48% in southern Africa, and 56% in western Africa (UNECA 2018).

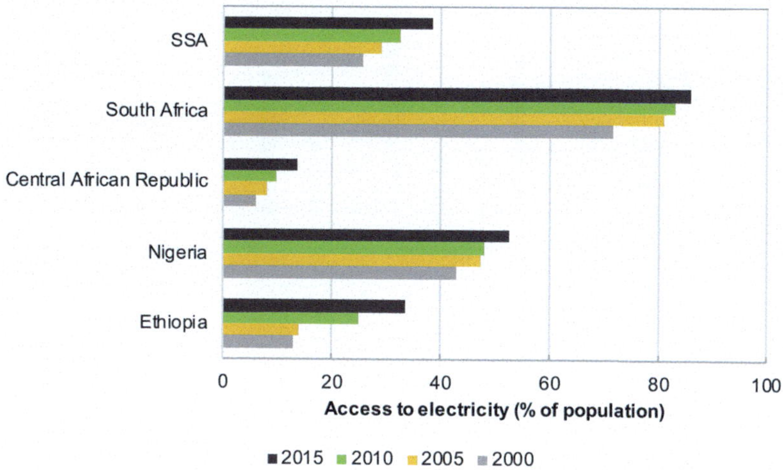

Fig. 1.8 Access to electricity in SSA and selected countries (in % of the population). Source: (World Bank 2018a)

(average of 57.9% across the continent), most of this renewable energy comes from traditional biomass used in the household sector (UNECA 2018).

It is doubtful that the current and future electricity demand will be met without significant changes in the national energy systems (Leimbach et al. 2018). Improving access to electricity would require an additional investment of USD 27–33 billion per annum until 2030, which is beyond the capacity of many countries (Dagnachew et al. 2018). However, it might be possible to meet some of this demand through domestic resources as many SSA countries have enormous renewable energy potential, especially hydropower and solar energy (da Silva et al. 2018; IEA 2019). However, most SSA countries are unable to expand renewable energy production effectively due to the combined challenges posed by funding constraints, institutional constraints and the lack of capacity and a good understanding of available potential (IEA 2019; UNECA 2018). For instance, Cameroon has been unable to tap into its high renewable energy potential, primarily due to the lack of clear policies, skilled labour, infrastructure, as well as the high prevalence of corruption and mismanagement (Mboumboue and Njomo 2016). Similarly, many Western African countries fail to tap their vast potential for small-scale hydropower due to the absence of reliable hydro-climatic data, low awareness, financial and technical limitations and policy and institutional constraints (Stanzel et al. 2018; Mohammed et al. 2013). Some scholars are concerned that the major needed shift to renewables could increase electricity costs, and thus put a further burden to poor consumers (Dagnachew et al. 2018). Therefore, affordability would be another challenge related to the sustained adoption of renewable energy options in many SSA countries.

Actually transforming the domestic energy systems across SSA would require many deep changes including, among others, (a) extensive fuel switch from solid

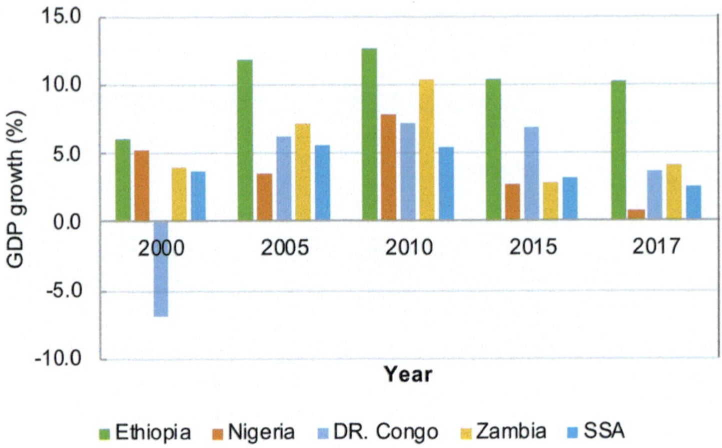

Fig. 1.9 Annual GDP growth for SSA and selected countries (in %). Source: (World Bank 2018a)

biomass fuels to more diversified energy portfolios in the domestic sector and (b) improved energy generation and efficiency (Leimbach et al. 2018). However, many challenges prohibit the transition of the current energy systems and the increased penetration of modern energy technologies, including the inadequate and unsupportive energy policies, poor technological diffusion, underdeveloped/crumbling energy infrastructure, low awareness about the benefits of modern energy technologies, last-mile distribution constraints, low end-user affordability and a series of other context-specific cultural, technical and environmental factors (ESMAP 2019; Davies 2018; Mohammed et al. 2013; Schwerhoff and Sy 2017; Karanja and Gasparatos 2019; IEA 2019).

1.2.8 SDG 8: Decent Work and Economic Growth

Many SSA countries have had robust economic growth since the 1990s, expanding substantially their national economies and moving millions of people out of poverty (Sect. 1.2.1) (AfDB 2019a; Arndt et al. 2016). However, this economic growth has decelerated substantially in the last decade in many parts of the continent (World Bank 2018b) (Fig. 1.9). However, unemployment rates remain quite high, hovering around 7% across the continent (*SDG Centre for Africa* 2019), with unemployment rates being disproportionately higher for females, youth and other vulnerable groups (AUC, AfDB, and UNDP 2016; UN 2019). Furthermore, labour productivity is abysmally low in SSA, with the average economic output per worker increasing only by 0.3% in 2018, compared to the global average of 2.1% (UN 2019).

Sustaining high economic growth over long periods of time is both necessary and difficult due to many internal (e.g. corruption, conflicts) and external (e.g. economic recessions in other parts of the world) challenges (Lange and Klasen 2017).

Similarly, it is challenging to boost formal employment generation for the many millions occupied in informal sectors across the continent, and especially the youth and women (UN 2019). Actually weak economic growth could cause the lower availability of financial resources, attraction of investment opportunities and generation of employment, all of which could undermine the realization of all other SDGs (Lange and Klasen 2017; *SDG Centre for Africa* 2019).

Catalyzing an economic diversification from primary economic sectors such as agriculture and mining is a particularly difficult challenge for many SSA countries (AfDB 2019b; Alobo Loison 2015) (Sect. 1.1). For example, the recent impressive economic growth of Botswana allowed government investment in development projects, but the acute price decline for its export commodities (e.g. diamonds) and the growing income inequalities (UN-DESA 2018b) took a significant toll on economic growth. Similarly, volatile international commodity markets and trade barriers tend to affect substantially the economies of countries that overly depend on commodity crops, such as Burkina Faso, Swaziland and Malawi (Vitale 2018; Terry and Ogg 2017; Jarzebski et al., 2020a). Poor infrastructure, restrictive policies and limited financial and technical capacity are some of the challenges facing local industries, which put serious obstacles to economic diversification (Morris and Fessehaie 2014) (Sect. 1.2.9). The increased spending for (and improved access to) infrastructure such as electricity, roads and Information and Communication Technologies (ICTs) could assist economic diversification, and eventually foster economic growth in SSA (Kodongo and Ojah 2016; AfDB 2019b). Nevertheless, the shortage of finance, policy constraints and poor infrastructure are some of the challenges in the region that still need attention.

Combating acute youth unemployment is another major challenge in many SSA countries, especially those that are highly dependent on agriculture and extractable natural resources (Ackah-Baidoo 2016; ILO 2012). Restrictive land policies, corruption, capital constraints and lack of entrepreneurial skills are some of the challenges faced by young people attempting to open their own businesses (Ackah-Baidoo 2016; Gossel 2018). While a lack of skills is a common hurdle for many young people seeking to enter the labour market, there is a pervasive mismatch between the skills demanded for formal employment and the skills of many young people in the continent (ILO 2012). There are concerns that the inability to effectively integrate youth in economic activities could contribute to conflicts, illegal migration and early marriage (Ackah-Baidoo 2016), all of which can affect negatively progress for other SDGs.

Another major challenge to achieve sustained economic growth in many SSA countries are the high trade imbalances, while often stem from their limited capacity to compete in a globalized world, and reduces their ability to reap benefits from international trade (Zahonogo 2016; Moussa 2016; UNCTAD 2018). For example, many countries in the region heavily rely on imported consumable goods, which they could have produced domestically to boost their manufacturing sector (Signé 2018; Mendes et al. 2014) (Sect. 1.2.9).

The sustained attraction of Foreign Direct Investment (FDI) and Overseas Development Assistance (ODA) could help address many of the above challenges and boost economic growth. Nonetheless, the flow of FDIs is notably low to many SSA countries, mainly due to corruption and weak institutions, among others (Gossel 2018). Similarly, many SSA countries do not use efficiently ODA, or even do not use it for the intended purposes due to corruption or technical incapability, thus discouraging donors to continue their much-needed support (Ndikumana and Pickbourn 2017).

Finally, apart from sustaining economic growth and creating decent employment, it is also important to ensure the fair distribution of economic benefits and employment opportunities. However, this is a major challenge for most countries in the region (Arndt et al. 2016). In fact, the high prevailing economic inequalities are one of the major reasons behind the growth constraints facing many national economies (World Bank 2016) (see Sect. 1.2.10 for more details). Institutional failures related to corruption and inadequate policy frameworks have often been blamed as serious challenges for curbing economic growth through increasing inequality (Sect. 1.2.10).

1.2.9 SDG 9: Industry, Innovation and Infrastructure

With a few exceptions, most SSA countries have not developed a strong industrial and manufacturing base (AfDB 2019b). The share of manufacturing in the regional GDP is low (just under 10% in 2017, and has hardly increased in decades). While SSA lags far behind the rest of the world (including most developing countries) in terms of manufacturing value added (MVA) and manufacturing exports (Signé 2018). More troubling is the fact that the proportion of medium–high and high-tech MVA in the total MVA is one of the lowest in the world and has actually declined between 2000 and 2016 (UN 2019). This is reflected by the fact that most SSA countries have also rather poor innovation performance,[6] usually ranking in the lowest parts of global innovation indices (WIPO 2019). Similarly, infrastructure is rather underdeveloped, costly, unreliable and prone to significant disruptions in many SSA countries (World Bank 2019b). As already mentioned, water, sanitation and energy infrastructure are rather underdeveloped and face significant challenges across the region (Sects. 1.2.6 and 1.2.7). Similarly, transportation networks are underdeveloped (with only a quarter of roads paved), airfreight and air travel unpopular and rail transportation still in its infancy (UNECA 2017).

At the same time, internet use is quite low for global standards, with less than 25% of the population having internet access in two-thirds of SSA countries, and practically all countries falling below the global average (*SDG Centre for Africa* 2019). However, the continent has also made impressive improvements over the past few years. For instance, the proportion of the population covered by 3G mobile

[6]South Africa is the innovation leader in the continent, but only ranks 63 out of 129 countries globally (WIPO, 2019).

networks has reportedly increased from 25% to 65% between 2010 and 2015, possibly enhancing financial inclusion by facilitating virtual access to financial services (UNECA 2017). There are also multiple examples of successful innovation across the continent (World Bank 2015). In fact when looking at innovation relative to level of development, 6 out of the 18 innovation achievers are SSA countries (the most out of each region) (WIPO 2019).

Clearly, achieving SDG9 would require coordinated action and massive investments for industrialization, research and infrastructure development, as it has been estimated that infrastructure and innovation constraints reduce industrial productivity by around 40% (UN 2015). According to the African Development Bank (AfDB), all these aspects are quite linked and highlighted in its top priorities for promoting industrialization, namely: (a) foster successful industrial policies, (b) attract and channel funding into infrastructure and funding projects, (c) grow liquid and effective capital markets, (d) promote and drive infrastructure development, (e) promote strategic partnerships and (f) develop efficient industry clusters (AfDB 2019b). Furthermore, many SSA countries have great capacity for leapfrogging, taking advantage of many tested technologies and approaches both from other parts of the world and within SSA (World Bank and China Development Bank 2017).

However, there are many challenges for effectively tackling SDG9 that are deeply cross-cutting and relate to multiple other SDGs. For example, the availability of funding and credit services (and their effective utilization) is the major constraint for fostering innovation, developing infrastructure and expanding the industrial sector (and especially small and medium enterprises) (UN 2019). Similarly, the lack of capacity and appropriate skillsets, combined with large youth unemployment and inability to access capital (Sects. 1.2.4 and 1.2.8) deeply affects the ability to attract the highly skilled personnel to drive innovation and industrialization (WIPO 2019 UNECA 2017). Related to the above is the very low level of funding invested in research that stands at 0.42% for the entire region (one of the lowest in the world) (UN 2019).

Another major challenge is the lack of ability to capitalize on emerging technologies and markets. This is again a cross-cutting issue, as apart from the lack of tech platforms and innovation capacity (UN 2019; WIPO 2019) there is a lack of robust institutional frameworks, skillsets, ability to attract global trade and investment opportunities, and a thriving demand environment (WEF 2018).

1.2.10 SDG 10: Reduced Inequalities

Despite the economic growth of the past decades (Sect. 1.2.8), SSA has remained one of the poorest and most unequal regions in the world (Sect. 1.2.1). Seven out of the ten countries with the highest income inequality are located in SSA, with South Africa topping the global inequality list (World Bank 2019a). However, there are very large regional disparities with countries in southern Africa having

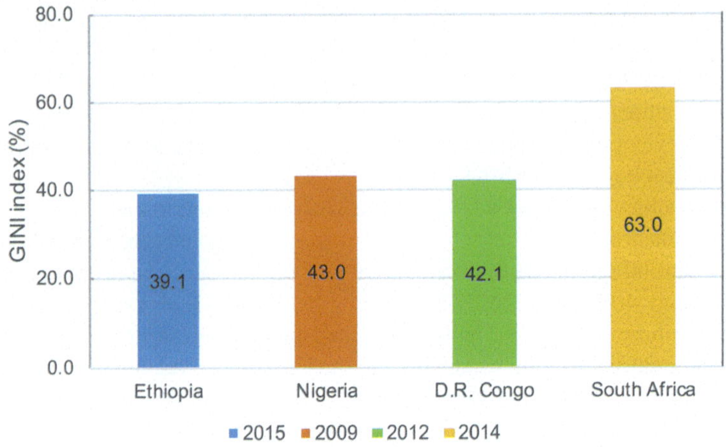

Fig. 1.10 Gini Index for selected SSA countries. Source: (World Bank 2018a)

rather large income inequalities, and countries in other regions recording medium- to low-income inequality (UNDP 2017) (Fig. 1.10). More importantly long-term inequality trends vary significantly between countries, with some countries experiencing a steady increase, others a steady decline, and others different U-shape trends (UNDP 2017). However, countries in southern and central Africa, and countries whose economies are dominated by the oil and mining sector have experienced increasing inequality (UNDP 2017). Apart from income inequality, there are substantial social inequalities in terms of, among others, access to health, education, and employment, both within and between countries, with those trends varying substantially between countries (Heaton et al. 2016).

Considering these rather different inequality trends, it is not easy to pinpoint an exact set of challenges that drive inequality within and between SSA countries. In fact, the perpetuation of inequalities in income, health and educational outcomes possibly emanates from a complex combination of discriminatory social norms (especially for women) and a skewed distribution of service provision. For example, some of the most commonly associated challenges perpetuating inequality in SSA include (a) dualistic economic structure (e.g. a minority 'labour elite' in the government, multinational companies and the resource extraction sector vs. the majority of labourers in informal sectors or subsistence agriculture); (b) concentration of physical/human capital, and land in elites; (c) resource curse; (d) urban bias of public policy; (e) limited or failed redistributive efforts from national governments; (f) prevailing ethnic and gender inequalities (UNDP 2017; *SDG Centre for Africa* 2019).

Ongoing processes such as population growth, urbanization, conflicts and environmental change often accentuate these challenges. For example, the low level of service delivery in rural areas compared to urban areas is often due to the sparse settlements, which complicates infrastructure expansion (Frankema 2014). In turn, the inadequate infrastructure and other outcomes associated with inequality (e.g. political instability) are major constraints for the expansion of investments and economic diversification (AUC, AfDB and UNDP 2016). Furthermore, despite measures to reduce income inequality such as tax reforms, the fair redistribution of the gains from economic growth has remained challenging throughout the region due to the lack of strong political commitment and the limited capacity (Masiya 2017).

Finally, as for most other SDGs, there is a critical lack of stratified and comparative datasets to appreciate the current extent of inequality within and between countries. In fact, there are major data gaps for most of the indicators for SDG10 (*SDG Centre for Africa* 2019), which have prevented its inclusion in most efforts to track SDG progress in the continent (e.g. UNECA 2017, 2018). This points to the ongoing need for improved data collection and statistical capacity building, especially in the poorer SSA countries (*SDG Centre for Africa* 2019).

1.2.11 SDG 11: Sustainable Cities and Communities

SSA is currently the least urbanized region in the world. Although only around 40% of its population is considered urban, it has the fastest urban growth rate globally standing at 4.1% (World Bank 2018a; UNECA 2018). Furthermore, many cities and towns in the region are poorly planned and constructed but account for about 55% of the regional GDP (UNECA 2015a). Large populations live in the continent's proliferating slums, which although declining moderately in terms of proportion, it remains quite high in absolute terms (Fig. 1.11) (UNECA 2018). Urban population in SSA lacks reliable access to many services such as transport. Only 18% of the residents in major SSA cities have access to reliable transport network, with

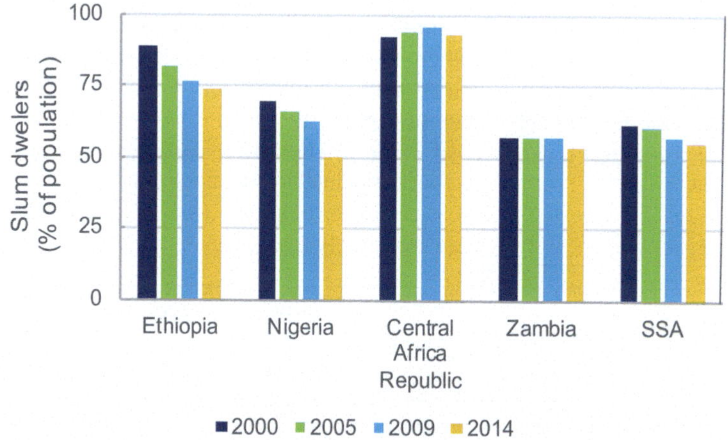

Fig. 1.11 Population living in urban slums (in % of urban population). Source: (World Bank 2018a)

transport expenses accounting for a large fraction of their household budget (UN 2019; UNECA 2018). Furthermore, most cities in the continent (and especially secondary cities) do not have proper sanitation and waste management systems[7] (Sects. 1.2.6 and 1.2.12), and experience acute pollution (UNECA 2018).

The changing demographic landscape and the increasingly urban future have enormous implications for human development, economic transformation and sustained economic growth, among others (Sect. 1.1). Considering the high rates of urbanization and the large number of slum dwellers, it will become an even greater challenge to provide adequate housing, social services and infrastructure (UNECA 2018). For example, when it comes to housing, a major challenge is both housing availability and also affordability. In other words, there is low delivery of appropriate housing at prices that most urban residents cannot afford without access to secure lines of credit (UNECA 2018). At the same time, the expansion of unplanned and informal housing undermines efficient land utilization, obstructs the development of the necessary infrastructure and sometimes poses security problems (Cobbinah et al. 2015). For example, urban green areas in most SSA cities are shrinking due to the ever-increasing competition for land for other uses (mainly housing) and poor urban planning and management (Cobbinah and Darkwah 2016).

As for other SDGs the lack of political will and the weak financial and institutional capacities often undermine the realization of SDG11 in SSA. For example, limited finance and the lack of skilled labour and modern technology likely impede the realization of interventions to tackle multiple SDGs in urban areas (Pieterse et al. 2018).[8] Urban planning agencies across the continent struggle to manage the many urban sustainability issues due to the limited skilled personnel and logistical capacities (Cobbinah and Aboagye 2017). Currently, only 13 countries have formulated (and 21 countries are in the process of implementing) national urban policies, while only a handful of countries are implementing national disaster risk-reduction strategies in line with the Sendai framework (*SDG Centre for Africa* 2019). Ensuring meaningful public participation in urban planning, inadequate legal enforcement and the prevailing customary land tenure systems are some of the key underlying challenges for urban planning and governance across SSA (Kleemann et al. 2017).

Finally, again, there is a critical lack of datasets with sufficient resolution to appreciate patterns at the city level. In fact there are major data gaps for most of the indicators for SDG11 (*SDG Centre for Africa* 2019), including disaster losses

---

[7]Estimates suggest that only 52% of municipal waste in terms of mass has been collected in SSA between 2010–2018 (up from 32% in 2001–2010) (UNECA 2018). Although this signifies a huge increase, it still remains the lowest in the world.

[8]For example, the biggest waste to energy power plant in SSA is built in Addis Ababa (National Planning Commission of Ethiopia 2017). This facility is expected to generate 50 MW of electricity (30% of the city's energy need) by incinerating 140 tons of dry waste per day (80% of Addis Ababa's dry waste). Even though this project would help upon completion address, multiple problems like waste disposal, power shortages, and job creation (contributing to multiple SDGs), it experiences substantial delays mainly due to financial constraints and technical issues.

(Osuteye et al. 2017). Interestingly, SDG11 has not been included in most efforts to track SDG progress in SSA (e.g. UNECA 2017, 2018).

1.2.12 SDG 12: Responsible Consumption and Production

Consumption and production patterns are shifting rapidly in SSA due to the underlying rapid socio-economic change discussed elsewhere in this chapter (e.g. Sects. 1.1 and 1.2.8). Even though the total domestic material consumption (DMC) remains one of the lowest in the world, it increased from 3.6 to 4.2 billion metric tons between 2010 and 2017 (UN 2019). Furthermore, despite the declining material use intensity in the past two decades, it still remains one of the highest in the world standing at 2.5 kg per USD of GDP, compared to the global average of 1.16 kg/USD (UN 2019). At the same time, food losses exceed 30% of total crop production in the continent, resulting in the loss of more than USD 4 billion each year (Sheahan and Barrett 2017). Most of this food loss comes from the production to the retailing part of the food system, rather than the consumer side (UNECA 2018).

Despite the lack of reliable statistics between countries, effective waste management is a major challenge across the continent, especially in the rapidly expanding cities (UNECA 2018). The lack of proper waste management facilities (and the underperformance of existing facilities) is compounded by the rapidly increasing and urbanizing population, funding constraints and lack of capacity and political will to create a complex set of challenges for most countries in the region (UNECA 2018). In Tanzania, for instance, local governments' efforts to provide solid waste management services to local communities have been severely hampered by the rapidly increasing human population (Hallaf 1999; Kassim and Ali 2006). However, apart from municipal waste, the management of other forms of hazardous waste (e.g. e-waste) is increasingly becoming a major sustainability challenge in the region. In particular, even though SSA still accounts for only 5% of global e-waste generation (having the lowest per capita generation rates), e-waste generation increases, partly due to illegal transboundary flows from other parts of the world (Baldé et al. 2017). For example, Western African countries such as Ghana, Nigeria, Benin, Ivory Coast and Liberia have become major e-waste destinations from other parts of the world (SBC 2011). Around 70% of all Ghanaian imports in 2009 were used electronic equipment (Amoyaw-Osei et al. 2011), with 30% being non-functioning (therefore e-waste), and much of the rest either repaired/sold locally or being un-repairable (SBC 2011).

The low political priority of SDG12 proves to be a major challenge for meaningful progress (UNECA 2018). This lack of political priority is evident both at different levels of governance (e.g. regional, national) and thematic areas (e.g. e-waste). For example, few SSA countries are signatories of all relevant Multilateral Environmental Agreements (MEAs), while SDG12 does not feature in some of the main documents related to SDG implementation in SSA, such as the African Union's Agenda 2063 and other key documents related to SDG progress (UNECA 2018;

SDG Centre for Africa 2019). Furthermore, most countries in SSA (and practically all countries in southern Africa) do not have any established legal framework for e-waste (ITU, 2017). An example has been the attempted ban on the imports of used air conditioners, television sets and fridges in Ghana 2012, which failed to receive public support.

Finally, as with other SDGs, the lack of data is a major obstacle for achieving meaningful progress in the implementation of SDG12 (*SDG Centre for Africa* 2019). This lack of reliable and comparable data prevents even the understanding of the baseline conditions of the current consumption/production patterns and waste generation levels, and thus the starting points and the tasks at hand. As mentioned throughout this chapter, this points to the underlying lack of capacity and available funds to develop technical expertise in these domains, as well as appropriate data collection, storing and sharing mechanisms.

1.2.13 SDG 13: Climate Action

SSA countries have a rather low contribution to global GHG emissions compared to other regions of the world (Hogarth et al., 2015; Adzawla et al. 2019). In fact, with per capita GHG emissions standing at 2.7 or 3.9 tonnes CO2eq (depending on whether land-use change and forestry is taken into account), SSA has the lowest GHG emissions in the world (Hogarth et al., 2015). Only South Africa ranks among the top 20 global GHG emitters (*SDG Centre for Africa* 2019), with most countries emitting less than 2 tCO2eq/cap (JRC 2017). However, despite its low contribution to global emissions, SSA is one of the regions expected to be hit the hardest by climate change (and associated climatic hazards) due to the comparatively higher risk of exposure and poor adaptive capacity (IPCC 2014; Engelbrecht et al. 2015; Müller et al. 2011; Kendon et al. 2019; Serdeczny et al. 2017). Despite evidence that the strength and periodicity of these hazards are likely to increase throughout the continent (IPCC 2014), the fraction of the population that is expected to be affected by such hazards varies substantially between countries (Fig. 1.12). Most countries have ratified the Paris Agreement [9] and have put in place a series of Intended Nationally Determined Contributions (INDCs) and National Adaptation Plans (NAPs) (UNFCCC 2018). The pledged INDCs and NAPs vary by country in their strategies, targets, levels of ambition and priority sectors (e.g. see Table 1.2).

One of the major challenges for successfully adapting to climate change would be reducing the vulnerability of agrarian communities and urban areas (Davis et al. 2017; Kahsay and Hansen 2016; Middleton and Sternberg 2013). In both cases, climate change can have cascading effects on multiple SDGs, but through rather different pathways. For instance, frequent recurrent droughts in rural areas have caused water scarcity, crop failure, livestock death and eventually collapse of

[9]Except for Angola, Eritrea, Liberia, Guinea-Bissau, Equatorial Guinea and South Sudan.

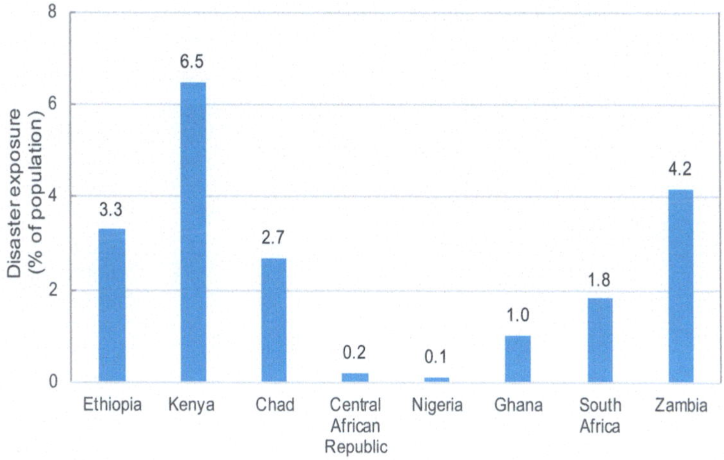

Fig. 1.12 Population affected by droughts, floods and extreme temperature (in % of the population, average 1990–2009). Source: (World Bank 2018a)

livelihoods and food insecurity in some countries (National Planning Commission of Ethiopia 2017; Shiferaw et al., 2014). Such cascading effects were very prevalent through the 2015 El Nino event that resulted in severe droughts and water shortages (Gizaw and Gan, 2017), affecting agricultural systems throughout the region. Conversely climate-related hazards and disasters such as floods and heat waves can have substantial effects in urban centres (Herslund et al., 2016), due to the high population density, poor urban planning and lack of disaster management and preparedness in most SSA cities (Sects. 1.2.6 and 1.2.11). Furthermore, such events could increase the risk of communicable disease outbreaks such as cholera (Wu et al., 2016), which can be particularly devastating in highly dense urban areas.

Adapting agriculture, cities and other vulnerable sectors to climate change would undoubtedly require innovative practices and policies backed by smart technologies, skilled labour and financial resources, all of which are in short supply in the region (Henderson et al. 2017). For example, the lack of operational meteorological infrastructure and accurate meteorological information (especially seasonal forecasts) poses major barriers for effective adaptation (NAP-GSP 2017). Furthermore, many organizations have pointed out the need for sustainable funding to enable adaptation actions throughout the continent (UNDRR 2018). However, despite these challenges, there have been many examples of successful climate change adaptation throughout SSA (UNDP 2018b), with many countries managing to attract sufficient funding and international support for their NAPs (UNFCCC 2018).

Finally, a particularly difficult challenge to solve would be to ensure policy coherence, especially by integrating and mainstreaming climate change adaptation into developmental and financial planning across sectors and scales (NAP-GSP 2017; UNDRR 2018). Achieving this would require high-level political support, institutional coordination and effective engagement with a wide range of

Table 1.2 Summary of INDCs for eight SSA countries

Country	Emission targets	Target type (year)	Baseline emissions (MtCO2eq)	Emission reduction compared to baseline (%)	Adaptation strategies for selected sectors
Ethiopia	Limit GHG emissions to 145 MtCO2e or lower in 2030	Fixed level target (2030)	400	64%	– *Agriculture* Agroforestry, climate-smart agriculture, food security, irrigation, land and soil management – *Cross-cutting areas* No specified actions
Kenya	Not specified	Baseline scenario target	143	30%	– *Agriculture* Climate-smart agriculture – *Cross-Cutting Areas* Capacity-building, knowledge transfer, climate risk management, climate services
D.R. Congo	Not specified	Baseline scenario target	430	17%	– *Agriculture* Climate-smart agriculture – *Cross-cutting areas* No specified actions
Central African Republic	Not specified	Baseline scenario target (2030)	NA	5%	– *Agriculture* Agroecology, food security, livestock – *Cross-cutting areas* Climate risk management
Nigeria	Not specified	Baseline scenario target (2030)	NA	45%	– *Agriculture* Climate-smart agriculture, fisheries/aquaculture, irrigation, land and soil management, livestock – *Cross-cutting areas* Capacity-building, knowledge transfer, climate services

(continued)

Table 1.2 (continued)

Country	Emission targets	Target type (year)	Baseline emissions (MtCO2eq)	Emission reduction compared to baseline (%)	Adaptation strategies for selected sectors
Ghana	Reduce GHG emission by 27% and 45% in 2025 and 2030, respectively	Baseline scenario target (2030)	NA	45%	– *Agriculture* Climate-smart agriculture, food security – *Cross-cutting areas* No specified actions
Zambia	Not specified	Baseline scenario target (2030)	NA	NA	– *Agriculture* Climate-smart agriculture, ps, fisheries/ aquaculture – *Cross-cutting areas* Capacity-building, knowledge transfer, climate services
South Africa	Limit GHG emissions in 2025–2030 to 398–614 Mt. CO2eq	Trajectory target (2030)	NA	NA	– *Agriculture* No specified actions – *Cross-cutting areas* Capacity-building. Knowledge transfer, climate risk management

Note: This table includes adaptation strategies for only two sectors, namely agriculture and crosscutting areas. Source: https://www.climatewatchdata.org/

stakeholders, including the private sector and vulnerable communities (NAP-GSP 2017). However, this is particularly difficult considering the challenges in forming effective partnerships on the continent (Sect. 1.2.17).

1.2.14 SDG 14: Life Below Water

SSA contains 38 coastal and island states, which collectively oversee 13 million km^2 of exclusive economic zones and have a combined coastline stretching over 47,000 km (UNECA 2016b). Most major coastal and marine habitats throughout the continent face significant degradation due to habitat loss/change, overexploitation, climate and invasive species, among others (IPBES 2018). For example, countries such as Kenya, Tanzania and Mozambique have lost 15–30% of their mangroves, while coral reefs in the Indian Ocean have also declined by as much as 15% (IPBES 2018). However, most coastal countries have very few marine protected areas,[10] which are collectively well below the expected goal of protecting 10% of coastal and marine areas (*SDG Centre for Africa* 2019; UNEP-WCMC 2016). Overall, the coverage of marine conservation areas in SSA is only 2.6%, which is one of the lowest in the world, and far below the global average of 8.4% (UNECA 2017). Furthermore, there are some indications of overfishing, with the oceanic areas bordering SSA having a similar proportion of marine fish stocks that are within biologically sustainable limits as the rest of the world (67%, in 2015, but declining from 90% in 1974, and stabilizing since 2008) (UN 2019). Similar patterns are also evident for water pollution in coastal areas, with water quality designated as mostly clean to average, apart from some equatorial zones (UN 2019).

Coastal and marine areas are increasingly identified as a major element of African development at the local, national and regional levels. For example, according to the Africa Agenda 2063, the transition to a Blue Economy could contribute substantially to continental transformation, development and growth (UNECA 2016b). However, there are many major challenges for meeting SDG14 in SSA, including (a) unsustainable coastal/marine management practices, (b) urbanization and unregulated trade, (c) lack of political commitment and (d) lack of data.

First, many coastal communities rely directly and indirectly on seascapes for their livelihoods and well-being (IPBES 2018; Owuor et al. 2017; Chaigneau et al. 2019; FAO 2014; Belhabib et al. 2015). However, many of these communities adopt unsustainable coastal/marine management practices due to the lack of capacity and resources and institutional lock-in (IPBES 2018). This is often aggravated by the fact that traditional top-down management approaches in coastal and marine contexts do not allow for the development of local management approaches that are contextually sensitive, and often preclude stakeholder participation and sense of ownership (UNECA 2016b; IPBES 2018).

Second, coastal cities are some of the most rapidly expanding urban areas in the continent, and host an increasing number of ports and industrial facilities. These ports play a very important role in trade and overall development, as more than 90% of SSA exports and imports are transported through them (compared to the global average of approximately 80%) (UNECA 2017). However, coastal cities, industrial zones and ports are major polluters, including hazardous waste and plastic pollution (UNECA 2016b; IPBES 2018). Undoubtedly, sustained investment would be needed to expand maritime facilities and enable related innovations to meet devel-

[10]Some island nations and countries with major coastlines such as Angola, Comoros, Mauritius, Nigeria, Seychelles and Sao Tome and Principe do not have any designated marine protected areas (*SDG Centre for Africa* 2019).

opment priorities, but this would need to take into account the substantial environmental implications for the coastal and marine environment (UNECA 2017).

Political instability and lack of political capacity plague many aspects related of coastal and marine management. For example, instability in some countries in eastern and western Africa has led to increased piracy in some areas (e.g. the Gulf of Guinea, off the coast of Somalia). Lack of capacity has precluded the coordination between and within coastal SSA countries to develop and implement integrated coastal zones and ocean management programmes (UNECA 2016b, 2017). Similarly institutional constraints have contributed to coastal and marine degradation through, for example, the inability to internalize the cost of 'clean' ballast water in ship design and operation, implement relevant multi-lateral environmental agreements and promote cleaner plastic production and use practices (UNDP 2016). A major overarching challenge, as in other parts of SSA, is the lack of resources and capacity to deal with the multi-faceted nature of coastal and marine environments (UNEP-WCMC 2016).

Finally, similar to other SDGs, there are major data gaps and a lack in methodological definitions for most SDG14 targets. For example, only targets 14.4.4 and 14.4.5 have indicators for which there is readily available data and an agreed methodology (*SDG Centre for Africa* 2019).

1.2.15 SDG 15: Life on Land

SSA contains a high diversity of endemic species and unique ecosystems that cater for multiple human needs (IPBES 2018). For example, natural resources from forests and agro-ecosystems support most of the livelihoods and economic output across the continent. In particular, more than 70% of the population depends on forests and woodlands for their livelihoods through multiple ecosystem services such as timber, fuelwood, wild food and medicinal plants, among others (IPBES 2018; CIFOR 2005). Many countries have experienced extensive net loss of forest area between 2010 and 2015, with four of the top ten countries located in SSA (i.e. Nigeria, Tanzania, Zimbabwe, DRC) (FAO 2015). Such high deforestation rates could possibly be due to the low proportion of forest area under long-term management plans (UNECA 2018). However, despite the prevailing terrestrial ecosystem loss and degradation, there has been some appreciable improvement in slowing down deforestation and increasing protected area coverage across SSA (FAO 2010). On average, 46.1% of the important terrestrial biodiversity sites are protected (one of the highest rates in the world), with 44% of SSA countries already protecting more than half of their important terrestrial areas (*SDG Centre for Africa* 2019).

Yet, despite the large fraction of protected terrestrial ecosystems, many areas of critical biodiversity importance remain outside of protected areas. Unsustainable land use change and management practices in such areas are driven by interconnected economic activities and demographic processes related to agriculture, mining, livestock production and urbanization, which collectively pose a major challenge to meeting SDG15. For example, such processes cause the annual loss of USD 195 billion of natural capital through illegal financial flows, mining, logging, wildlife trade, unregulated fishing and environmental degradation (AMCEN 2019). Apart from negatively affecting the achievement of SDG15, these resources could have contributed to meeting multiple other SDGs. These underlying economic and demographic processes are particularly multi-faceted as explained throughout this chapter (see Sects. 1.1, 1.2.8, 1.2.9 and 1.2.11 among others) and, as a result, are bound to pose substantial challenges in effectively protecting terrestrial ecosystems, biodiversity and ecosystem services (IPBES 2018).

There is a large array of different policy instruments and practical solutions that can curb biodiversity loss and ecosystem degradation, especially outside protected areas. These can range from community-based conservation and other participatory land/resource management/planning practices (Duguma et al. 2018; Galvin et al. 2018; Dyer et al. 2014) to the promotion of good production practices (UNDP 2011; Mijatović et al. 2018), and market-based conservation mechanisms (Brownlie et al., 2017). Such diverse instruments can increase the potential to link environmental conservation outside of protected areas, to policies and efforts related to other SDGs, environmental issues and broader economic, agricultural, poverty and trade policies (IPBES 2018). However, the proper implementation and effectiveness of these instruments is often compromised by institutional failures (e.g. corruption, elite capture), market failures (e.g. environmental externalities) and legal/policy failures (e.g. distorted subsidies, lack of secure property rights), essentially posing significant challenges in promoting effective measures to meet SDG15 (IPBES 2018).

Finally, as for many other SDGs, the lack of resources, data, knowledge and capacity in enforcing existing regulations, designing new innovative instruments and harnessing synergies between existing instruments can compromise the positive progress in meeting SDG15 throughout SSA (IPBES 2018; O'Connell et al. 2019; UNEP-WCMC 2016; Siddig 2019; *SDG Centre for Africa* 2019; UNECA 2018). However, on the positive side, there is an increasing commitment of ODA flows for biodiversity conservation in most sub-regions with the exception of southern Africa (UNECA 2018).

1.2.16 SDG 16: Peace, Justice and Strong Institutions

Many SSA countries have experienced protracted armed conflicts following colonial rule, often lasting for decades. Even though the number of conflict- and terrorism-related deaths in SSA is much lower compared to Northern Africa, it is still very high in some countries such as Somalia, South Sudan and the Central African Republic (*SDG Centre for Africa* 2019). However, the vast majority of SSA countries (more than 90% of countries) register fewer than the global average of 1.9 deaths per 100,000 inhabitants (*SDG Centre for Africa* 2019). Violent crimes have become more prevalent across most sub-regions, with the continent registering an increasing share of homicides, from 25% in 2000 to 33% in 2017 (UN 2019). Countries in southern Africa have some of the highest homicide rates globally, with Lesotho topping the list at 38 homicides per 100,000 people (World Bank 2017). At the same time, SSA countries register some of the highest perceived corruption rates from the private sector (Transparency International 2018). SSA has also by far the lowest rates of child registration globally, with fewer than half (46%) of all children under 5 years of age registered across the continent, and even fewer recorded in some countries such as Somalia, Malawi and Ethiopia (UN 2019).

Many scholars have suggested that failure to effectively meet SDG16 might have ramifications and for other SDGs as effective governance systems, peace and security are critical elements for sustainable development (Bolaji-Adio 2015). As discussed below, various interconnected challenges prevent progress for SDG16, including constraints in developing appropriate governance systems/frameworks, weak institutions, institutional failures, inequality and the lack of data and clear definitions of SDG16 indicators.

Establishing appropriate governance systems is not always straightforward in many parts of the continent, both at the local and national level. Often the promoted ideals of good governance according to SDG16 language have predisposition towards a certain ideological model of good governance, which does not necessarily reflect local realities, capabilities and possibly needs (Bolaji-Adio 2015). For example, in many parts of the continent seemingly participatory and representative systems of governance have sometimes heightened polarization and failed to prevent or contain communal violence (Bolaji-Adio 2015). At the same time, the prevailing and deeply ingrained racial/ethnic/religious/*cultural divisions* and differences pose serious challenges in establishing good governance systems, and often underlie conflicts and violence at different scales (Raleigh 2014).

Weak institutions and institutional failures are prevalent through the region, providing a fertile ground for bribery and corruption that characterizes many business activities such as import licensing and construction permits, among others (World Bank 2017; ACCA 2014).

The pervasive inequality is another major challenge for meeting SDG16, which has been thoroughly discussed throughout this chapter. The lack of access to justice, information, education and other fundamental freedoms in many SSA countries are only some of the major factors related to how low inequality can influence the attainment of SDG16 (Sects. 1.2.4, 1.2.5 and 1.2.10). Inequality also stems from low birth registration (one of the major targets of SDG16), which is often constrained by low capacity and specific local cultural characteristics, beliefs and taboos (UN 2019).

Finally, there are also major challenges related to the definition and measurement of SDG16. Apart from the lack of defined indicators and data for some targets (*SDG Centre for Africa* 2019), there are grave concerns that most of the underlying language and measures of SDG16 are framed in a way that will not eventually encourage governance improvements and the achievement of peace and security in the region (Bolaji-Adio 2015).

1.2.17 SDG 17: Partnerships for the Goals

SDG17 has three major elements namely finance, technology and systemic issues. In terms of finance, despite the gradual increase in ODA flows over the past decades, there has been a leveling and slight decline since 2016 (net ODA decreased by 2.7% between 2016 and 2018) (UN 2019). However, comparatively less aid was sent to SSA countries, with the net flows falling by 4% in the same period (UN 2019). In terms of technology, despite significant expansion in the number of broadband connections and internet users across the continent (Sect. 1.2.9), the overall access for most SSA countries fall well below global averages (Mahler et al. 2018; ITU, 2019). When it comes to systemic issues, in 2015 more than half of SSA countries had a national statistical plan that was fully funded and under implementation[11] (*SDG Centre for Africa* 2019). Despite this fraction being lower compared to global standards, it shows positive signs in terms of statistical development and capacity, especially considering that the number of SSA countries with adequate data for some SDG indicators increased significantly between 2000 and 2015 (UNECA, 2018). A recently developed composite measure of the quality of national statistical systems and related capacity across three domains (i.e. methodology, source data, periodicity) suggests that 21 SSA countries rank above the global average (*SDG Centre for Africa* 2019).

Some of the key elements for catalysing meaningful progress for SDG17 include (a) the development of strong and effective partnerships, (b) the ability to mobilize and use funds effectively (both domestic and international) and (c) the development and implementation of proper tracking mechanisms. However, as for many other SDGs, an interconnected set of challenges related to high corruption, low capacity and a lack of resources pose major barriers for achieving the aforementioned elements.

Regarding (a) it has been argued that effective partnerships for SDG17 should link multiple stakeholders from the national/local governments, international community, civil society and the private sector, among others (Haywood et al. 2018). This includes public–private partnership (PPP) that can play a multi-faceted role in supporting economic growth and delivering infrastructure solutions (Dykes and Jones 2016) (Sects. 1.2.6–1.2.9). However, weak institutions, conflicts and other institutional failures often put obstacles in the formation of trust and an enabling environment for meaningful collaboration between such stakeholders (Sect. 1.2.16). For example, public mistrust (El-Gohary et al. 2006), corruption (Babatunde et al. 2015) and other social conflicts pose major challenges that can derail otherwise useful development projects developed through PPPs (Dykes and Jones 2016).

Regarding (b), corruption and lack of capacity pose major challenges for the mobilization and effective use of domestic and international funds. For example, they can put obstacles in the development and effective implementation of national

[11]This includes 10 out of 13 East African countries, 6 out of 15 West African countries and 9 out of 12 Southern African countries (*SDG Centre for Africa* 2019).

budgets, with four-fifths of SSA countries always falling more than 5% below budget (World Bank 2017). Corruption and low capacity also contribute to the lack of confidence from potential investors and donors in the ability of national governments to effectively manage funds and target the intended uses/beneficiaries (Sects. 1.1, 1.2.8 and 1.2.9).

Regarding (c), the development and implementation of appropriate statistical systems are a backbone for tracking progress for the SDGs (SDSN and TReNDS 2019), especially by generating reliable data that are missing for many individual SDG targets as discussed throughout this chapter (e.g. Sects. 1.2.1, 1.2.6, 1.2.7, 1.2.11, 1.2.12, 1.2.14 and 1.2.16 among others) (*SDG Centre for Africa* 2019). However, developing, implementing and managing consistent and good quality statistical data would require substantial investments over time to both attract capable personnel and also build the infrastructure for data collection, storage, analysis and sharing. Currently the lack of resources, capacity constraints and lack of political will are some of the major challenges to develop such systems in the region (SDSN and TReNDS 2019).

1.3 Structure and Outline of the Edited Volumes

This two-volume edited series seeks to discuss in more depth some of the main sustainability challenges facing SSA countries in the context of the SDGs. In particular, each individual chapter tackles different sustainability challenges across the full spectrum of the SDGs. Furthermore, each chapter pays special attention in introducing the respective sustainability challenges, and the policy and practice implications in the context of the SDGs. Notably, in each chapter, there is a clear paragraph in the introduction outlining the sustainability challenge explored, as well as a dedicated sub-section about the main policy and practice implications and recommendations. In most cases, these sub-sections are linked directly to the SDGs.

The two volumes contain five main sections divided across geographical considerations. Thematically the chapters cover a breadth of different sustainability challenges ranging from agriculture and food security to energy security, deforestation, mining and urban sustainability, among others. Volume 1 contains 10 chapters, including this introductory chapter (Chap. 1 Vol. 1), four chapters focusing on the entire continent (Chap. 2–5 Vol. 1), three chapters focusing on Western Africa (Chap. 6–8 Vol. 1) and two chapters focusing on Central Africa (Chap. 9–10 Vol. 1). Volume 2 contains ten chapters, of which four focus on Eastern Africa (Chap. 1–4 Vol. 2), four focus on Southern Africa (Chap. 5–8 Vol. 2) and two are synthesis chapters (Chap. 9–10 Vol. 2). Figure 1.13 highlights the regional distribution of the chapters from Western, Central, Eastern and Southern Africa.

The 'Continental Perspectives' section contains four individual chapters. Chapter 2 (Vol. 1) discusses the main sustainability challenges associated with current bioenergy pathways, and some of the key priority areas for facilitating large-scale bioenergy transitions in SSA (Johnson et al., 2020). Chapter 3 (Vol. 1)

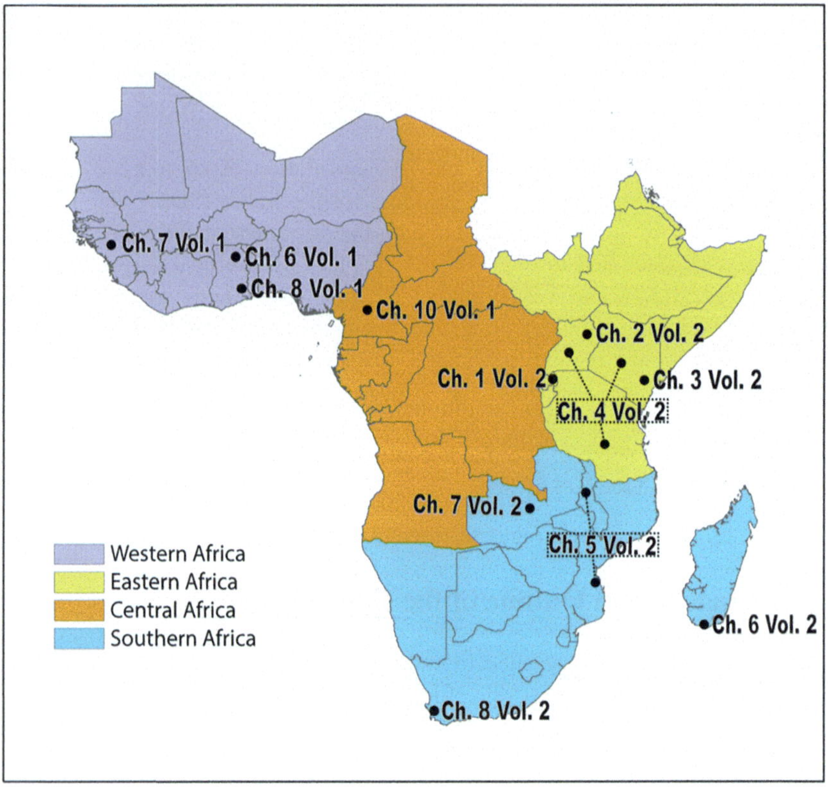

Fig. 1.13 Geographical scope of each chapter

outlines the history and drivers of industrial crop production across SSA and how it intersects with food security (Jarzebski et al., 2020b). Chapter 4 (Vol. 1) uses the case of Italian investors to highlight some of the main patterns, drivers and outcomes of large-scale land acquisitions in SSA, and how corporate social responsibility can mediate some of the negative outcomes (Antonelli et al., 2020). Chapter 5 (Vol. 1) outlines patterns in ODA and FDI flows in SSA as they relate to different SDGs, and the main challenges and opportunities for improving the effective utilization of these flows (Lopes et al., 2020).

The 'Western Africa' section contains three individual chapters. Chapter 6 (Vol. 1) outlines how local communities in two areas of Northern Ghana perceive their resilience to climatic hazards, especially droughts and floods (Boafo et al., 2020). Chapter 7 (Vol. 1) discusses how different rural livelihood options influence fuelwood procurement practices in rural Guinea, and how this intersects with mangrove degradation (Balde et al., 2020). Chapter 8 (Vol. 1) outlines how academic institutions in Ghana engage in research collaborations with the private sector, and what are the barriers and solutions for improving such partnerships (Mensah and Gordon, 2020).

The 'Central Africa' section contains two chapters. Chapter 9 (Vol. 1) highlights the natural and anthropogenic drivers affecting forest and savanna cover in Central Africa over the past millennia (Alleman and Fayolle, 2020). Chapter 10 (Vol. 1) focuses on the competition between traditional and imported agricultural innovations in forest–agriculture landscapes of central Cameroon, and how traditional ecological knowledge can enhance biodiversity conservation in such landscapes (Mala et al., 2020).

The 'Eastern Africa' section contains four chapters. Chapter 1 (Vol. 2) focuses on how nutritional interventions in rural areas of Rwanda can reduce under-nutrition, especially for children (Sekiyama et al., 2020). Chapter 2 (Vol. 2) explores how climatic hazards can affect the consumption and food security of rural and urban households in Uganda (Akampumuza et al., 2020). Chapter 3 (Vol. 2) tracks some of the main livelihood and food security issues in different areas of Kenya affected by climate change, and how traditional and local knowledge (TLK) can provide relevant options to adapt to climate change (Ndalilo et al., 2020). Chapter 4 (Vol. 2) uses a sustainability science lens to identify the main challenges for effectively promoting sanitation interventions in eastern Africa, and the lessons learnt from some successful community-led initiatives (Gabrielsson et al., 2020).

The 'Southern Africa' section contains four chapters. Chapter 5 (Vol. 2) discusses how sugarcane ethanol can provide a feasible fuel for transport and cooking in SSA, and identifies some of the key lessons learnt from production and use sites in Malawi and Mozambique, respectively (Nyambane et al., 2020). Chapter 6 (Vol. 2) highlights how these hedges of an alien invasive plant can create corridors for species movements in agricultural landscapes of Madagascar, and provide multiple ecosystem services to local communities (Andriamparany et al., 2020). Chapter 7 (Vol. 2) adopts a resilience lens to explain the history of collapse of a mining town in Zambia (and its outcomes), and provides insights for the development of appropriate management frameworks (Mfune et al., 2020). Chap. 8 (Vol. 2).

The 'Synthesis' section contains two individual chapters. Chapter 9 (Vol. 2) distils and critically discusses certain vital aspects for strengthening the science–policy interface in SSA, mainly using as an example the experience gained through involvement in the processes of the Intergovernmental Science–Policy Platform on Biodiversity and Ecosystem Services (IPBES) (von Maltitz, 2020). Chapter 10 (Vol. 2) synthesizes the main findings of the two edited volumes, especially focusing on how sustainability science can catalyse progress for meeting the SDGs in SSA (Gasparatos et al., 2020).

Acknowledgements We acknowledge the support of the Japan Science and Technology Agency (JST) for project FICESSA, the Japan International Cooperation Agency (JICA) for project USiA and the Japan Society for the Promotion of Science (JSPS) for the Grant-in-Aid for Young Scientists A (Project 17H05037) and the Core-to-core project 'Establishment and Advancement of a Global Meta-Network on Sustainability Science' (Project 23001). Merle Naidoo, Alice Karanja and Abubakari Ahmed received Monbukagakusho PhD scholarships offered by the Japanese Ministry of Education, Culture, Sports, Science and Technology (MEXT) through the Graduate Program in Sustainability Science – Global Leadership Initiative (GPSS-GLI). Jie Su assisted with the formatting of the chapter.

References

ACCA (2014) Leadership and governance in Africa. Association of Chartered Certified Accountants (ACCA), London

Ackah-Baidoo P (2016) Youth unemployment in resource-rich sub-Saharan Africa: a critical review. The Extractive Industries and Society 3(1):249–261. https://doi.org/10.1016/j.exis.2015.11.010

Adzawla W, Sawaneh M, Yusuf AM (2019) Greenhouse gasses emission and economic growth nexus of sub-Saharan Africa. Scientific African 3:e00065. https://doi.org/10.1016/j.sciaf.2019.e00065

AfDB (2015a) Agenda 2063, the Africa we want. African Development Bank (AfDB), Addis Ababa

AfDB (2015b) Water Supply & Sanitation in Africa: findings, lessons and good practices to improve delivery. African Development Bank (AfDB), Addis Ababa

AfDB (2016) Catalyzing growth and development through effective natural resources management. African Development Bank (AfDB), Addis Ababa

AfDB (2018) African economic outlook 2018. African Development Bank (AfDB), Addis Ababa

AfDB (2019a) African economic outlook 2019. African Development Bank (AfDB), Addis Ababa

AfDB (2019b) Industrialise Africa. African Development Bank (AfDB), Addis Ababa

Alagidede P, Alagidede AN (2016) The public health effects of water and sanitation in selected west African countries. Public Health 130:59–63. https://doi.org/10.1016/j.puhe.2015.07.037

Alobo Loison S (2015) Rural livelihood diversification in sub-Saharan Africa: a literature review. J Dev Stud 51(9):1125–1138. https://doi.org/10.1080/00220388.2015.1046445

Akampumuza P, Ggombe KM, Matsuda H (2020) Weather Shocks, Gender and Household Consumption: evidence from the urban households in Teso sub-region, Uganda. In: Gasparatos A, Naidoo M, Ahmed A, Karanja A, Fukushi K, Saito O, Takeuchi K (eds) Sustainability challenges in sub-Saharan Africa II: insights from eastern and southern Africa. Springer, Berlin

Aleman JC, Fayolle A (2020) Long-term vegetation change in central Africa: the need for an integrated management framework for forests and savannas. In: Gasparatos A, Ahmed A, Naidoo M, Karanja A, Fukushi K, Saito O, Takeuchi K (eds) Sustainability challenges in sub-Saharan Africa I: continental perspectives and insights from western and central Africa. Springer, Berlin

Andriamparany R, Lundberg J, Pyykönen M, Wurz S, Elmqvist T (2020) The effect of introduced Opuntia (Cactaceae) species on landscape connectivity and ecosystem service provision in southern Madagascar. In: Gasparatos A, Naidoo M, Ahmed A, Karanja A, Fukushi K, Saito O, Takeuchi K (eds) Sustainability Challenges in sub-Saharan Africa II: insights from eastern and southern Africa. Springer, Berlin

Antonelli M, Bracco S, Turvani ME, Vicario A (2020) Large-scale land acquisitions in sub-Saharan Africa and corporate social responsibility (CSR): insights from Italian investments. In: Gasparatos A, Ahmed A, Naidoo M, Karanja A, Fukushi K, Saito O, Takeuchi K (eds) Sustainability challenges in sub-Saharan Africa I: continental perspectives and insights from western and central Africa. Springer, Berlin

AMCEN (2019) Action for environmental sustainability and prosperity in Africa: embracing policy delivery and implementation through the African ministerial conference on the environment. African Ministerial Conference on the Environment (AMCEN), Addis Ababa

Amoah PA, Phillips DR (2018) Health literacy and health: rethinking the strategies for universal health coverage in Ghana. Public Health 159:40–49. https://doi.org/10.1016/j.puhe.2018.03.002

Amoyaw-Osei Y, Opoku Agyekum O, Pwamang JA, Mueller E, Fasko R, Schluep M (2011) Ghana e-waste country assessment. GreenAd, Accra

Arndt C, Mahrt K, Hussain MA, Tarp F (2018) A human rights-consistent approach to multidimensional welfare measurement applied to sub-Saharan Africa. World Dev 108:181–196. https://doi.org/10.1016/j.worlddev.2018.03.022

Arndt C, McKay A, Tarp F (eds) (2016) Growth and poverty in sub-Saharan Africa. Oxford University Press, Oxford

AUC, AfDB, UNDP (2016). MDGs to agenda 2063/SDGs transition report 2016. African Union Commission (AUC), African Development Bank (AfDB), & United Nations Development Program (UNDP) Addis Ababa

Babatunde S, Perera S, Zhou L, Udeaja C (2015) Barriers to public private partnership projects in countries: a case of Nigeria. Eng Constr Archit Manag 22(6):669–691. https://doi.org/10.1108/ECAM-12-2014-0159

Baldé CP, Forti V, Gray V, Kuehr R, Stegmann P (2017) The global E-waste monitor—2017. United Nations University (UNU), International Telecommunication Union (ITU) & International Solid Waste Association (ISWA), Bonn/Geneva/Vienna

Balde BS, Karanja A, Kobayashi H, Gasparatos G (2020) Linking rural livelihoods and fuelwood demand from mangroves and upland forests in the coastal region of Guinea. In: Gasparatos A, Ahmed A, Naidoo M, Karanja A, Fukushi K, Saito O, Takeuchi K (eds) Sustainability challenges in sub-Saharan Africa I: continental perspectives and insights from western and central Africa. Springer, Berlin

Belhabib D, Sumaila UR, Pauly D (2015) Feeding the poor: contribution of west African fisheries to employment and food security. Ocean & Coastal Management 111:72–81

Bluhm R, de Crombrugghe D, Szirmai A (2018) Poverty accounting. Eur Econ Rev 104:237–255. https://doi.org/10.1016/j.euroecorev.2018.03.003

Boafo, Y. A., Lam, R. D., Juju, D. B., Saito, O. (2019). The role of sustainability science in addressing Africa's sustainable development challenges and priorities: a bibliometric survey. In J. C.-C. C. Aguirre-Bastos & and S. Aricò (Eds.), Co-designing science in Africa: first steps in assessing the sustainability science approach on the ground (pp. 38–64). UNESCO, Paris

Boafo YA, Saito O, Jasaw GS, Yiran GAB, Lam RD, Mohan G, Kranjac-Berisavljevic G (2020) Perceived community resilience to floods and droughts induced by climate change in semi-arid Ghana. In: Gasparatos A, Ahmed A, Naidoo M, Karanja A, Fukushi K, Saito O, Takeuchi K (eds) Sustainability challenges in sub-saharan Africa I: continental perspectives and insights from western and central Africa. Springer, Berlin

Bolaji-Adio A (2015) The challenge of measuring SDG 16: what role for African regional frameworks? European Centre for Development Policy Management (ECDPM), Maastricht

Brinkel J, Krämer A, Krumkamp R, May J, Fobil J (2014) Mobile phone-based mHealth approaches for public health surveillance in sub-Saharan Africa: a systematic review. Int J Environ Res Public Health 11:11559–11582

Brownlie S, von Hase A, Botha M, Manuel J, Balmforth Z, Jenner N (2017) Biodiversity offsets in South Africa–challenges and potential solutions. Impact Assessment and Project Appraisal 35 (3):248–256

Budds J, McGranahan G (2003) Are the debates on water privatization missing the point? Experiences from Africa, Asia and Latin America. Environ Urban 15:87–114. https://doi.org/10.1177/095624780301500222

Bayliss K (2003) Utility privatisation in sub-Saharan Africa: a case study of water. J Mod Afr Stud 41(4):507–531

Canning D, Raja S, Yazbeck AS (eds) (2015) Africa's demographic transition: dividend or disaster? World Bank, Washington DC

Chaigneau T, Brown K, Coulthard S, Daw TM, Szaboova L (2019) Money, use and experience: identifying the mechanisms through which ecosystem services contribute to wellbeing in coastal Kenya and Mozambique. Ecosyst Serv 38:100957. https://doi.org/10.1016/j.ecoser.2019.100957

Challinor AJ, Watson J, Lobell DB, Howden S, Smith D, Chhetri N (2014) A meta-analysis of crop yield under climate change and adaptation. Nat Clim Chang 4:287–291. https://doi.org/10.1038/nclimate2153

CIFOR (2005) Contributing to Africa's development through forests strategy for engagement in sub-Saharan Africa. Center for International Forestry Research (CIFOR), Bogor

Cobbinah PB, Aboagye HN (2017) A Ghanaian twist to urban sprawl. Land Use Policy 61:231–241. https://doi.org/10.1016/j.landusepol.2016.10.047

Cobbinah PB, Darkwah RM (2016) African urbanism: the geography of urban greenery. Urban Forum 27(2):149–165. https://doi.org/10.1007/s12132-016-9274-z

Cobbinah PB, Erdiaw-Kwasie MO (2016). Urbanization in Ghana (pp. 82–104). doi:https://doi.org/10.4018/978-1-5225-0187-9.ch005

Cobbinah PB, Erdiaw-Kwasie MO, Amoateng P (2015) Africa's urbanisation: implications for sustainable development. Cities 47:62–72. https://doi.org/10.1016/j.cities.2015.03.013

da Silva PP, Cerqueira PA, Ogbe W (2018) Determinants of renewable energy growth in sub-Saharan Africa: evidence from panel ARDL. Energy 156:45–54. https://doi.org/10.1016/j.energy.2018.05.068

Dagnachew AG, Lucas PL, Hof AF, van Vuuren DP (2018) Trade-offs and synergies between universal electricity access and climate change mitigation in sub-Saharan Africa. Energy Policy 114:355–366. https://doi.org/10.1016/j.enpol.2017.12.023

Davies G (2018) Clean energy product markets in sub-Saharan Africa: complex market devices and power asymmetries. Energy Res Soc Sci 42:80–89. https://doi.org/10.1016/j.erss.2018.03.009

Davis B, Di Giuseppe S, Zezza A (2017) Are African households (not) leaving agriculture? Patterns of households' income sources in rural sub-Saharan Africa. Food Policy 67:153–174. https://doi.org/10.1016/j.foodpol.2016.09.018

de Vries SG, Visser BJ, Nagel IM, Goris MGA, Hartskeerl RA, Grobusch MP (2014) Leptospirosis in sub-Saharan Africa: a systematic review. Int J Infect Dis 28:47–64. https://doi.org/10.1016/j.ijid.2014.06.013

Deloitte (2015) Reducing food loss along African agricultural value chains. Deloitte, London. https://www2.deloitte.com/content/dam/Deloitte/za/Documents/consumer-business/ZA_FL1_ReducingFoodLossAlongAfricanAgriculturalValueChains.pdf

Demie G, Bekele M, Seyoum B (2016) Water accessibility impact on girl and women's participation in education and other development activities: the case of Wuchale and Jidda Woreda, Ethiopia. Environ Syst Res 5(1):11. https://doi.org/10.1186/s40068-016-0061-6

dos Santos R, Gupta J (2017) Pro-poor water and sanitation: operationalising inclusive discourses to benefit the poor. Curr Opin Environ Sustain 24:30–35

Dreibelbis R, Greene LE, Freeman MC, Saboori S, Chase RP, Rheingans R (2013) Water, sanitation, and primary school attendance: a multi-level assessment of determinants of household-reported absence in Kenya. Int J Educ Dev 33(5):457–465. https://doi.org/10.1016/j.ijedudev.2012.07.002

Duguma LA, Atela J, Ayana AN, Alemagi D, Mpanda M, Nyago M et al (2018) Community forestry frameworks in sub-Saharan Africa and the impact on sustainable development. Ecol Soc 23(4):21. https://doi.org/10.5751/ES-10514-230421

Dyer J, Stringer LC, Dougill AJ, Leventon J, Nshimbi M, Chama F et al (2014) Assessing participatory practices in community-based natural resource management: experiences in community engagement from southern Africa. J Environ Manag 137:137–145. https://doi.org/10.1016/j.jenvman.2013.11.057

Dykes BJ, Jones CD (2016) Public-private partnerships in Africa: challenges and opportunities for future management research. Afr J Manage 2(3):381–393. https://doi.org/10.1080/23322373.2016.1206806

Ekpa O, Palacios-Rojas N, Kruseman G, Fogliano V, Linnemann AR (2018) Sub-Saharan African maize-based foods: technological perspectives to increase the food and nutrition security impacts of maize breeding programmes. Glob Food Sec 17:48–56

El-Gohary N, Osman H, El-Diraby T (2006) Stakeholder management for public private partnerships. Int J Proj Manag 24(7):595–604. https://doi.org/10.1016/j.ijproman.2006.07.009
Engelbrecht F, Adegoke J, Bopape MJ, Naidoo M, Garland R, Thatcher M, Gatebe C (2015) Projections of rapidly rising surface temperatures over Africa under low mitigation. Environ Res Lett 10:085004. https://doi.org/10.1088/1748-9326/10/8/085004
Ericsson M, Löf O (2019) Mining's contribution to national economies between 1996 and 2016. Miner Econ 32:223–250. https://doi.org/10.1007/s13563-019-00191-6
ESMAP (2019) The energy progress report. Energy Sector Management Assistance Program (ESMAP), Washington DC
Fan S, Brzeska J (2016) Sustainable food security and nutrition: demystifying conventional beliefs. Glob Food Sec 11:11–16. https://doi.org/10.1016/j.gfs.2016.03.005
Fan S, Brzeska J, Keyzer M, Halsema A (2013) From subsistence to profit: transforming smallholder farms. International Food Policy Research Institute (IFPRI), Washington, DC
FAO (2010) Global Forest Resources Assessment 2010. Food and Agriculture Organization of the United Nations (FAO), Rome
FAO (2014) The value of African fisheries. Food and Agriculture Organization of the United Nations (FAO), Rome
FAO (2015) The global forest resources assessment. Food and Agriculture Organization of the United Nations (FAO), Rome
FAO (2016) 2015–2016 El Niño, early action and response for agriculture, food security and nutrition. Food and Agriculture Organization of the United Nations (FAO), Rome
FAO, UNECA (2018) Regional overview of food security and nutrition. Addressing the threat from climate variability and extremes for food security and nutrition. Food and Agriculture Organization of the United Nations (FAO) and United Nations Economic Cooperation for Africa (UNECA), Accra
Federal Ministry of Education (2016) Nigeria education indicators 2016. Federal Ministry of Education, Abuja
Frankema E (2014) Africa and the green revolution a global historical perspective. NJAS – Wageningen J Life Sci 70–71:17–24
Galvin KA, Beeton TA, Luizza MW (2018) African community-based conservation: a systematic review of social and ecological outcomes. Ecol Soc 23(3):39. https://doi.org/10.5751/ES-10217-230339
Gasparatos A, Ahmed A, Naidoo M, Karanja A, Fukushi K, Saito O, Takeuchi K (2020) Sustainability challenges in sub-Saharan Africa: trade-offs, opportunities and priority areas for sustainability science. In: Gasparatos A, Naidoo M, Ahmed A, Karanja A, Fukushi K, Saito O, Takeuchi K (eds) Sustainability challenges in sub-Saharan Africa II: insights from eastern and southern Africa. Springer, Berlin
Gabrielsson S, Huston A, Gaskin S (2020) Reframing the challenges and opportunities for improved sanitation services in eastern Africa through sustainability science. In: Gasparatos A, Naidoo M, Ahmed A, Karanja A, Fukushi K, Saito O, Takeuchi K (eds) Sustainability challenges in sub-Saharan Africa II: insights from eastern and southern Africa. Springer, Berlin
Gizaw MS, Gan TY (2017) Impact of climate change and El Niño episodes on droughts in sub-Saharan Africa. Clim Dyn 49(1–2):665–682
GIZ (2019) Access to water and sanitation in sub-Saharan Africa. Deutsche Gesellschaft für, Internationale Zusammenarbeit (GIZ), Berlin
Gossel SJ (2018) FDI, democracy and corruption in sub-Saharan Africa. J Policy Model 40:647–662. https://doi.org/10.1016/j.jpolmod.2018.04.001
Hakura D, Hussain M, Newiak M, Thakoor V, Yang F (2016) Inequality, gender gaps and economic growth: comparative evidence for sub-Saharan Africa. International Monetary Fund (IMF), Washington, D.C.

Hallaf MB (1999) Innovative ways for solid waste management in Dar Es Salaam: towards stakeholder partnerships. Habitat Int 23:351–361. https://doi.org/10.1016/S0197-3975(98)00057-5

Haywood LK, Funke N, Audouin M, Musvoto C, Nahman A (2018) The sustainable development goals in South Africa: investigating the need for multi-stakeholder partnerships. Dev South Afr 39:555–569. https://doi.org/10.1080/0376835X.2018.1461611

Heaton TB, Crookston B, Pierce H, Amoateng AY (2016) Social inequality and children's health in Africa: a cross sectional study. Int J Equity Health 15:92. https://doi.org/10.1186/s12939-016-0372-2

Herslund LB, Jalayer F, Jean-Baptiste N, Jørgensen G, Kabisch S, Kombe W, Lindley S, Karlsson Nyed P, Pauleit S, Printz A, Vedeld T (2016) A multi-dimensional assessment of urban vulnerability to climate change in sub-Saharan Africa. Nat Hazards 82(S2):149–172

Henderson JV, Storeygard A, Deichmann U (2017) Has climate change driven urbanization in Africa? J Dev Econ 124:60–82. https://doi.org/10.1016/j.jdeveco.2016.09.001

Hogarth JR, Haywood C, Whitley S (2015) Low-carbon development in sub-Saharan Africa: 20 cross-sector transitions. Overseas Development Institute (ODI), London

ICSU (2017) A guide to SDG interactions: from science to implementation. International Council for Science (ICSU), Paris

IEA (2019) Africa energy outlook 2019. International Energy Agency, Paris

ILO (2012) Youth employment interventions in Africa a mapping report of the employment and labour sub-cluster of the regional coordination mechanism (RCM) for Africa. International Labour Organisation (ILO), Geneva

ILO (2017) Global employment trends for youth 2017 paths to a better working future. International Labour Organisation (ILO), Geneva

ITU (2017) Global E-waste Monitor 2017. International Telecommunication Union (ITU), Geneva

ITU (2019) Measuring digital development: facts and figures 2019. International Telecommunication Union (ITU), Geneva

IPBES (2018) The IPBES regional assessment report on biodiversity and ecosystem services for Africa. In: Archer E, Dziba L, Mulongoy KJ, Maoela MA, Walters M (eds) . Secretariat of the Intergovernmental Science-Policy Platform on Biodiversity and Ecosystem Services, Bonn

IPCC (2014). Africa. In: Climate Change 2014: Impacts, Adaptation, and Vulnerability. Part B: Regional Aspects. Contribution of Working Group II to the Fifth Assessment Report of the Intergovernmental Panel on Climate Change. Cambridge University Press, Cambridge, pp. 1199–1265.

JRC (2017) Fossil CO2 and GHG emissions of all world countries. Joint Research Centre (JRC), Ispra

Jarzebski MP, Ahmed A, Boafo YA, Balde BS, Chinangwa L, Saito O, von Maltitz G, Gasparatos A (2020a) Food security impacts of industrial crop production in sub-Saharan Africa: a systematic review of the impact mechanisms. Food Security 12(1):105–135

Jarzebski MP, Ahmed A, Karanja A, Boafo YA, Balde BS, Chinangwa L, Degefa S, Dompreh EB, Saito O, Gasparatos A (2020b) Linking industrial crop production and food security in sub-Saharan Africa: local, national and continental perspectives. In: Gasparatos A, Ahmed A, Naidoo M, Karanja A, Fukushi K, Saito O, Takeuchi K (eds) Sustainability challenges in sub-Saharan challenges and potential: continental perspectives and insights from western and central Africa. Springer, Berlien

Johnson FX, Batidzirai B, Iiyama M, Ochieng CA, Olsson O, Mofor L, Gasparatos A (2020) Enabling sustainable bioenergy transitions in sub-Saharan Africa: strategic issues for achieving climate-compatible development. In: Gasparatos A, Ahmed A, Naidoo M, Karanja A, Fukushi K, Saito O, Takeuchi K (eds) Sustainability challenges in sub-Saharan Africa I: continental perspectives and insights from western and central Africa. Springer, Berlin

Kahsay GA, Hansen LG (2016) The effect of climate change and adaptation policy on agricultural production in eastern Africa. Ecol Econ 121:54–64. https://doi.org/10.1016/j.ecolecon.2015.11.016

Karanja A, Gasparatos A (2019) Adoption and impacts of clean bioenergy cookstoves in Kenya. Renew Sust Energ Rev 102:285–306. https://doi.org/10.1016/j.rser.2018.12.006
Kassim S, Ali M (2006) Solid waste collection by the private sector: households' perspective findings from a study in Dar Es Salaam city, Tanzania. Habitat Int 30:769–780
Kendon EJ, Stratton RA, Tucker S, Marsham JH, Berthou S, Rowell DP, Senior CA (2019) Enhanced future changes in wet and dry extremes over Africa at convection-permitting scale. Nat Commun 10:1794. https://doi.org/10.1038/s41467-019-09776-9
Kleemann J, Inkoom JN, Thiel M, Shankar S, Lautenbach S, Fürst C (2017) Peri-urban land use pattern and its relation to land use planning in Ghana, West Africa. Landsc Urban Plan 165:280–294. https://doi.org/10.1016/j.landurbplan.2017.02.004
Kodongo O, Ojah K (2016) Does infrastructure really explain economic growth in sub-Saharan Africa? Rev Dev Finance 6(2):105–125. https://doi.org/10.1016/j.rdf.2016.12.001
Kuivanen KS, Alvarez S, Michalscheck M, Adjei-Nsiah S, Descheemaeker K, Mellon-Bedi S, Groot JCJ (2016) Characterising the diversity of smallholder farming systems and their constraints and opportunities for innovation: a case study from the northern region, Ghana. NJAS – Wageningen J Life Sci 78:153–166. https://doi.org/10.1016/j.njas.2016.04.003
Lange S, Klasen S (2017) How the new international goal for child mortality is unfair to sub-Saharan Africa (again). World Dev 90:128–146. https://doi.org/10.1016/j.worlddev.2016.09.002
Leimbach M, Roming N, Schultes A, Schwerhoff G (2018) Long-term development perspectives of Sub-Saharan Africa under climate policies. Ecol Econ 144:148–159. https://doi.org/10.1016/j.ecolecon.2017.07.033
Lopes J, Somanje AN, Velez E, Dam Lam R, Saito O (2020) Determinants of foreign investment and international aid for meeting the Sustainable Development Goals in Africa: a visual cognitive review of the literatur. In: Gasparatos A, Ahmed A, Naidoo M, Karanja A, Fukushi K, Saito O, Takeuchi K (eds) Sustainability Challenges in sub-Saharan Africa I: continental perspectives and insights from western and central Africa. Springer, Berlin
Lucci P, Bhatkal T, Khan A (2018) Are we underestimating urban poverty? World Dev 103:297–310. https://doi.org/10.1016/j.worlddev.2017.10.022
Lundgren CJ, Thomas AH, York RC (2013) Boom, bust, or prosperity? Managing Sub-Saharan Africa's natural resource wealth. International Monetary Fund (IMF), Washington, DC
Mahler DG, Montes J, Newhouse D (2018) Internet access in sub-Saharan Africa. World Bank, Washington DC
Mainali B, Luukanen J, Silveira S, Kaivo-Oja J (2018) Evaluating synergies and trade-offs among sustainable development goals (SDGs): explorative analyses of development paths in South Asia and sub-Saharan Africa. Sustainability 10:1–25
Mala WA, Geldenhuys CJ, Prabhu R, Essouma FM (2020) Forest-agriculture in the Center-South region of Cameroon: how does traditional knowledge inform integrated management approaches? In: Gasparatos A, Ahmed A, Naidoo M, Karanja A, Fukushi K, Saito O, Takeuchi K (eds) Sustainability challenges in sub-Saharan Africa I: continental perspectives and insights from western and central Africa. Springer, Berlin
Marson M, Savin I (2015) Ensuring sustainable access to drinking water in sub Saharan Africa: conflict between financial and social objectives. World Dev 76:26–39
Masiya M (2017). Taxation, inequality and poverty in Sub-Saharan Africa: a panel data approach. Available at SSRN: doi: https://doi.org/10.2139/ssrn.3082747
Mboumboue E, Njomo D (2016) Potential contribution of renewables to the improvement of living conditions of poor rural households in developing countries: Cameroon's case study. Renew Sust Energ Rev 61:266–279. https://doi.org/10.1016/j.rser.2016.04.003
Mendes APF, Bertella MA, Teixeira RFAP (2014) Industrialization in sub-Saharan Africa and import substitution policy. Revista de Economia Política 34:120–138. https://doi.org/10.1590/S0101-31572014000100008
Mfune O, Kunda-Wamuwi CF, Chansa-Kabali T, Chisola MN, Manchisi J (2020) The legacy of mine closure in Kabwe, Zambia: what can resilience thinking offer to the mining sustainability

discourse? In: Gasparatos A, Naidoo M, Ahmed A, Karanja A, Fukushi K, Saito O, Takeuchi K (eds) Sustainability challenges in sub-Saharan Africa II: insights from eastern and southern Africa. Springer, Berlin

Mensah AM, Gordon C (2020) Strategic partnerships between universities and non-academic institutions for sustainability and innovation: Insights from the University of Ghana. In: Gasparatos A, Ahmed A, Naidoo M, Karanja A, Fukushi K, Saito O, Takeuchi K (eds) Sustainability challenges in sub-Saharan Africa I: continental perspectives and insights from western and central Africa. Sprineger, Berelin

Middleton NJ, Sternberg T (2013) Climate hazards in drylands: a review. Earth Sci Rev 126:48–57. doi:10.1016/j.earscirev.2013.07.008

Mijatović D, Sakalian M, Hodgkin T (2018) Mainstreaming biodiversity in production landscapes. United Nation Environment Programme, Nairobi

Mohale H, Sweet L, Graham K (2017) Maternity health care: the experiences of sub-Saharan African women in sub-Saharan Africa and Australia. Women Birth 30:298–307

Mohammed YS, Mustafa MW, Bashir N (2013) Status of renewable energy consumption and developmental challenges in sub-Sahara Africa. Renew Sust Energ Rev 27:453–463. https://doi.org/10.1016/j.rser.2013.06.044

Morris M, Fessehaie J (2014) The industrialisation challenge for Africa: towards a commodities based industrialisation path. J Afr Trade 1(1):25–36. https://doi.org/10.1016/j.joat.2014.10.001

Moussa N (2016) Trade and current account balances in sub-Saharan Africa: stylized facts and implications for poverty. United Nations Conference on Trade and Development (UNCTAD), Geneva

Mugagga F, Nabaasa BB (2016) The centrality of water resources to the realization of sustainable development goals (SDG). A review of potentials and constraints on the African continent. Int Soil Water Conserv Res 4:215–223

Müller C, Cramer W, Hare WL, Lotze-Campen H (2011) Climate change risks for African agriculture. Proc Natl Acad Sci U S A 108:4313–4315. https://doi.org/10.1073/pnas.1015078108

Mutunga C, Zulu E, De Souza RM (2012) Population dynamics and sustainable development in Africa. Population Action International (PAI) and African Institute for Development Policy (AFIDEP), Washington, DC and Nairobi

Nakakana UN, Akite JE, Mungu BE, Uche IV, Ahmed IM (2015) Mass vaccination campaigns in Nigeria and sub Saharan Africa: a review of three vaccines. IOSR J Dent Med Sci 14(9). https://doi.org/10.9790/0853-14956063

Ndalilo L, Wekesa C, Mbuvi MTE (2020) Indigenous and Local Knowledge practices and innovations for enhancing food security under climate change: examples from Mijikenda communities in coastal Kenya. In: Gasparatos A, Naidoo M, Ahmed A, Karanja A, Fukushi K, Saito O, Takeuchi K (eds) Sustainability challenges in sub-Saharan Africa II: insights from eastern and southern Africa. Springer, Berlin

NAP-GSP (2017) Regional briefing on National Adaptation Plans: Africa in focus. National Adaptation Plan Global Support Programme (NAP-GSP), New York

National Planning Commission of Ethiopia (2017) The 2017 voluntary National Reviews on SDGs of Ethiopia: government commitments, National Ownership and performance trends. National Planning Commission of Ethiopia, Addis Ababa

Ndikumana L, Pickbourn L (2017) The impact of foreign aid allocation on access to social services in sub-Saharan Africa: the case of water and sanitation. World Dev 90:104–114. https://doi.org/10.1016/j.worlddev.2016.09.001

Nhamo G, Nhamo S, Nhemachena C (2018) What gets measured gets done! Towards an afro-barometer for tracking progress in achieving sustainable development goal 5. Agenda 32 (1):60–75. https://doi.org/10.1080/10130950.2018.1433365

O'Connell MJ, Nasirwa O, Carter M, Farmer KH, Appleton M, Arinaitwe J et al (2019) Capacity building for conservation: problems and potential solutions for sub-Saharan Africa. Oryx 53:273–283. https://doi.org/10.1017/S0030605317000291

Omisore AG (2018) Attaining sustainable development goals in sub-Saharan Africa; the need to address environmental challenges. Environ Dev 25:138–145. https://doi.org/10.1016/j.envdev.2017.09.002

Onyutha C (2018a) African crop production trends are insufficient to guarantee food security by 2050 owing to persistent poverty. Food Secur 10(5):1203–1219. https://doi.org/10.1007/s12571-018-0839-7

Onyutha C (2018b) African food insecurity in a changing climate: the roles of science and policy. Food Energy Secur:e00160. https://doi.org/10.1002/fes3.160

Ortega DL, Tschirley D (2017) Demand for food safety in emerging and developing countries: a research agenda for Asia and sub-Saharan Africa. J Agribus Dev Emerg Econ 7:21–34

Osuteye E, Johnson C, Brown D (2017) The data gap: an analysis of data availability on disaster losses in sub-Saharan African cities. Int J Disaster Risk Reduct 26:24–33. https://doi.org/10.1016/j.ijdrr.2017.09.026

Owuor MA, Icely J, Newton A, Nyunja J, Otieno P, Tuda AO, Oduor N (2017) Mapping of ecosystem services flow in Mida Creek, Kenya. Ocean Coast Manag 140:11–21. https://doi.org/10.1016/j.ocecoaman.2017.02.013

Parfitt J, Barthel M, Macnaughton S (2010) Food waste within food supply chains: quantification and potential for change to 2050. Philos Trans R Soc B: Biol Sci 365(1554):3065–3081. https://doi.org/10.1098/rstb.2010.0126

Parnell S, Walawege R (2011) Sub-Saharan African urbanisation and global environmental change. Glob Environ Chang 21:S12–S20

Patel Z, Marrengane N, Smit W, Anderson P (2020) Knowledge co-production in Sub-Saharan African cities: building capacity for the Urban Age. In: Gasparatos A, Naidoo M, Ahmed A, Karanja A, Fukushi K, Saito O, Takeuchi K (eds) Sustainability challenges in sub-Saharan Africa II: insights from eastern and southern Africa. Springer, Berlin

Pauleit S, Coly A, Fohlmeister S, Gasparini P, Jorgensen G, Kabisch S, Kombe WJ, Lindley S, Simonis I, Yeshitela K (eds) (2015) Urban vulnerability and climate change in Africa: a multidisciplinary approach. Springer, Berlin

Pieterse E, Parnell S, Haysom G (2018) African dreams: locating urban infrastructure in the 2030 sustainable developmental agenda. Area Dev Policy 3:149–169. https://doi.org/10.1080/23792949.2018.1428111

Raleigh C (2014) Political hierarchies and landscapes of conflict across Africa. Polit Geogr 42:92–103. https://doi.org/10.1016/j.polgeo.2014.07.002

Rembold F, Kerdiles H, Lemoine G, Perez-Hoyos A (2016) Impact of El Niño on agriculture in southern Africa for the 2015/2016. Joint Research Centre (JRC), Ispra

Rosche D (2016) Agenda 2030 and the sustainable development goals: gender equality at last? An Oxfam perspective. Gend Dev 24(1):111–126. https://doi.org/10.1080/13552074.2016.1142196

Save the Children International (2013) Ending the hidden exclusion: learning and equity in education post-2015. Save the Children International, London

SBC (2011) Where are Weee in Africa?: findings from the Basel convention E-waste Africa programme. Secretariat of the Basel Convention (SBC), Geneva

Schwerhoff G, Sy M (2017) Financing renewable energy in Africa—key challenge of the sustainable development goals. Renew Sust Energ Rev 75:393–401. 10.1016/j.rser.2016.11.004

SDG Centre for Africa (2019) Africa 2030: sustainable development goals three year reality check. Sustainable Development Goal (SDG) Centre for Africa, Kigali

SDSN, TReNDS (2019) A roadmap for governments to achieve modern data systems for sustainable development. Sustainable Development Solution Network (SDSN) and Thematic Research Network on Data and Statistics (TRenDS), New York

Sekiyama M, Matsuda H, Mohan G, Yanagisawa A, Sudo N, Amitani Y, Caballero Y, Matsuoka T, Imanishi H, Sasaki T (2020) Tackling child malnutrition by strengthening the linkage between agricultural production, food security and nutrition in rural Rwanda. In: Gasparatos A,

Naidoo M, Ahmed A, Karanja A, Fukushi K, Saito O, Takeuchi K (eds) Sustainability challenges in sub-Saharan Africa II: insights from eastern and southern Africa. Springer, Berlin
Sembene D (2015) Poverty, growth, and inequality in sub-Saharan Africa: did the walk match the talk under the PRSP approach? International Monetary Fund (IMF), Washington DC
Serdeczny O, Adams S, Baarsch F, Coumou D, Robinson A, Hare W et al (2017) Climate change impacts in sub-Saharan Africa: from physical changes to their social repercussions. Reg Environ Chang 17:1585–1600. https://doi.org/10.1007/s10113-015-0910-2
Shiferaw B, Tesfaye K, Kassie M, Abate T, Prasanna BM, Menkir A (2014) Managing vulnerability to drought and enhancing livelihood resilience in sub-Saharan Africa: technological, institutional and policy options. Weather and Climate Extremes 3:67–79
Sheahan M, Barrett CB (2017) Review: food loss and waste in sub-Saharan Africa. Food Policy 70:1–12
Siddig AAH (2019) Why is biodiversity data-deficiency an ongoing conservation dilemma in Africa? J Nat Conserv 50:125719. https://doi.org/10.1016/j.jnc.2019.125719
Signé L (2018) The potential of manufacturing and industrialization in Africa: trends, opportunities, and strategies. Brookings Institution, New York
Ssozi J, Amlani S (2015) The effectiveness of health expenditure on the proximate and ultimate goals of healthcare in sub-Saharan Africa. World Dev 76:165–179
Stanzel P, Kling H, Bauer H (2018) Climate change impact on west African rivers under an ensemble of CORDEX climate projections. Clim Serv 11:36–48
Sørensen K, Van den Broucke S, Fullam J, Doyle G, Pelikan J, Slonska Z, Brand H (2012) Health literacy and public health: a systematic review and integration of definitions and models. BMC Public Health 12(1):80
Terry A, Ogg M (2017) Restructuring the Swazi sugar industry: the changing role and political significance of smallholders. J South Afr Stud 43(3):585–603. https://doi.org/10.1080/03057070.2016.1190520
Transparency International (2018) Corruption perception index 2018. Transparency International, Berlin
Tsinda A, Abbott P, Pedley S, Charles K, Adogo J, Okurut K, Chenoweth J (2013) Challenges to achieving sustainable sanitation in informal settlements of Kigali, Rwanda. Int J Environ Res Public Health 10:6939–6954. https://doi.org/10.3390/ijerph10126939
UN (2015) The millennium development goals report. United Nations, New York
UN (2018) The World's cities in 2018: data booklet. United Nations, New York
UN (2019) The sustainable development goals report 2019. United Nations, New York
UNCTAD (2018) Export diversification and employment in Africa. United Nations Conference on Trade and Development (UNCTAD), Geneva
UN-DESA (2018a) World urbanization prospects: the 2018 revision. United Nations Department of Economic and Social Affairs (UN-DESA), New York
UN-DESA (2018b) 2017 voluntary National Reviews: synthesis report. United Nations Department of Economic and Social Affairs (DESA), New York
UNDP (2011) Protecting biodiversity by working with agribusiness supply chains. United Nations Development Programme (UNDP), New York
UNDP (2016) Africa human development report 2016: accelerating gender equality and Women's empowerment in Africa. United Nations Development Programme (UNDP), New York
UNDP (2017) Income inequality trends in sub-Saharan Africa: divergence, determinants and consequences. United Nations Development Programme (UNDP), New York
UNDP (2018a) Africa sustainable development report: towards a transformed and resilient continent. United Nations Development Programme (UNDP), New York
UNDP (2018b) Climate Change Adaptation in Africa: UNDP Synthesis of Experiences and Recommendations United Nations Development Programme (UNDP), Nairobi.
UNDP, World Bank (2016) Transitioning from the MDGs to the SDGs. United Nations Development Programme (UNDP) and World Bank, New York and Washington DC

UNDRR (2018) Building disaster resilience to natural hazards in sub-Saharan African regions, countries and communities. UN Office for Disaster Risk Reduction (UNDRR), Geneva

UNECA (2015a) Africa regional report on the sustainable development goals. United Nations Economic Commission for Africa (UNECA), Addis Ababa

UNECA (2015b) Sustainable development goals for the eastern Africa sub-region. United Nations economic Commission for Africa (UNECA), Addis Ababa

UNECA (2016a) The demographic profile of African countries. United Nations Economic Commission for Africa (UNECA), Addis Ababa

UNECA (2016b) Africa's blue economy: a policy handbook. United Nations Economic Commission for Africa (UNECA), Addis Ababa

UNECA (2017) 2017 Africa sustainable development report: tracking Progress on agenda 2063 and the sustainable development goals. United Nations Economic Commission for Africa (UNECA), Addis Ababa

UNECA (2018) 2018 Africa sustainable development report: towards a transformed and resilient continent. United Nations Economic Commission for Africa (UNECA), Addis Ababa

UNEP-WCMC (2016) The state of biodiversity in Africa: a mid-term review of progress towards the Aichi biodiversity. Targets. UN Environment Programme World Conservation Monitoring Centre (UNEP-WCMC), Cambridge

UNESCO (2015) Rethinking education: towards a global common good? United Nations Educational, Scientific and Cultural Organization (UNESCO), Paris

UNFCCC (2018) National Adaptation Plans 2018: Progress in the process to formulate and implement National Adaptation Plans. United Nations Framework Convention on Climate Change (UNFCCC), Bonn

UN-HABITAT (2016) World cities report 2016: urbanization and development: emerging futures. UN-Habitat, Nairobi

Vitale J (2018) Economic importance of cotton in Burkina Faso. Food and Agriculture Organization, Rome

von Maltitz G (2020) Harnessing science-policy interface processes to tackle sustainability challenges in sub-Saharan Africa. In: Gasparatos A, Naidoo M, Ahmed A, Karanja A, Fukushi K, Saito O, Takeuchi K (eds) Sustainability challenges in sub-Saharan Africa II: insights from eastern and southern Africa. Springer, Berlin

Webber H, Gaiser T, Ewert F (2014) What role can crop models play in supporting climate change adaptation decisions to enhance food security in sub-Saharan Africa? Agric Syst 127:161–177

WEF (2018) Readiness for the future of production report 2018. World Economic Forum (WEF), Cologny

WHO (2017) World malaria report 2017. World Health Organization (WHO), Geneva

WHO (2018a) The state of health in the WHO African region: an analysis of the status of health, health services and health systems in the context of the sustainable development goals. World Health Organisation Regional Office for Africa, Brazzaville

WHO (2018b) Guidelines on sanitation and health. World Health Organization (WHO), Geneva

WHO, UNICEF (2018) Water and sanitation coverage. Drinking water, sanitation and hygiene database. World Health Organization (WHO) and United Nations Children Fund (UNICEF), Geneva. Available at: https://data.humdata.org/dataset/water-sanitation-drinking-water-sanitation-and-hygiene-database

WIPO (2019) Global innovation index 2019 – 12th edition. World Intellectual Property Organisation (WIPO), Geneva

World Bank (2014) Improving gender equality in Africa. World Bank, Washington DC

World Bank (2016) Poverty in a rising Africa. World Bank, Washington DC

World Bank (2017) World development report 2017: governance and the law. World Bank, Washington DC

World Bank (2018a) Poverty and equity data portal. World Bank, Washington DC. Available at http://povertydata.worldbank.org/poverty/home/

World Bank (2018b) World development report 2018 (WDR 2018). World Bank, Washington DC

World Bank (2019a) Accelerating poverty reduction in Africa. World Bank, Washington DC
World Bank (2019b) Lifelines: The Resilient Infrastructure Opportunity. Sustainable Infrastructure
World Bank (2015) Driving innovation in Africa. World Bank, Washington DC
World Bank and China Development Bank (2017) Leap frogging: the key to Africa's development? World Bank, Washington, DC
Xie H, You L, Wielgosz B, Ringler R (2014) Estimating the potential for expanding smallholder irrigation in sub-Saharan Africa. Agric Water Manag 131:183–193. https://doi.org/10.1016/j.agwat.2013.08.011
Wu X, Lu Y, Zhou S, Chen L, Xu B (2016) Impact of climate change on human infectious diseases: empirical evidence and human adaptation. Environ Int 86:14–23
Zahonogo P (2016) Trade and economic growth in developing countries: evidence from sub-Saharan Africa. Journal of African Trade 3(1–2):41–56. https://doi.org/10.1016/j.joat.2017.02.001

Chapter 2
Enabling Sustainable Bioenergy Transitions in Sub-Saharan Africa: Strategic Issues for Achieving Climate-Compatible Developments

Francis X. Johnson, Bothwell Batidzirai, Miyuki Iiyama, Caroline A. Ochieng, Olle Olsson, Linus Mofor, and Alexandros Gasparatos

2.1 Introduction

Bioenergy accounts for 10% of primary energy supply, which is more than the combined contribution from all other renewable energy sources and nuclear power (IEA 2018). However, most of the biomass used for energy in developing countries is in the form of firewood, charcoal, agricultural residues and animal dung for cooking and heating purposes, which is not so different from the way biomass has been used for thousands of years (Mattick et al. 2010). In fact, almost 3 billion persons worldwide use biomass in this manner, including the overwhelming

F. X. Johnson (✉) · C. A. Ochieng · O. Olsson
Stockholm Environment Institute (SEI), Stockholm, Sweden
e-mail: francis.johnson@sei.org; caroline.ochieng@sei.org; olle.olsson@sei.org

B. Batidzirai
University of Cape Town, Cape Town, South Africa
e-mail: bbatidzirai@gmail.com

M. Iiyama
World Agroforestry Centre (ICRAF), Nairobi, Kenya

Japan International Research Center for Agricultural Sciences (JIRCAS), Tsukuba, Japan
e-mail: miiyama@affrc.go.jp

L. Mofor
United Nations Economic Commission for Africa (UNECA), Addis Ababa, Ethiopia
e-mail: mofor@un.org

A. Gasparatos
Institute for Future Initiatives (IFI), The University of Tokyo, Tokyo, Japan
e-mail: gasparatos@ifi.u-tokyo.ac.jp

A. Gasparatos et al. (eds.), *Sustainability Challenges in Sub-Saharan Africa I*,
Science for Sustainable Societies, https://doi.org/10.1007/978-981-15-4458-3_2

majority of households in the least developed countries (LDCs) of sub-Saharan Africa (SSA) (ESMAP 2018).

The supply and share of different types of bioenergy fuels and products vary tremendously by region (Table 2.1) and is closely linked to levels of urbanisation, economic development/growth, standard of living, income, availability of affordable alternative energy sources and national policies (IEA 2018; Bildirici and Ersin 2015). For example, in OECD countries and some emerging economies, there is a diverse mix of modern biomass-based fuels and applications available on the market (e.g. liquid biofuels, solid biomass, biogas) (Table 2.1). This is largely due to policy-driven changes during the past three decades arising from concerns over energy security and climate change (Chum et al. 2011). By contrast, in SSA, more than 90% of the biomass used for energy is in traditional forms (e.g. charcoal, fuelwood) and is mainly used for household cooking and heating, with urban charcoal use driving most of the growth in bioenergy demand during the past two decades (IEA 2018). Charcoal is preferred to fuelwood due to its higher energy density, cleaner burning characteristics and easier storage (Smeets et al. 2012). Electricity, kerosene and liquefied petroleum gas (LPG) offer cleaner alternatives, but generally require substantial subsidies to become affordable for the poor in SSA (Takama et al. 2012; Mudombi et al. 2018b) (Chap. 5 Vol. 2).

The lack of extensive modern bioenergy adoption in SSA lies in stark contrast to its potential, including for liquid biofuels and for other modern gaseous and solid fuels. For example, agricultural harvesting and processing residues in SSA have

Table 2.1 Trends in global bioenergy use (in PJ)

	Biomass type	Africa	Non-OECD Asia	Non OCED Americas	Non-OECD Europe/Eurasia/ Middle East	OECD	World
1990	Municipal waste	<1	3	<1	<1	562	565
	Solid biomass	8206	19,510	3053	731	5658	37,159
	Biogas	<1	<1	<1	<1	64	64
	Liquid biofuels	<1	<1	224	<1	<1	224
2000	Municipal waste	<1	27	<1	131	1046	1204
	Solid biomass	10,466	20,730	2889	54	6284	40,906
	Biogas	<1	49	<1	<1	236	285
	Liquid biofuels	<1	3	269	<1	148	420
2010	Municipal waste	3	241	<1	172	1413	1828
	Solid biomass	13,806	19,491	4173	699	7522	45,692
	Biogas	<1	314	1.4	1.6	520	837
	Liquid biofuels	<1	111	642	8	1679	2220
2015	Municipal waste	3	324	<1	211	1632	2169
	Solid biomass	15,800	19,887	4312	781	7904	48,685
	Biogas	<1	374	9	6565	899	1289
	Liquid biofuels	2	251	933	21	2093	3300

Source: IEA (2018)

been estimated to possibly provide energy equivalent to 4.2 EJ, i.e. more than one-fourth of current solid biomass use in SSA (IRENA 2015). Five SSA countries (Ghana, Mozambique, Nigeria, South Africa, Uganda) have a combined bioenergy potential equal to 117% and 190% of their projected energy needs in 2050 for transport and heat/power, respectively (IRENA 2017).[1] Furthermore, there is substantial potential in SSA to harness biogas both in commercial/industrial settings and at the household level. It is estimated that 18.5 million African households have sufficient dung and water for biogas production (IRENA 2015).

Considering the high bioenergy potential in SSA and the multiple significant sustainability impacts from traditional biomass, the transition to modern bioenergy is important in achieving the long-term development goals embodied in the sustainable development goals (SDGs) and the African Union Vision 2063 (AUC 2015) (Chap. 1 Vol. 1). It is possible that the speed and nature of bioenergy transitions are connected to many other sustainability issues, such that improved understanding of these connections can inform the design of appropriate programmes and policies.

However, catalysing and achieving sustainable bioenergy transitions pose major challenges for most SSA countries. Indeed there have not been major developments in modern bioenergy pathways outside of selected cases and countries such as ethanol in Malawi or bagasse cogeneration in Mauritius and South Africa (Johnson and Matsika 2006; Batidzirai and Johnson 2012; Gasparatos et al. 2015) (see Chap. 3 Vol. 1; Chap. 5 Vol. 2). A host of reasons such as low levels of technology adoption, immature markets and widespread poverty pose major barriers for the transition to modern bioenergy. However, there is a need to make some basic distinctions before considering how to catalyse such a transition.

First, the distinction between "traditional" and "modern" bioenergy is sometimes mistakenly assumed to be a technical issue. In fact, this distinction primarily relates to the improved energy services obtained and new applications developed (Bazilian et al. 2010; Chum et al. 2011; Smeets et al. 2012). Traditional biomass only provides heat or light that is difficult to regulate, often in open fires or simple household stoves that result in high levels of incomplete combustion and the release of indoor air pollutants and greenhouse gases (GHGs). Conversely, modern bioenergy offers higher quality energy services across different carriers (i.e. solid, liquid, gas, electricity) that can be better matched to end-user needs (Faaij 2006; Macqueen and Korhaliller 2011). Nevertheless, the efficiency improvements of modern bioenergy become quite significant when considering the entire supply chain. This is because the same amount of raw materials can provide much higher amounts of useful energy, thus reducing environmental and economic costs. Some fuels and applications, such as improved fuelwood/charcoal stoves and small-scale biogas systems, can be seen as an intermediate stage between traditional and modern energy, in that

[1]It is worth noting that apart from contributing to energy security, well-developed biofuel crop systems such as those based on sugarcane can offer poverty reduction benefits and create long-term livelihood opportunities within rural landscapes that otherwise might not have other major economic opportunities (Mudombi et al. 2018a; von Maltitz et al. 2019) (Chap. 3 Vol. 1; Chap. 5 Vol. 2).

there is an improvement in the quality of energy services, although the efficiency or flexibility might be lower (Barnes and Floor 1996; Foell et al. 2011; Dresen et al. 2014).

Second, the shift from traditional to modern bioenergy can be direct such as, for example, the shift from fuelwood to gaseous or liquid cooking fuels. However, such a shift is more likely to occur indirectly, alongside broader industrialisation pathways and economic development processes, as energy production shifts away from the informal sector and energy use moves outside the household itself (Leach 1992; Silveira and Johnson 2016). As different bioenergy products become more standardised or "commoditised", they also become more flexible in terms of transport and trade (Junginger et al. 2011; Olsson and Johnson 2014). As shown in Table 2.2, bioenergy applications extend across all carriers (i.e. solid, liquid, gas, heat, electricity) and all end-use sectors (i.e. transport, residential, commercial, industry).

Third, bioenergy use transitions in SSA occur at the same time as vulnerability to climate change increases (Olsson and Johnson 2014) (Chap. 1 Vol. 1). At the same time, the continued reliance on bioenergy can put further pressure on land-intensive livelihoods in rural areas, especially those associated with high dependence on traditional biomass and subsistence farming (Chap. 7 Vol. 1; Chap. 2–3 Vol. 2). In this sense, the twin challenges of improving energy access and adapting to climate change can be approached simultaneously if proper support is mobilised for the necessary investment, infrastructure and management systems (Suckall et al. 2015).

It is projected that the demand for traditional biomass will increase in absolute terms in SSA due to the increasing population, accelerated urbanisation and the lack

Table 2.2 Bioenergy options by carrier, end-use sector and market orientation

	Primarily domestic options	Export options	Options not yet widely commercialised
Liquid biofuels (for transport or other uses)	• Unrefined oils • Ethanol • Methanol	• Refined oils • Ethanol	• Pyrolysis oils • Biobutanol • Biogasoline
Solid biofuels (for heat and power)	• Wood pellets • Wood chips • Briquettes	• Wood pellets • Chips	• Torrefied biomass
Solid biofuels (for domestic and institutional uses)	• Charcoal • Agricultural residues • Fuelwood	• Charcoal • Wood pellets • Wood chips	• Biochar
Gaseous biofuels	• Biogas • Synthesis gas	• Biogas (if cleaned and if exportable to gas grid	• Pyrolysis gas
Feedstocks, carriers and co-products	• Agricultural residues • Municipal solid waste • Black liquor • Waste oils	• Oilseeds or oils for refining into biodiesel • High-quality or processed wastes	• Lignin (by-product of lignocellulosic ethanol) • Carbon-rich chains • Bio-hydrogen

Source: adapted from (FAO/UNEP 2011; FAO 2004)

of mass adoption of alternative energy options (Smeets et al. 2012) (Chap. 1 Vol. 1). Under most scenarios, traditional biomass will still cater for most energy needs in SSA in 2050 (IEA 2014). Given that this dependency on traditional biomass will also remain a major sustainability challenge into the near-to-medium future, improvements in household energy services would be a key part of the transition towards sustainable bioenergy supply. However, through appropriate strategies and policy interventions this trajectory could be altered to accelerate the transition to modern energy systems and the transformation of rural economies (IRENA 2015).

This chapter aims to outline some of the critical aspects that can enable sustainable bioenergy transitions in SSA. In particular, it highlights four interlinked strategic aims related to the modernisation of biomass for energy (and other uses) that are relevant for climate-compatible bioenergy development. These interlinked strategic issues or aims are to (a) identify and strengthen positive linkages across the different SDGs associated within modern bioenergy transitions in SSA (Sect. 2.2); (b) choose the most appropriate markets and production modes for modern bioenergy (Sect. 2.3); (c) promote integrated landscape approaches for biomass and bioenergy feedstock production to improve resource efficiency and climate resilience, and at the same time reduce land competition (Sect. 2.4); (d) foster synergies between climate mitigation and adaptation (Sect. 2.5). Section 2.6 discusses these priorities in connection to the policy implications and governance requirements for sustainable bioenergy transitions across the continent.

2.2 Identify and Strengthen Positive SDG Inter-Linkages in Bioenergy Transitions

Traditional bioenergy production and use have been linked to multiple sustainability impacts such as forest degradation, GHG emissions, health, poverty, food security, and gender inequality, to mention just a few (Karanja and Gasparatos 2019; Iiyama et al. 2014; van de Ven et al. 2019) (Chap. 7 Vol. 1; Chap. 5 Vol. 2). As a result, the heavy reliance of households in SSA on traditional biomass fuels can complicate the achievement of many different sustainable development goals (SDGs) (Nerini et al. 2018; McCollum et al. 2018).

For example, poor and vulnerable populations, particularly women and girls in rural areas, spend considerable amount of time gathering fuelwood (Karanja and Gasparatos 2019). Households also spend a significant portion of their income to purchase traditional biomass fuels such as charcoal (Takama et al. 2012; Masera et al. 2015). This has been linked with different direct and opportunity costs, as well as energy poverty (Karanja and Gasparatos 2019). Furthermore, biomass dependence can affect cooking habits and dietary choices, which may further directly affect household nutrition and food security (Sola et al. 2016). Indoor air pollution from biomass fuel use is currently one of the leading risks for human health in SSA, and has been linked to high mortality across the continent (Lim et al. 2012; Lakshmi et al. 2010; Lam et al. 2012). The above mechanisms suggest some important linkages between SDGs in the context of traditional biomass fuel use, including especially SDG1, 2, 3, 5 and 7.

In urban areas of SSA, charcoal remains the fuel of choice although of course it is sourced from rural areas (Sect. 2.1). Even though charcoal supply is smaller compared to firewood, charcoal production in SSA may be unsustainable or "non-renewable" and contribute to net GHG emissions, especially in eastern Africa (Bailis et al. 2015). About 20% of harvested woodfuel in SSA (which often involves cutting live hardwood trees) is converted to charcoal (IRENA 2015). This has led to deforestation and land degradation around densely populated peri-urban and urban areas (Ndegwa et al. 2016; Kiruki et al. 2017; Jagger and Kittner 2017). Inefficiencies across the charcoal supply chains and the tendency to use whole trees for charcoal production result in much higher wood consumption compared to direct fuelwood use (World Bank 2009; Smeets et al. 2012; Chidumayo and Gumbo 2013). Furthermore, fuel combustion in inefficient stoves and charcoal kilns contributes significantly to outdoor air pollution and GHG emissions (Shindell et al. 2012; Anenberg et al. 2013; Bailis et al. 2003).[2] The above mechanisms suggest important linkages between multiple SDGs in the context of traditional biomass use, including SDG 7, 12, 13 and 15.

However, it is difficult to halt charcoal production due to the lack of alternative livelihoods across the value/supply chain and/or the affordability of other fuels by users (World Bank 2009; Zulu 2010; Smith et al. 2015; Taylor et al. 2019). So far, the attempts to impose sustainable feedstock sourcing and to formalise and control the charcoal market have had little success in SSA due to the combined effects of poor law enforcement, prevailing land ownership/tenure rules, poor socioeconomic conditions and the high reliance of rural households on charcoal earnings (IEA 2014; Smith et al. 2015; Wanjiru et al. 2016; Taylor et al. 2019). In fact, charcoal contributes significantly to livelihoods in many areas across SSA (Jones et al. 2016; Zulu and Richardson 2013).[3] The above mechanisms suggest some important linkages between multiple SDGs in the context of traditional biomass use such as SDG 1, 8, 9, 12 and 15.

Considering the aforementioned linkages and impacts, transitioning to modern bioenergy production and sustained use can create multiple trade-offs between SDGs through a multitude of different pathways and mechanisms (Table 2.3). Often these pathways relate to multiple SDGs. For example, the transition to modern bioenergy for cooking can have positive health effects (SDG 3) but also contribute to energy access and climate change mitigation and adaptation, goals (related to SDG7 and

[2] It is worth noting that the rate of increase in charcoal use is normally much higher than the rate of urbanisation itself (e.g. due to demographic factors such as the smaller size of urban households compared to rural households) (Hosier et al. 1993). Thus, rapid urbanization and/or commercialisation can result in significantly higher forest degradation from charcoal demand (Santos et al. 2017).

[3] Charcoal production in some dryland areas can also provide a socio-economic adaptation approach when agricultural livelihood opportunities are impacted by climate change (Ochieng et al. 2014).

Table 2.3 Impact pathways of modern bioenergy development and use in the household sector

Impact category	Key relevant SDGs	Impact pathway
Improved energy services	1, 2, 7	– Higher quality of energy services, and overall higher levels of human Well-being, poverty reduction and food security.
Health	3, 11, 12	– Reduced risk of disease from indoor air pollution due to biomass use for cooking, and kerosene for cooking and lighting
Rural development	1, 5, 15	– Employment and income generation (both gains and losses) from the stove sector and bioenergy feedstock production, transport, processing and sales – Income diversification for rural households – Increased time availability (especially for women) to engage in income-generating activities and development initiatives (e.g. self-help groups)
Education	4, 5, 8	– Increased time availability (especially for women and girls) to engage in education and other gainful ventures – Improved conditions (e.g. lighting) to allow better studying
Ecosystem protection	2, 15	– Reduced fuelwood requirement reduces rates of deforestation and forest degradation, with positive outcomes for the provision of ecosystem services and biodiversity conservation
Climate change mitigation	7, 13, 15	– Reduced loss of carbon stocks from deforestation and forest degradation for fuelwood – Reduced emission of black carbon and other GHGs from charcoal production and biomass combustion.
Climate change adaptation	1, 2, 13, 15	– Reduced deforestation and forest degradation improves the availability of natural resources used directly or indirectly to help cope with climatic events (e.g. forest products). – Provision of energy for adaptation measures such as water pumping (e.g. for drinking, irrigation), food processing and storage and medicine storage.

Note: Modern bioenergy fuels can be either used directly in the household sector as alternative fuels for cooking (e.g. ethanol or biogas) or converted to electricity or heat for local use

13, respectively) (Cameron et al. 2016). Conversely, improved access to modern energy services (related to SDG 7) can simultaneously promote climate adaptation and mitigation, and broader development goals related to multiple SDGs such as SDG1 and SDG13 (Suckall et al. 2015) (see also Sect. 2.5).

However, even though most of the outcomes of modern bioenergy transitions are expected to be positive for attaining the SDGs, there could also be some negative or uncertain outcomes. For example, modern bioenergy transitions can cause, in some cases, the loss of employment and income along charcoal value chains, which implies negative trade-offs with SDG8 (Karanja and Gasparatos 2019) (Table 2.3). Other, uncertain outcomes could, for example, relate to climate change mitigation and be linked to the significant emissions associated with land use change (e.g. Chap. 5 Vol. 2) and the difficulty in estimating the emissions of traditional biomass in SSA, due to its informal nature, lifecycle accounting complications and the common practice of using multiple fuels (i.e. fuel stacking) (Masera et al. 2000;

Lee et al. 2013; Cerutti et al. 2015) (Sects. 2.3 and 2.4). This is compounded by the high prevalence of subsistence agriculture often using slash-and-burn methods in rural SSA, which is characterised by low productivity and high GHG emissions (Palm et al. 2013; Johnson and Jumbe 2013). This suggests some important uncertainties at the interface of SDG2, 13 and 15.

Some trade-offs might also emerge due to institutional and/or cultural factors. For example, improving energy access (or similarly reducing energy poverty) is a key enabler of economic development (Sovacool 2012), and at the same time a major possible outcome of clean bioenergy transitions. With increasing income or wealth, households and businesses can switch to higher quality fuels, following the so-called energy ladder, which leads to better energy services (Leach 1992). As the low access to modern energy services in SSA leads to high reliance on the lowest rungs of the energy ladder for cooking and heating, it has been suggested that the thrust of the efforts seeking to catalyse bioenergy transitions should be on accelerating these shifts up the ladder (Bazilian et al. 2010; IEA 2014; Johnson and Diaz-Chavez 2018). However, strong policy incentives for moving up the energy ladder are not always appropriate or desirable, as such shifts also need to consider the prevailing cultural, practical and socio-economic factors (e.g. reliance on multiple fuels and stoves for flexibility at the household and community levels in meeting energy needs) (Masera et al. 2000; Takama et al. 2012).

Identifying such sustainability synergies and trade-offs would be necessary for informing different bioenergy transition pathways. This knowledge would undoubtedly provide a much-needed evidence base that can inform bioenergy transitions in SSA, not the least by allowing them to reach their full potential by maximising multiple positive sustainability outcomes. Integrated research approaches based on sustainability science or the ecosystem services approach have been shown to hold great potential in SSA contexts for synthesising current evidence, assessing the multiple impacts of bioenergy systems and identifying pathways to maximise the positive synergies (Gasparatos et al. 2011; von Maltitz et al. 2016; Baumber 2017; Johnson et al. 2018; Gasparatos et al. 2013, 2018).

2.3 Choose the Most Appropriate Scale, Markets and Production Modes for Modern Bioenergy Options

Until the past decade or so, bioenergy was considered to be primarily a local resource, with international trade being rather limited (Sect. 2.1). Some of the few exceptions were major biofuel programmes and markets in Brazil and United States, and solid biomass for heat and power in a few OECD countries. However, this perception has shifted considerably in the past decade, as the rapidly growing bioenergy demand has also boosted the international trade and commoditisation of liquid biofuels and solid bioenergy (e.g. wood pellets) (Junginger et al. 2011; Faaij et al. 2014; Olsson and Johnson 2014).

Whereas OECD countries invested in modern bioenergy long after phasing out traditional biomass fuels, most developing countries and emerging economies only started investing in modern bioenergy recently and alongside the traditional uses that dominate their energy systems. This has opened up different development pathways for modern bioenergy transitions (Johnson and Jumbe 2013; Johnson and Silveira 2014). For example, in Malawi and Ethiopia, ethanol production for fuel blending in the transport sector has developed over the past decades (see Chap. 3 Vol. 1; Chap. 5 Vol. 2), but traditional biomass still overwhelmingly dominates their domestic energy market (as in practically every other SSA country except for South Africa) (Sect. 2.1).

The growing demand for modern bioenergy (including at the household level) could influence SSA countries to develop both markets simultaneously, targeting both exports and domestic demand (Faaij et al. 2014). In this respect, international trade aspirations could also support domestic agro-industrial development (Batidzirai and Johnson 2012), while the resulting north–south and south–south relations could offer different impetus for trade, technology transfer and land investment in bioenergy, agriculture and forestry (Mathews 2007; Dauvergne and Neville 2009).

However, the actual feedstock type and mode of production can have significant interdependencies with scale economies and market orientation (Batidzirai and Johnson 2012; Gasparatos et al. 2015). Furthermore, the feasible scale of feedstock production and end use can vary considerably between areas. For example, the characteristics of the local economy can determine the availability of labour and agricultural inputs. Similarly, the logistics and economics of bioenergy production and/or conversion may constrain the sourcing of feedstock (e.g. feedstock production becomes uneconomic outside of a certain radius from the conversion facility) (FAO/UNEP 2011).

Figure 2.1 outlines some of the major bioenergy production and use alternatives according to a simple bimodal division between markets (local vs. export) and scale of feedstock production (small vs. large). Small-scale bioenergy production and local use (Type 1) can in principle can have greater development benefits, although these benefits can only be realised when the economic viability is assured, either through public support (e.g. quotas or mandates) or strong local institutions (Gasparatos et al. 2015). On the contrary large-scale bioenergy production has mainly been associated with feedstock production for national and international markets (Type 4) (Gasparatos et al. 2015). Sometimes small-scale production can also be combined with national and/or export markets (Type 3), which has often been the case in some SSA countries for sugarcane production (Mudombi et al. 2018a; von Maltitz et al. 2019) (see Chap. 3 Vol. 1). We should note that both Fig. 2.1 and the examples outlined above are for liquid biofuels (Gasparatos et al. 2015). However, the underlying logic would not be much different for other bioenergy options available in SSA such as solid biomass for heat and power production, biogas or multi-product biorefineries.

Regardless of the scale of bioenergy production and use, there is a need for substantial investments in infrastructure and institutions for enabling bioenergy

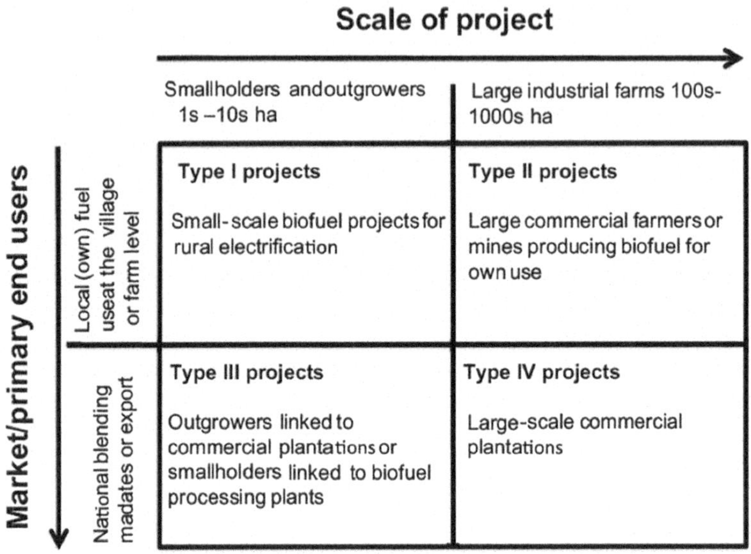

Fig. 2.1 Alternative configurations for bioenergy production and use according to scale and market. Source: (Gasparatos et al. 2015)

transitions (IRENA 2015, 2017; Silveira and Johnson 2016) (Chap. 1 Vol. 1). Such investments would be instrumental for funding the different stages of bioenergy systems, from production (e.g. production systems, ancillary infrastructure), to providing incentives to producers and end users (Souza et al. 2015; da Maia 2018) (see below for more details). An interesting example was the case of jatropha that attracted different types of financing and investment, ranging from foreign direct investments (FDIs) for jatropha-related large-scale land acquisitions (see Chap. 3–4 Vol. 1) to social investments emphasising local benefits through small-scale production and use (Liu et al. 2013; von Maltitz et al. 2014; Gasparatos et al. 2015) (Chap. 5 Vol. 1).

For those bioenergy transitions geared towards meeting domestic demand, it is important to note that, generally speaking, the scale of bioenergy systems is modest compared to those of fossil fuels and nuclear power. Furthermore, the scale can also vary considerably depending on the applications, end-use markets (e.g. local or national) and feedstocks. However, it is this modest scale that makes bioenergy projects and investments well-suited to most SSA countries, especially considering the institutional risks associated with large bioenergy-related investments (e.g. see experience from Clean Development Mechanism projects in SSA) (Lee and Lazarus 2013; Burian and Arens 2014). Smaller scale models do have risks but can also possibly offer greater social benefits in terms of the impacts identified in Sect. 2.2 (Gasparatos et al. 2015). Furthermore, their lower capital investment needs may make it easier to overcome barriers to financing.

For bioenergy transitions geared towards exports, it is important to achieve economies of scale and high economic efficiency for bioenergy production and export. This would require the adoption of at least some level of large-scale production models that can possibly lead to faster transitions and have broader impacts. However, such approaches can also be risky in terms of negative impacts and their effectiveness being curtailed by multiple factors (von Maltitz et al. 2014; Ahmed et al. 2019; Iiyama et al. 2014). The collapse of the jatropha sector across SSA was a painful reminder of the multiple factors that can affect negatively the viability of bioenergy production for exports (von Maltitz et al. 2014; Ahmed et al. 2019). In such models, value addition and export potential could be greater if bioenergy pathways are based on international commodities such as palm oil or sugar/ethanol that are already well-established in terms of agronomic knowledge and international markets (Batidzirai and Johnson 2012; Johnson and Seebaluck 2012; Faaij et al. 2014).

In a sense this issue of choice between local and global markets is inherently reflected in the wide range of global bioenergy potential estimates.[4] At the low end of the spectrum, these estimates correspond to bioenergy catering for less than 10% of the forecasted global energy demand in the year 2050, while at the high end, bioenergy could supply more than the entire global energy demand in 2050 (IPCC 2014). It can thus be argued that the lower end estimates reflect a view of bioenergy as a largely local resource, whereas the higher end estimates view bioenergy as a global market commodity. This divergence implies the emergence of two schools of thought concerning bioenergy in a climate and development context, with the first advocating considerable caution for the possible ecological/environmental impacts of a major global expansion (Beringer et al. 2011), and the other focusing on the possible considerable energy, socioeconomic and environmental benefits of such an expansion (Souza et al. 2015).

Emphasising the domestic use of bioenergy in SSA (rather than export markets) is sound in principle. This is especially true when considering the potential synergies with agricultural development and the significant economic and environmental benefits of shifting away from traditional biomass (Sect. 2.2). However, the scale of energy demand is also low in most SSA countries, and is compounded by the lack of infrastructure and investment options (Chap. 1 Vol. 1), making it difficult to attract sufficient and stable investments to reach economies of scale and/or centres of demand (IEA 2018). Consequently, focusing solely on domestic markets to achieve modern bioenergy transitions can be a lengthy process. In the meantime, the prevailing business-as-usual patterns of traditional biomass production and use can further deepen the cycle of poverty and resource degradation (Sect. 2.2). On the

[4]Modeling results suggest that the global bioenergy potential is largely situated in Latin America and SSA mainly due to climatic and demographic factors (Hoogwijk et al. 2005; Smeets et al. 2007; WGBU 2009; Haberl et al. 2010; van Vuuren et al. 2009; Beringer et al. 2011; Chum et al. 2011; IPCC 2014). A common starting point of these modelling studies is that "food/fibre" should be prioritised, with sustainable bioenergy potential calculated after accounting for the land needed for food production and also excluding deforestation (IPCC 2014; Batidzirai et al. 2016).

other hand, stronger linkages with the larger export markets in the EU and elsewhere could stimulate technology transfer and investment in SSA countries that would otherwise not materialise (Mathews 2007; Johnson 2011; Johnson and Mulugetta 2017) (Chap. 4 Vol. 1). However, strengthening institutions and investment scrutiny would be also needed to increase the long-term viability of modern bioenergy investments, as evidenced from the collapse of the jatropha sector throughout SSA (von Maltitz et al. 2014; Ahmed et al. 2019).

In this regard, the development of bioenergy systems that are flexible enough to cater to both domestic and export markets could be valuable. For example, bioenergy feedstocks such as wood pellets and bioethanol could offer this flexibility (Table 2.2), whereas feedstocks such as biogas and some types of waste (e.g. municipal waste) can be more appropriate for domestic markets for logistical and economic reasons.

In any case, the rural poor must be involved in bioenergy transitions in terms of energy demand and land use if modern bioenergy options are to reach domestic markets in SSA (Johnson and Diaz-Chavez 2018). Subsistence farmers and the rural poor in SSA are extremely constrained in terms of cash and often have almost no disposable income for investing in modern energy options after meeting basic needs (Takama et al. 2012; Mudombi et al. 2018a). Yet they play a major role in their respective national economies through informal markets, especially those related to food and energy (Leach 1992; Sola et al. 2016) (Chap. 5 Vol. 1). The shift from fuelwood to charcoal is a prominent example of a shift from non-cash to a cash economy that occurs partly through urbanisation. This shift has important environmental ramifications (Sect. 2.2) depending on the extent to which charcoal markets are regulated (Zulu 2010).[5]

2.4 Promote Integrated Landscape Approaches for Feedstock Production

Traditional bioenergy production systems can have substantial negative impacts on terrestrial ecosystems in SSA. For example, charcoal production is often associated with various negative environmental impacts such as deforestation and land degradation, particularly in semi-arid areas (IPBES 2018) (Chaps. 1 and 7 Vol. 1). For example, land degradation from unsustainable charcoal production in Kenya and other eastern African areas threatens local livelihoods through declining yields, biodiversity loss and other environmental impacts (Kiruki et al. 2017; Ndegwa et al. 2016). However, the actual links between bioenergy and land degradation

[5]Despite its negative environmental impacts, charcoal production and trade can improve rural livelihoods in terms of cash income (Openshaw 2010; Smith et al. 2015; Karanja and Gasparatos 2019). However, charcoal production does not necessarily reduce poverty in SSA, as revealed by multi-dimensional poverty indicators that incorporate health, housing and other fundamental indicators of well-being (Vollmer et al. 2017).

are rather complex, with charcoal production often being a by-product of other livelihood activities such as land clearing for agriculture (Iiyama 2013; IPBES 2018) (Sect. 2.2). Some countries have attempted to criminalise charcoal trade, but this has been largely unsuccessful due to the lack of affordable energy alternatives and enforcement challenges (Zulu 2010; Smith et al. 2015) (Chap. 1 Vol. 1). On the other hand, fuelwood collection in rural areas is rather different than charcoal use in that it can often be environmentally sustainable (Swemmer et al. 2019), although not necessarily socially desirable in terms of development goals (Sect. 2.2).

However, as discussed above modern bioenergy production can also be rather land-intensive compared to other energy options, especially for Type 2 and 4 modes of production (Sect. 2.3) (Fthenakis and Kim 2009; Emberson et al. 2012; Gasparatos et al. 2017). Land competition between bioenergy feedstock production and food crop production has emerged as a major concern for bioenergy expansion in SSA, especially for first generation liquid biofuels sourced from food crops (Rosillo-Calle and Johnson 2010; WGBU 2009; Gasparatos et al. 2015) (see Chap. 3 Vol. 1; Chap. 5 Vol. 2).[6] Furthermore, bioenergy feedstocks and food crops can compete for water, nutrients and other resources or agricultural inputs, having thus multiple linkages to food security (Wiggins et al. 2015; Jarzebski et al. 2020) (see Chap. 3 Vol. 1).[7] At the same time, there can also be complementarities and co-benefits when food and energy crops are produced and/or used across common systems or landscapes (Johnson and Virgin 2010; Bogdanski 2012; Souza et al. 2015; Kline et al. 2016; Mudombi et al. 2018a).

Landscape approaches across scales, sectors and/or markets can potentially address the competition for land, water and other resources, and help break down the, sometimes unnecessary, distinction between traditional and modern bioenergy. Landscape integration approaches can create opportunities to exploit synergies between food, fibre and fuel production (Dale et al. 2013).[8] Such synergies can occur through common supply chains and infrastructure development. Economic linkages in inputs and outputs can offer complementarities with food production, in terms of the flexibility afforded to producers to adjust over time the production of food, fuel, feed and fibre according to market signals (Bogdanski 2012; Rosillo-Calle and Johnson 2010; Kline et al. 2016).

[6]This has included in some cases the issue of indirect land use change. Indirect land use change (ILUC) can occur when non-food (e.g. bioenergy) production expands onto agricultural land and displaces food production, which then leads to additional land use elsewhere for food production to compensate the shortfall; ILUC cannot be measured empirically but instead is estimated through assumptions and modelling (Berndes et al. 2013; Finkbeiner 2014; Wicke et al. 2015).

[7]It is worth noting that modern bioenergy systems normally include multiple co-products or waste streams such as bagasse and molasses, respectively, in the case of sugarcane ethanol. The use of such co-products and waste streams can increase land and water efficiency and reduce competition with food (Ackom et al. 2013).

[8]Integrated food-energy systems are a particular class of such systems that can be very important in some SSA countries as they offer both synergies and complementarities between food and bioenergy production (Bogdanski 2012).

Agro-forestry offers another possible approach to reduce the negative impacts arising from land competition between bioenergy production systems and ecosystem services (Duguma et al. 2014; Mbow et al. 2014). In agro-forestry systems, farming practices are adapted to incorporate the multi-functional use of inputs and soil to support tree growth on farms, including ecosystem services such as biological nitrogen fixation (Nair et al. 2009). Feedstock production through agro-forestry systems can be combined with improved stoves to reduce pressure on forests and put fuelwood consumption on a more sustainable path (Iiyama et al. 2014).

Apart from reducing land competition, landscape approaches can also improve bioenergy value chains by emphasising the utilisation of downstream products and factoring them into the initial design of integrated systems (Dale et al. 2013). Such approaches might incorporate broader bioeconomy and land use management perspectives when planning programmes and supporting investments to facilitate transitions away from traditional biomass and subsistence agriculture (Johnson 2017; van de Ven et al. 2019). Furthermore, combining conservation efforts with income-generating activities across integrated landscapes can further offer co-benefits and shift practices away from slash and burn agriculture Rosenzweig and Tubiello 2007; Palm et al. 2013).

When using a landscape lens, bioenergy transitions essentially become a cross-sectoral issue where linkages, synergies and conflicts across agriculture, forestry and bioenergy systems must be addressed (Dale et al. 2013; Johnson and Jumbe 2013; Iiyama et al. 2014). Landscape approaches can incentivise the adoption of various good production practices that can facilitate the useful synergies and reduce the environmental and food security trade-offs of bioenergy production (Milder et al. 2008; Ackom et al. 2013; Kline et al. 2016; see Table 2.4). It must also be noted that the competition for land and biomass between different needs (i.e. food, feed, fibre, fuel) is not necessarily negative. On the contrary it can have positive impact by improving the overall land and resource utilisation efficiency towards a sustainable bioeconomy (Johnson and Virgin 2010; Johnson 2017). The issue is thus not to prevent land use competition but rather to ensure that such competition does not unduly impact the more vulnerable segments of society.

2.5 Foster Synergies between Climate Change Mitigation and Adaptation

As discussed above, modern bioenergy transitions entail multiple processes across different scales and sectors, which collectively have diverse sustainability impacts (Sects. 2.2 and 2.4). Similarly, bioenergy transitions can have important ramifications for climate change mitigation and adaptation in SSA. Although such synergies between climate change adaptation and mitigation could offer an incentive to further promote modern bioenergy transition in the continent, they have, so far, been relatively underappreciated in the SSA context. In this sense, in those contexts that

Table 2.4 Good bioenergy production practices using landscape approaches

	Climate	Ecosystems and biodiversity	Socio-economic	Food security	Energy security
Land use	• Incentives to use degraded or marginal land where feasible	• Enforce high biodiversity exclusion zones	• Land reform considerations in bioenergy programmes • Employment creation efforts in bioenergy programmes	• Food security assessments	• Integrated food-energy systems • Integrated provision of fibre and energy
Water	• Adaptation schemes for reduced water availability	• Protection of wetlands and watersheds • Effluent capture methods	• Systems for enabling water access of smallholders	• Evaluate availability of water for food crops • Advanced irrigation methods	• Bioenergy use for water pumping • Water-efficient energy crops
Soils and nutrients	• Low tillage practices	• Conservation agriculture • Nutrient recovery systems	• Technical support to smallholders for sustainable land management, and agro-chemical/ fertiliser use	• Capacity-building for soil nutrient impact assessments	• Guidelines for removal of agricultural residues
Forests	• Measures to safe-guard high carbon stock areas	• Protection of ecologi-cally sensitive forests	• Organisation of wood and char-coal markets	• Include methods for more efficient woody biomass use in bioenergy programmes	• Energy access evaluation • Incentives for fuel-switching

Source: Adapted from (FAO/UNEP 2011)

it is feasible, strategies can be aimed at win–win–win measures to pursue simultaneously adaptation, mitigation and basic development goals (Suckall et al. 2015).

In terms of climate change mitigation, many studies have noted the high GHG emission savings potential of some bioenergy pathways (Chum et al. 2011; Popp et al. 2011; Albanito et al. 2016). However, the estimated GHG emissions savings can vary widely between different bioenergy pathways due to factors as diverse as the feedstock, mode of production, end use and the different policies and practices governing bioenergy production, use and trade (Smith et al. 2014; Creutzig et al. 2015; Hurlbert et al. 2019). Modern bioenergy transitions can have substantial mitigation benefits if they succeed in curbing the use of traditional biomass fuels such as charcoal, considering the high GHG emissions associated with their production and use (Sect. 2.2).

However, as SSA countries are generally not expected to contribute to large-scale climate change mitigation efforts due to their low overall GHG emissions (Chap. 1 Vol. 1), the main challenge for sustainable bioenergy transitions is how to phase out traditional biomass (and/or use it more effectively and efficiently), rather than maximise emission reductions (Smeets et al. 2012; Karlberg et al. 2015). There are nevertheless many opportunities to achieve large-scale climate change mitigation from bioenergy pathways in SSA, particularly in some agro-industries such as sugarcane where agricultural residues are readily available (Batidzirai and Johnson 2012; da Maia 2018). In this sense, climate change mitigation from bioenergy transitions in SSA could be a valuable co-benefit to attract climate funding to assist the transitions themselves (Lee and Lazarus 2013).

Conversely, the links between climate change adaptation and bioenergy transitions can be less obvious and indirect in SSA. Below we attempt to outline some key, but rather underappreciated, aspects at the interface of modern bioenergy transitions and climate change adaptation in SSA. In particular, we focus on the (a) mechanisms linking bioenergy transitions and climate change adaptation and the (b) possible measures for addressing the adaptation of the bioenergy sector.

One of the most important mechanisms linking modern bioenergy transitions and climate change adaptation is the reduced reliance on climate-induced fuel scarcity. Many rural communities in SSA are highly vulnerable to climate change (especially precipitation changes), as it affects vegetation growth patterns, and thus agricultural productivity and woody biomass availability (Chaps. 1 and 9 Vol. 1; Chap. 2 Vol. 2). In this respect, as such climatic factors affect rural livelihoods and contribute to fuelwood scarcity (Karlberg et al. 2015), then a decreased reliance on traditional biomass fuels through improved energy access could have substantial adaptation benefits (Lambe and Johnson 2009). Another mechanism relates to the reduced reliance on centralised energy systems that are vulnerable and/or prone to disruption. For example, locally available small-scale renewable energy systems (Type 1 systems, Fig. 2.1) could reduce such dependencies, while also offering useful synergies between adaptation and development (Venema and Rehman 2007; Batidzirai and Johnson 2012; Gasparatos et al. 2015).

However, using the same logic as above it is also important to keep in mind that the bioenergy sector is vulnerable to climate change. This because most bioenergy feedstocks in SSA originate from either the agricultural or forestry sectors, which are highly exposed to (and affected by) climate change (IPCC 2014) (Chap. 1 Vol. 1). The impacts of climate change on the bioenergy sector (as well as its prospects for successful adaptation) depend substantially on actual implementation factors such as production site conditions, crop choices, management systems and supply chain structures (Field et al. 2014; Kongsager et al. 2016).

For example, there are significant disparities between the adaptation (and mitigation) potential of different bioenergy feedstocks. Annual agricultural crops (e.g. corn/maize, soybean, rapeseed) used for first generation liquid biofuels may have a negative effect on climate adaptation goals,[9] as they are vulnerable to erosion and drought, which are likely to become more serious in SSA due to ongoing climate change (Rosenzweig and Tubiello 2007; Nguyen and Tenhunen 2013; Smith and Olesen 2010). In contrast, perennial bioenergy crops (e.g. sugarcane, switchgrass, miscanthus) and trees for woody biomass are more resilient to climatic disturbances (thus offering greater adaptation potential), as they can enhance soil stability, reduce erosion risk and improve water retention in soils (Anderson-Teixeira et al. 2009; Wright and Wimberly 2013). It is also worth noting that such feedstocks have generally higher energy yields and GHG emission savings (Fazio and Barbanti 2014; Pugesgaard et al. 2015), offering thus valuable synergies between adaptation and mitigation (Smith and Olesen 2010).

Further, mitigation and adaptation synergies can be leveraged through the adoption of sustainable feedstock production practices. Agro-forestry and other integrated landscape approaches can offer perhaps the greatest potential, despite some negative adaptation and mitigation examples (Table 2.5). Other promising production practices include: (a) landscape management approaches that integrate livestock for biogas production[10] and (b) feedstock production practices that use timber damaged by insects to partially offset forest ecosystem degradation and reduce fire risks by creating incentives to remove dead trees (Lamers et al. 2014).

2.6 Implications for Policy and Governance

Through the different pathways outlined in Table 2.3, modern bioenergy transitions can contribute to multiple SDGs, including SDG 1, 2, 3, 7, 8, 11, 12, 13 and 15 (Sect. 2.2). Indeed, modern bioenergy transitions can become integral parts of climate-compatible development that "minimises the harm caused by climate impacts, while

[9]At the same time, these crops may require large amounts of agricultural inputs (e.g. fertiliser, agrochemicals, fuels), while their yields can be moderate, thus only having modest lifecycle GHG emission savings compared to fossil fuel alternatives (Fazio and Barbanti 2014; Pugesgaard et al. 2015). Implementing best practices could nevertheless facilitate improved scenarios and greater competitiveness for the use of annual crops as bioenergy feedstocks (Souza et al. 2015).

[10]For similar reasons, biogas has become a major part of national adaptation strategies in some SSA countries facing significant land scarcity such as Malawi (Johnson and Jumbe 2013).

Table 2.5 Positive and negative agro-forestry practices for climate change mitigation and adaptation

		Mitigation	
		Positive	Negative
Adaptation	Positive	• Soil carbon sequestration • Improved water holding capacity • Diversification of commercial products • Reduced nitrogen fertiliser use and fertiliser substitution with manure • Fire management	• High dependence on biomass for energy • Overexploitation of ecosystem services • Increased use of mineral fertilisers • Poor management of nitrogen and manure • Emphasis on non-timber forest products
	Negative	• Protection of forest reserves • Forest plantation excluding harvest • Large-scale biofuels export only through international carbon finance	• Use forest fires for pastoral management • Tree exclusion in farmland • Increased reliance on urban charcoal use without land tenure for rural production

Source: Adapted from (Mbow et al. 2014)

maximising the many human development opportunities presented by a low emission, more resilient future" (Mitchell and Maxwell 2010).[11] The goal in this case is to catalyse win–win–win situations rather than focusing separately on development, mitigation and adaptation goals (Suckall et al. 2015). Below, we discuss some critical governance and policy aspects to catalyse the effective integration of bioenergy pathways in climate-compatible development in SSA.

First, it would be necessary to ensure complementarities in national policy frameworks by avoiding the tendency of separating programmes aiming at phasing out traditional biomass from programmes aiming at promoting modern bioenergy. Instead there should be a nested approach in that climate-compatible bioenergy development should emerge from overall development objectives, and then integrate climate change adaptation and biomass promotion strategies across different sectors and scales (Fig. 2.2). In some respect, the missing link is how to better understand the strategic value of modern bioenergy in terms of how a reduction in traditional biomass use can free up biomass for more productive uses (Johnson and Jumbe 2013; Souza et al. 2015). In this sense, the higher productivity of modern bioenergy production (compared to traditional biomass) can thereby contribute significantly to climate-compatible development. Thus, the transition away from traditional biomass fuels in SSA countries would not necessarily mean that biomass use for energy will be reduced in aggregate terms. Rather it means that biomass needs to be used more

[11]There is a wide scope for strategies incorporating climate-compatible and/or "low carbon resilient" development in the context of a green economy. Such strategies focus on innovation and improved management in sectors that have significant climate implications such as agriculture, forestry and transport (Fisher 2013; Stringer et al. 2014; Kongsager et al. 2016).

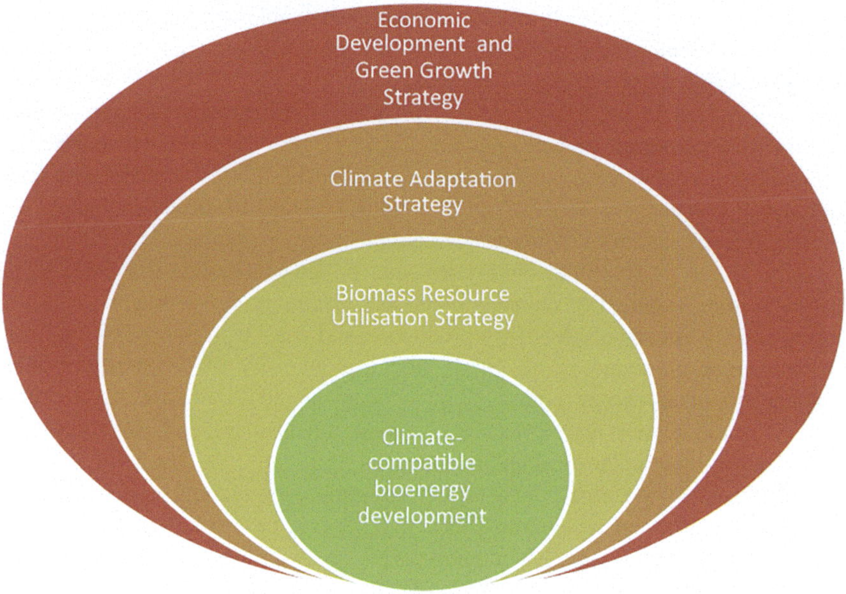

Fig. 2.2 Embedding climate-compatible bioenergy development within broader strategic policy goals

effectively, efficiently and synergistically across its many different uses for food, feed, fuel and fibre (Johnson and Virgin 2010).

Second, there should be concerted effort in national policy frameworks to incorporate mitigation and adaptation measures into broader sectoral policies, particularly in agriculture, forestry and other land-based activities. This would offer a more effective means of implementing climate policies than pursuing specific climate measures per se (Klein et al. 2005). Sectoral approaches to bioenergy development are especially relevant for SSA countries that lack fossil fuel resources, but have sufficient land and water availability. Such countries can benefit from expanding modern bioenergy, while at the same time phasing out traditional biomass, modernising their agricultural sectors and improving forest management. By integrating and coordinating climate policy with agricultural development and forest management, it is possible to create useful synergies for catalysing modern bioenergy transitions, not the least by expanding sustainable biomass supply for both food and fuel, as well as bio-based materials (Johnson and Virgin 2010; Davis 2012; Johnson 2017).

Third, apart from ensuring coherence and complementarity in national policy frameworks, it would be necessary to also consider issues related to national and regional markets (Arndt et al. 2019). The shift of modern bioenergy demand to China, India and other large emerging economies has created new South–South dynamics in technology transfer and energy trade (Dauvergne and Neville 2009). At the same time, the increasing prominence of non-state actors and transnational governance systems in the climate regime and sustainable certification has further complicated the integration of development strategies with national and local priorities (van Asselt et al. 2015).[12] The lack of appropriate cross-level governance systems can lead to the exploitation of precisely those groups that the biofuels expansion is purported to help, namely the rural poor (Dauvergne and Neville 2009). Thus, strengthening national and regional institutions in concert with local governance mechanisms in developing countries would be needed to allow the sustainable exploitation of their considerable bioenergy potential (Sect. 2.1).

Fourth, there are multiple biophysical and policy constraints that need to be navigated in this context of climate-compatible bioenergy development when promoting specific modes of bioenergy production and use. A prominent example are the challenges presented by the high land use intensity of bioenergy systems compared to other energy options (Emberson et al. 2012; Fritsche et al. 2017) when choosing the most appropriate combination of feedstocks, end uses and market orientation in a particular local and national setting (Sect. 2.3). Depending on such factors, bioenergy systems can be either supportive or disruptive in relation to climate mitigation and adaptation (Sect. 2.4). It is thus crucial that the choices made at the policy and implementation levels are well-informed and take these complexities into account.

Fifth, the direct transition route for some household cooking options is rather difficult in practice. The logistical challenges in SSA suggest that there are some advantages for portable and tradeable fuels such as bioethanol. However, the introduction of new fuels and stove technologies is rather complicated, with many factors influencing its effective large-scale uptake, e.g. as witnessed through bioethanol cookstoves promotion in Ethiopia, Kenya and Mozambique (Box 1). Other direct transition routes such as biogas offer similarly clean renewable options and benefits related to land-use management. Such options can indeed offer the most promising pathways for the transition away from traditional biomass, but there are many practical implementation issues that would need to be addressed (van de Ven et al. 2019).

Box 1 Policy Lessons from Bioethanol Promotion for Household Cooking in SSA

Ethiopia has had a long experience promoting ethanol as a cooking fuel. Following its initial introduction in refugee camps, there was a concerted effort to commercialise ethanol fuel through the introduction of highly efficient stoves (Stokes and Ebbeson 2005). However, there has been a competition for bioethanol feedstock (molasses) with other sectors, as well as

(continued)

[12]A prominent example comes from the EU, where the biofuels targets and sustainability criteria have had repercussions globally for markets and policies related to bioenergy, forest and agriculture (Johnson 2011; Pacini et al. 2013; Johnson and Mulugetta 2017) (Chap. 4 Vol. 1).

Box 1 (continued)
competition for the fuel itself with the transport sector. The Ethiopian government has tended to prioritise the transport sector for energy security reasons (Chap. 3 Vol. 1), posing a major barrier for the development of a household bioethanol market, as consumers want a fuel whose availability is assured (Rogers et al. 2013).

Ethanol for cooking was introduced in Maputo (Mozambique) in the early 2010s to divert some of the rapidly increasing charcoal demand (Chap. 5 Vol. 2). This has been the only successful large-scale promotion of ethanol stoves in SSA (Karanja and Gasparatos 2019). The initial success of the large-scale introduction was due to a favourable policy environment, with adoption rates increasing fairly rapidly until supply constraints prevented further expansion (Mudombi et al. 2018b) (Chap. 5 Vol. 2). However, the collapse of the domestic supply for the Cleanstar project, compared with technical and market-related problems also reduced some of the original motivation for ethanol market development as it was intended to boost local production (Chap. 5 Vol. 2).

Kenya has a high national ethanol production capacity that can potentially meet a large share of the domestic household energy demand. However, this bioenergy potential is hampered by unfavourable policies. For example, ethanol is treated as an alcoholic beverage regardless of its end use levying heavy taxes (Karanja and Gasparatos 2019) (Chap. 3 Vol. 1). At the same time, the largest sugarcane factory in Kenya has an annual production capacity of 22 mL of ethanol, but it is not fully utilised. Even though the acceptability and potential of ethanol as a cooking fuel has been strongly demonstrated in pilot studies in Western Kenya, the slow policy progress has prevented uptake of ethanol for household energy use. Instead, this ethanol is used for potable applications or industrial processes, targeting both the local and European markets. The elimination of taxes could make ethanol price competitive to charcoal or kerosene, and possibly contribute to its long-term adoption for household energy use (Karanja and Gasparatos 2019).

Finally, effective bioenergy transitions in SSA must include meaningfully the household sector. If this does not happen then bioenergy transition cannot be effective due to the overwhelming household dependence on traditional biomass and the significant sustainability impacts of this dependence. At the same time, the small scale of the household sector and its informal nature present barriers to the overall bioenergy transitions. The informal nature of the fuelwood and charcoal markets presents considerable sustainability and governance challenges that have created substantial barriers for effective bioenergy transitions. In this sense, transition pathways emphasising fuel-switching are likely to be more effective (van de Ven et al. 2019).

2.7 Conclusions

This chapter discussed some of the key critical aspects that can facilitate sustainable bioenergy transitions in SSA. In particular, it outlined the importance of (a) identifying and strengthening positive linkages across the different SDGs associated with bioenergy transitions in SSA; (b) choosing the most appropriate scales, markets and production modes for modern bioenergy; (c) promoting integrated landscape approaches for feedstock production and (d) fostering synergies between climate mitigation and adaptation.

It must be noted that the choice of these critical aspects has been somewhat selective, emphasising especially how biomass is utilised in the evolving context of land-use change, climate change and development in SSA. Other important aspects, such as water resource management and food security, were somewhat less emphasised, but were highlighted where appropriate. Even though the focus has been on transitions and pathways over time (as opposed to spatial aspects or sustainability assessments at fixed points in time), some important aspects have been highlighted in relation to how sustainability is evaluated and assessed in existing bioenergy policies and related frameworks.

Modernising bioenergy systems is critical for achieving many of the SDGs in SSA. Yet, it must be recognised that it is not simply a local and national issue, but also a regional and international issue. Tradable and environment-friendly bioenergy commodities must be developed across the continent. This could increase their competitiveness with charcoal, which is practically the only widely available current bioenergy commodity. Thus, modern bioenergy markets require deeper international linkages and trade, as much as they require deeper local engagement. This is the dual nature of the bioenergy transition facing SSA.

Bioenergy modernisation can in turn contribute to climate-compatible development, having both environmental and economic benefits. Thus the modernisation process does not have to entail a conflict with ecological or equity goals, but can be rather based on the best combination of local knowledge and global capital. In this respect, the bioenergy transition is not just about meeting the SDGs per se, but also about modernising economies in SSA by using their tremendous natural resource base in a sustainable manner. This virtuous pathway could reduce the tendency observed in many SSA countries to export raw materials (regardless of whether they are renewable or non-renewable). Instead, it could be a starting point for creating value-added knowledge-based products in the pursuit of a sustainable bioeconomy for all.

Acknowledgements This chapter was partly based on a previous research synthesis funded through institutional programme support provided to Stockholm Environment Institute (SEI) by the Swedish International Development Cooperation Agency (SIDA) within the SEI Reducing Climate Risk programme under the leadership of Richard J.T. Klein. However, SIDA was not involved in the choice of research topics or questions. The opinions expressed in the chapter are solely those of the authors.

References

Ackom EK, Alemagi D, Ackom NB, Minang PA, Tchoundjeu Z (2013) Modern bioenergy from agricultural and forestry residues in Cameroon: potential, challenges and the way forward. Energy Policy 63:101–113

Ahmed A, Campion B, Gasparatos A (2019) Towards a classification of the drivers of jatropha collapse in Ghana elicited from the perceptions of multiple stakeholders. Sustain Sci 14:315–339

Albanito F, Beringer T, Corstanje R, Poulter B, Stephenson A, Zawadzka J, Smith P (2016) Carbon implications of converting cropland to bioenergy crops or forest for climate mitigation: a global assessment. GCB Bioenergy 8(1):81–95

Anderson-Teixeira KJ, Davis SC, Masters MD, Delucia EH (2009) Changes in soil organic carbon under biofuel crops. GCB Bioenergy 1(1):75–96

Anenberg S, Balakrishnan K, Jetter J, Masera O, Mehta S, Moss J, Ramanathan V (2013) Cleaner cooking solutions to achieve health, climate, and economic co-benefits. Environ Sci Technol 47:3944–3952

Arndt C, Henley G, Hartley F (2019) Bioenergy in southern Africa: an opportunity for regional integration? Dev South Afr 36(2):145–154

AUC (2015) AGENDA 2063: The Africa We Want—A Shared Strategic Framework for Inclusive Growth and Sustainable Development, FIRST TEN-YEAR IMPLEMENTATION PLAN, 2014–2023, African Union Commission (AUC), Addis Ababa, September 2015

Bailis R, Drigo R, Ghilardi A, Masera O (2015) The carbon footprint of traditional woodfuels. Nat Clim Chang 5:266–272

Bailis R, Ezzati M, Kammen D (2003) Greenhouse gas implications of household energy technologies in Kenya. Environ Sci Technol 37:2051–2059

Barnes DF, Floor WM (1996) Rural energy in developing countries: a challenge for economic development. Annu Rev Energy Environ 21:497–530

Batidzirai B, Johnson FX (2012) Energy security, agro-industrial development and international trade: the case of sugarcane in southern Africa. In: Gasparatos A, Stromberg P (eds) Socio-economic and environmental impacts of biofuels: evidence from developing nations. Cambridge University Press, London. (Chapter 12)

Batidzirai B, Junginger M, Klemm M, Schipfer F, Thrän D (2016) Chapter 5: biomass supply and trade opportunities of pre-processed biomass for power generation. In: Lamers P, Stichnothe H, Hess JR, Searcy EM (eds) Developing the global bioeconomy. Elsevier, Chennai

Baumber A (2017) Enhancing ecosystem services through targeted bioenergy support policies. Ecosyst Serv 26:98–110

Bazilian M, Sagar A, Detchon R, Yumkella K (2010) More heat and light. Energy Policy 38:5409–5412

Beringer T, Lucht W, Schaphoff S (2011) Bioenergy production potential of global biomass plantations under environmental and agricultural constraints. GCB Bioenergy 3:299–312

Berndes G, Ahlgren S, Börjesson P, Cowie AL (2013) Bioenergy and land use change—state of the art'. Wiley Interdisciplinary Reviews:. Energy Environ 2(3):282–303. https://doi.org/10.1002/wene.41

Bildirici M, Ersin Ö (2015) An investigation of the relationship between the biomass energy consumption, economic growth and oil prices. Procedia Soc Behav Sci 210:203–212

Bogdanski A (2012) Integrated food–energy systems for climate-smart agriculture. Agric Food Secur 1:1–9

Burian M, Arens C (2014) The clean development mechanism: a tool for financing low carbon development in Africa? Int J Clim Change Strategies Manage 6(2):166–191

Cameron C, Pachauri S, Rao ND, McCollum D, Rogelj J, Riahi K (2016) Policy tradeoffs between climate mitigation and clean cook stove access in South Asia. Nat Energy 1:15010

Cerutti PO, Sola P, Chenevoy A, Iiyama M, Yila J, Zhou W et al (2015) The socioeconomic and environmental impacts of wood energy value chains in sub-Saharan Africa: a systematic map protocol. Environ Evidence 4(1):1

Chidumayo EN, Gumbo DJ (2013) The environmental impacts of charcoal production in tropical ecosystems of the world: a synthesis. Energy Sustain Dev 17(2):86–94

Chum H, Faaij A, Moreira J, Berndes G, Dhamija P, Dong H, Gabrielle B, Goss Eng A, Lucht W, Mapako M, Masera Cerutti O, McIntyre T, Minowa T, Pingoud K (2011) Bioenergy. In: Edenhofer O, Pichs-Madruga R, Sokona Y, Seyboth K, Matschoss P, Kadner S, Zwickel T, Eickemeier P, Hansen G, Schlömer S, von Stechow C (eds) IPCC special report on renewable energy sources and climate change mitigation. Cambridge University Press, Cambridge, UK and New York

Creutzig F, Ravindranath NH, Berndes G, Bolwig S, Bright R, Cherubini F, Chum H, Corbera E, Delucchi M, Faaij A, Fargione J (2015) Bioenergy and climate change mitigation: an assessment. GCB Bioenergy 7(5):916–944

da Maia RC (2018) Methodology of assessment and mitigation of risks facing bioenergy investments in sub-Saharan Africa. In: Cortez LAB, Leal MRLV, Nogueira LAH (eds) Sugarcane bioenergy for sustainable development: expanding production in Latin America and Africa. Routledge, London, UK. Chapter 31

Dale VH, Kline KL, Kaffka SR, Langeveld JH (2013) A landscape perspective on sustainability of agricultural systems. Landsc Ecol 28(6):1111–1123

Dauvergne P, Neville KJ (2009) The changing north–south and south–south political economy of biofuels. Third World Q 30(6):1087–1102

Davis M (2012) Building a Low-carbon Future: Resource Constraints and Key Strategies to overcome them. Project report, Stockholm Environment Institute. http://www.sei-international.org/publications?pid=2120

Dresen E, DeVries B, Herold M, Verchot L, Müller R (2014) Fuelwood savings and carbon emission reductions by the use of improved cooking stoves in an Afromontane forest, Ethiopia. Land 3(3):1137–1157

Duguma LA, Minang PA, van Noordwijk M (2014) Climate change mitigation and adaptation in the land use sector: from complementarity to synergy. Environ Manag 54(3):420–432

Emberson L, He K, Rockström J, Amann M, Barron J, Corell R, Feresu S, Haeuber R, Hicks K, Johnson FX, Karlqvist A, Klimont Z, Mylvakanam I, Song WW, Vallack H, Qiang Z (2012) Energy and environment. In: Global energy assessment—toward a sustainable future. Cambridge University Press, Cambridge, UK and New York, NY, USA. and the International Institute for Applied Systems Analysis, Laxenburg, Austria (Chapter 3), pp 191–254

ESMAP (2018) SDG7 tracking: the energy progress report, World Bank, 2018

Faaij A (2006) Modern biomass conversion technologies. Mitig Adapt Strateg Glob Chang 11 (2):335–367

Faaij A, Junginger M, Goh CS (2014) A general introduction to international bioenergy trade. In: Junginger M, Goh CS, Faaij A (eds) International bioenergy trade. Springer, Netherlands, pp 1–15

FAO (2004) Unified Bioenergy Terminology: UBET. Food and Agriculture Organization of the United Nations. Forestry Department, Wood Energy Programme

FAO/UNEP (2011) Bioenergy decision support tool. Food and Agriculture Organization of the United Nations (FAO), Rome. Online: http://www.bioenergydecisiontool.org/

Fazio S, Barbanti L (2014) Energy and economic assessments of bio-energy systems based on annual and perennial crops for temperate and tropical areas. Renew Energy 69:233–241

Field CB, Barros VR, Mach K, Mastrandrea M (2014) Climate change 2014: impacts, adaptation, and vulnerability. Contribution of Working Group II to the Fifth Assessment Report of the Intergovernmental Panel on Climate Change

Finkbeiner M (2014) Indirect land use change–help beyond the hype? Biomass Bioenergy 62:218–221

Fisher S (2013) Low-carbon resilient development in the least developed countries: emerging issues and areas of research. IIED Issue Paper. IIED, London

Foell W, Pachauri S, Spreng D, Zerriffi H (2011) Household cooking fuels and technologies in developing economies. Energy Policy 39(12):7487–7496

Fritsche U, Berndes G, Cowie A, Dale VH, Kline KL, Johnson FX, Langeveld H, Sharma N, Watson H, Woods J (2017) Sustainable energy options and implications for land use, Global Land Outlook (GLO), UNCCD and IRENA

Fthenakis V, Kim HC (2009) Land use and electricity generation: a life-cycle analysis. Renew Sust Energ Rev 13(6–7):1465–1474

Gasparatos A, Doll C, Esteban M, Ahmed A, Olang T (2017) Biodiversity and renewable energy: implications for transitioning to a Green economy. Renew Sust Energ Rev 70:161–184

Gasparatos A, Lehtonen M, Stromberg P (2013) Do we need a unified appraisal framework to synthesize biofuel impacts? Biomass Bioenergy 50:75–80

Gasparatos A, Romeu-Dalmau C, von Maltitz G, Johnson FX, Shackleton C, Jarzebski MP, Jumbe C, Ochieng C, Mudombi S, Nyambane A, Willis KJ (2018) Mechanisms and indicators for assessing the impact of biofuel feedstock production on ecosystem services. Biomass Bioenergy 114:157–173

Gasparatos A, Stromberg P, Takeuchi K (2011) Biofuels, ecosystem services and human wellbeing: putting biofuels in the ecosystem services narrative. Agric Ecosyst Environ 142(3):111–128

Gasparatos A, von Maltitz GP, Johnson FX, Lee L, Mathai M, de Oliveira JP, Willis KJ (2015) Biofuels in sub-Sahara Africa: drivers, impacts and priority policy areas. Renew Sust Energ Rev 45:879–901

Haberl H, Beringer T, Bhattacharya SC, Erb KH, Hoogwijk M (2010) The global technical potential of bioenergy in 2050 considering sustainability constraints. Curr Opin Environ Sustain 2 (5):394–403

Hoogwijk M, Faaij A, Eickhout B, de Vries B, Turkenburg W (2005) Potential of biomass energy out to 2100, for four IPCC SRES land-use scenarios. Biomass Bioenergy 29(4):225–257

Hosier RH, Mwandosya MJ, Luhanga ML (1993) Future energy development in Tanzania: the energy costs of urbanization. Energy Policy 21(5):524–542

Hurlbert M, Krishnaswamy J, Davin E, Johnson FX, Mena CF, Morton J, Myeong S, Viner D, Warner K, Wreford A, Zakieldeen S, Zommers Z (2019) Emergent risks, decision-making and sustainable development (Chapter 7) in Climate Change and Land: an IPCC special report on climate change, desertification, land degradation, sustainable land management, food security, and greenhouse gas fluxes in terrestrial ecosystems. https://www.ipcc.ch/srccl-report-download-page/

IEA (2014) Energy access. World energy outlook. Int Energy Agency, Paris. www.iea.org

IEA (2018) Key world energy statistics 2014. International energy agency. http://www.iea.org/publications/freepublications/publication/KeyWorld2014.pdf

Iiyama M (2013) Charcoal: a driver of dryland forest degradation in Africa? (fact sheet). World Agrofor Centre, Nairobi, Kenya

Iiyama M, Neufeldt H, Dobie P, Njenga M, Ndegwa G, Jamnadass R (2014) The potential of agroforestry in the provision of sustainable woodfuel in sub-Saharan Africa. Curr Opin Environ Sustain 6:138–147

IPBES (2018) The IPBES regional assessment report on biodiversity and ecosystem services for Africa. In: Archer E, Dziba L, Mulongoy KJ, Maoela MA, Walters M (eds) . Secretariat of the Intergovernmental Science-Policy Platform on Biodiversity and Ecosystem Services, Bonn

IPCC (2014) Agriculture, forestry and other land use (AFOLU). Intergovernmental Panel on Climate Change (IPCC) Working Group III report, www.ipcc.ch

IRENA (2015) Africa 2030: roadmap for a renewable energy future. Int Renewable Energy Agency, Abu Dhabi. www.irena.org/

IRENA (2017) Biofuel potential in sub-Saharan Africa: raising food yields, reducing food waste and utilising residues. International Renewable Energy Agency, Abu Dhabi. www.irena.org/

Jagger P, Kittner N (2017) Deforestation and biomass fuel dynamics in Uganda. Biomass Bioenergy 105:1–9
Jarzebski MP, Ahmed A, Boafo YA, Balde BS, Chinangwa L, Saito O et al (2020) Food security impacts of industrial crop production in sub-Saharan Africa: a systematic review of the impact mechanisms. Food Security 12(1):105–135
Johnson F, Diaz-Chavez RA (2018) The role of the sugarcane bioeconomy in supporting energy access and rural development. In: Cortez LAB, Leal MRLV, Nogueira LAH (eds) Sugarcane bioenergy for sustainable development: expanding production in Latin America and Africa. Routledge, London, UK. Chapter 5
Johnson FX (2011) Regional-global linkages in the energy-climate-development policy nexus: the case of biofuels in the EU renewable energy directive. Renewable Energy Law and Policy Journal (RELP) 2:91–106
Johnson FX (2017) Biofuels, bioenergy and the bioeconomy in North and South. Ind Biotechnol 13 (6):289–291
Johnson FX, Jumbe C (2013) Energy Access and Biomass Transitions in Malawi. Policy Brief. Stockholm Environment Institute, Stockholm. Online www.sei.se
Johnson FX, Matsika E (2006) Bioenergy trade and regional development: the case of bioethanol in southern Africa. Energy Sustain Dev 10(1):42–53
Johnson FX, Mulugetta Y (2017) Biofuels for sustainable energy and mobility in the EU and Africa. In: Virgin I, Morris EJ (eds) Creating sustainable bio-economies: the bioscience revolution in Europe and Africa. Routledge, London. Chapter 11
Johnson FX, Nyambane A, von Maltitz G, Luhanga D, Jarzebski M, Balde BS, Gasparatos A (2018) Impacts of biofuel crop production in southern Africa: land use change, ecosystem services, poverty alleviation and food security. ESPA policy and practice brief, ESPA, Edinburgh. https://www.espa.ac.uk/publications/impacts-biofuel-production-southern-africa
Johnson FX, Seebaluck V (2012) Bioenergy for sustainable development and international competitiveness: the role of sugar cane in Africa. Routledge/Earthscan, London
Johnson FX, Silveira S (2014) Pioneer countries in the transition to alternative transport fuels: comparison of ethanol programmes and policies in Brazil, Malawi and Sweden. Environ Innov Soc Trans 11:1–24
Johnson FX, Virgin I (2010) Future trends in markets for food, feed, fibre and fuel. In: Rosillo-Calle F, Johnson FX (eds) Food vs. fuel: an informed introduction. ZED, London, pp 164–190
Jones D, Ryan CM, and Fisher J (2016) Charcoal as a diversification strategy: the flexible role of charcoal production in the livelihoods of smallholders in central Mozambique. doi:https://doi.org/10.1016/j.esd.2016.02.009
Junginger M, van Dam J, Zarrilli S, Ali Mohamed F, Marchal D, Faaij A (2011) Opportunities and barriers for international bioenergy trade. Energy Policy 39(4):2028–2042
Karanja A, Gasparatos A (2019) Adoption and impacts of clean bioenergy cookstoves in Kenya. Renew Sust Energ Rev 102:285–306
Karlberg L, Hoff H, Flores-López F, Goetz A, Matuschke I (2015) Tackling biomass scarcity—from vicious to virtuous cycles in sub-Saharan Africa. Curr Opin Environ Sustain 15:1–8
Kiruki HM, van der Zanden EH, Malek Ž, Verburg PH (2017) Land cover change and woodland degradation in a charcoal producing semi-arid area in Kenya. Land Degrad Dev 28:472–481
Klein R, Schipper L, Dessai S (2005) Integrating mitigation and adaptation into climate and development policy: three research questions. Environ Sci Policy 8(6):579–588
Kline KL, Msangi S, Dale VH, Woods J, Souza GM, Osseweijer P, Clancy JS, Hilbert JA, Johnson FX, McDonnell PC, Mugera HK (2016) Reconciling food security & bioenergy: priorities for action. GCB Bioenergy 9:557–576
Kongsager R, Locatelli B, Chazarin F (2016) Addressing climate change mitigation and adaptation together: a global assessment of agriculture and forestry projects. Environ Manag 57 (2):271–282
Lakshmi PVM, Virdi NK, Thakur JS, Smith KR, Bates MN, Kumar R (2010) Biomass fuel and risk of tuberculosis: a case–control study from Northern India. J Epidemiol Community Health. https://doi.org/10.1136/jech.2010.115840

Lam NL, Chen Y, Weyant C, Venkataraman C, Sadavarte P, Johnson MA et al (2012) Household light makes global heat: high black carbon emissions from kerosene wick lamps. Environ Sci Technol 46(24):13531–13538. https://doi.org/10.1021/es302697h
Lambe F, Johnson FX (2009) Energy access, climate and development: a paper contributed to the report of the Commission on Climate Change and Development. Stockholm Environment Institute
Lamers P, Junginger M, Dymond CC, Faaij A (2014) Damaged forests provide an opportunity to mitigate climate change. GCB Bioenergy 6(1):44–60
Leach G (1992) The energy transition. Energy Policy 20:116–123
Lee CM, Chandler C, Lazarus M, Johnson FX (2013) Assessing the climate impacts of Cookstove projects: issues in emissions accounting. Challenges in Sustainability 1(2):53–71
Lee CM, Lazarus M (2013) Bioenergy projects and sustainable development: which project types offer the greatest benefits? Clim Dev 5(4):305–317
Lim SS, Vos T, Flaxman AD, Danaei G, Shibuya K, Adair-Rohani H et al (2012) A comparative risk assessment of burden of disease and injury attributable to 67 risk factors and risk factor clusters in 21 regions, 1990–2010: a systematic analysis for the global burden of disease study 2010. Lancet 380(9859):2224–2260. https://doi.org/10.1016/S0140-6736(12)61766-8
Liu P, Koroma S, Arias P, Hallam D (2013) Trends and impacts of foreign investment in developing country agriculture – evidence from case studies. Food and Agriculture Organization of the United Nations (FAO), Rome. http://www.fao.org/economic/est/publications/trends/en/
Macqueen D, Korhaliller S (2011) Bundles of energy: the case for renewable biomass energy. Natural resource issues no. 24. International Institute for Environment and Development (IIED), London
Masera OR, Saatkamp DB, Kammen DM (2000) From linear switching to multiple cooking strategies: a critique and alternative to the energy ladder model. World Dev 28:2083–2103
Masera OR et al (2015) Environmental burden of traditional bioenergy use. Annu Rev Environ Resour 40(1):121–150
Mathews J (2007) Biofuels: what a biopact between north and south could achieve. Energy Policy 35:3550–3570
Mattick CS, Williams E, Allenby BR (2010) Historical trends in global energy consumption. Technol Soc Mag IEEE 29(3):22–30
Mbow C, Smith P, Skole D, Duguma L, Bustamante M (2014) Achieving mitigation and adaptation to climate change through sustainable agroforestry practices in Africa. Curr Opin Environ Sustain 6:8–14
McCollum DL, Echeverri LG, Busch S, Pachauri S, Parkinson S, Rogelj J, Riahi K (2018) Connecting the sustainable development goals by their energy inter-linkages. Environ Res Lett 13(3):033006
Milder JC, McNeely JA, Shames SA, Scherr SJ (2008) Biofuels and ecoagriculture: can bioenergy production enhance landscape-scale ecosystem conservation and rural livelihoods? Int J Agric Sustain 6(2):105–121. https://doi.org/10.3763/ijas.2008.0344
Mitchell T, Maxwell S (2010) Defining climate compatible development: policy brief. https://www.dfid.gov.uk/R4D/PDF/Outputs/CDKN/CDKN-CCD-DIGI-MASTER-19NOV.pdf . Accessed 20 July 2018
Mudombi S, Nyambane A, von Maltitz GP, Gasparatos A, Johnson FX, Chenene ML, Attanassov B (2018b) User perceptions on adoption and use of ethanol fuel and cookstoves in Maputo, Mozambique. Energy Sust Dev 44:97–108
Mudombi S, von Maltitz GP, Gasparatos A, Romeu-Dalmau C, Johnson FX, Jumbe C et al (2018a) Multi-dimensional poverty effects around operational biofuel projects in Malawi, Mozambique and Swaziland. Biomass Bioenergy 114:41–54
Nair PKR, Kumar MB, Nair VD (2009) Agroforestry as a strategy for carbon sequestration. J Plant Nutr Soil Sci 172(1):10–23, February, 2009 doi:. https://doi.org/10.1002/jpln.200800030
Ndegwa G, Nehren U, Grüninger F, Iiyama M, Anhuf D (2016) Charcoal production through selective logging leads to degradation of dry woodlands: a case study from Mutomo District, Kenya. J Arid Land 8:618–631

Nerini FF, Tomei J, To LS, Bisaga I, Parikh P, Black M, Borrion A, Spataru C, Broto VC, Anandarajah G, Milligan B (2018) Mapping synergies and trade-offs between energy and the sustainable development goals. Nat Energy 3(1):10

Nguyen TT, Tenhunen J (2013) Review of integrated ecological-economic analyses for bioenergy plants under climate change at local scale. Int J Clim Change Strategies Manage 5(3):324–343

Ochieng C, Juhola S, Johnson FX (2014) The societal role of charcoal production in climate change adaptation of the arid and semi-arid lands (ASALs) of Kenya. In: Inderberg TH, Eriksen S, O'Brien K, Sygna L (eds) Climate change adaptation and development: transforming paradigms and practices. Routledge. Chapter 3

Olsson O, Johnson FX (2014) Bioenergy Trade in a Changing Climate. NORDSTAR Working Paper 2014–01, Nordic Center of Excellence for Strategic Adaptation Research. www.nord-star.info/workingpapers

Openshaw K (2010) Biomass energy: employment generation and its contribution to poverty alleviation. Biomass Bioenergy 34(3):365–378

Pacini H, Assunção L, van Dam J, Toneto R Jr (2013) The price for biofuels sustainability. Energy Policy 59:898–903

Palm CA, Vosti SA, Sanchez PA, Ericksen PJ (2013) Slash-and-burn agriculture: the search for alternatives. Columbia University Press, New York

Popp A, Dietrich JP, Lotze-Campen H, Klein D, Bauer N, Krause M, Beringer T, Gerten D, Edenhofer O (2011) The economic potential of bioenergy for climate change mitigation with special attention given to implications for the land system. Environ Res Lett 6(3):034017

Pugesgaard S, Schelde K, Larsen SU, Lærke PE, Jørgensen U (2015) Comparing annual and perennial crops for bioenergy production–influence on nitrate leaching and energy balance. GCB Bioenergy 7(5):1136–1149

Rogers C, Sovacool BK, Clarke S (2013) Sweet nectar of the Gaia: lessons from Ethiopia's "project Gaia". Energy Sustain Dev 17(3):245–251

Rosenzweig C, Tubiello FN (2007) Adaptation and mitigation strategies in agriculture: an analysis of potential synergies. Mitig Adapt Strateg Glob Chang 12(5):855–873

Rosillo-Calle F, Johnson FX (2010) Food versus fuel: an informed introduction to biofuels. ZED Books, London

Santos MJ, Dekker SC, Daioglou V, Braakhekke MC, van Vuuren DP (2017) Modeling the effects of future growing demand for charcoal in the tropics. Front Environ Sci 5:1–12

Shindell D, Kuylenstierna JC, Vignati E, van Dingenen R, Amann M, Klimont Z et al (2012) Simultaneously mitigating near-term climate change and improving human health and food security. Science 335(6065):183–189

Silveira S, Johnson FX (2016) Navigating the transition to sustainable bioenergy in Sweden and Brazil: lessons learned in a European and international context. Energy Res Soc Sci 13:180–193

Smeets EM, Faaij AP, Lewandowski IM, Turkenburg WC (2007) A bottom-up assessment and review of global bio-energy potentials to 2050. Prog Energy Combust Sci 33(1):56–106

Smeets EM, Johnson FX, Ballard-Tremeer G (2012) Traditional and improved use of biomass for energy in Africa. In: Janssen R, Rutz D (eds) Bioenergy for sustainable development in Africa. Springer, Netherlands, pp 3–12

Smith HE, Eigenbrod F, Kafumbata D, Hudson MD, Schreckenberg K (2015) Criminals by necessity: the risky life of charcoal transporters in Malawi. For Trees Livelihoods 24 (4):259–274

Smith P, Bustamante M, Ahammad H, Clark H, Dong H, Elsiddig EA, Haberl H, Harper R, House J, Jafari M, Masera O, Mbow C, Ravindranath NH, Rice CW, Robledo Abad C, Romanovskaya A, Sperling F, Tubiello F (2014) Agriculture, forestry and other land use (AFOLU). In: Edenhofer O, Pichs-Madruga R, Sokona Y, Farahani E, Kadner S, Seyboth K, Adler A, Baum I, Brunner S, Eickemeier P, Kriemann B, Savolainen J, Schlömer S, von Stechow C, Zwickel T, Minx JC (eds) Climate change 2014: mitigation of climate change. Contribution of working group III to the fifth assessment report of the intergovernmental panel on climate change. Cambridge University Press, Cambridge, United Kingdom and New York, NY, USA, pp 811–922

Smith P, Olesen JE (2010) Synergies between the mitigation of, and adaptation to, climate change in agriculture. J Agric Sci 148(05):543–552
Sola P, Ochieng C, Yila J, Iiyama M (2016) Links between energy access and food security in sub Saharan Africa: an exploratory review. Food Secur 8(3):635–642
Souza GM, Victoria R, Joly C Verdade L (Eds), (2015). Bioenergy & Sustainability: bridging the gaps (Vol. 72, 779p.). SCOPE, UNESCO: Paris. ISBN 978–2–9545557-0-6, http://bioenfapesp.org/scopebioenergy/index.php
Sovacool BK (2012) The political economy of energy poverty: a re-view of key challenges. Energy Sustain Dev 16(3):272–282
Stokes H, Ebbeson B (2005) Project Gaia: commercializing a new stove and new fuel in Africa. Boiling Point 50:31–33
Stringer LC, Dougill AJ, Dyer JC, Vincent K, Fritzsche F, Leventon J, Falcão MP et al (2014) Advancing climate compatible development: lessons from southern Africa. Reg Environ Chang 14(2):713–725
Suckall N, Stringer LC, Tompkins EL (2015) Presenting triple-wins? Assessing projects that deliver adaptation, mitigation and development co-benefits in rural sub-Saharan Africa. Ambio 44:34–41
Swemmer AM, Mashele M, Ndhlovu PD (2019) Evidence for ecological sustainability of fuelwood harvesting at a rural village in South Africa. Reg Environ Chang 19(2):403–413
Takama T, Tsephel S, Johnson FX (2012) Evaluating the relative strength of product-specific factors in fuel switching and stove choice decisions in Ethiopia: a discrete choice model of household preferences for clean cooking alternatives. Energy Econ 34(6):1763–1773
Taylor R, Wanjiru H, Johnson O, Johnson FX (2019) Combining agent-based modelling and local stakeholder perspectives to analyse sustainable charcoal in Kenya. Environmental Innovation and Societal Transitions (in press: doi:https://doi.org/10.1016/j.eist.2019.10.001
van Asselt H, Mehling M, Kehler Siebert C (2015) The changing architecture of international climate change law. In: Van Calster G, Vandenberghe W, Reins L (eds) Research handbook on climate change mitigation law. Edward Elgar, Cheltenham
van de Ven DJ, Sampedro J, Johnson FX, Bailis R, Forouli A, Nikas A et al (2019) Integrated policy assessment and optimisation over multiple sustainable development goals in eastern Africa. Environ Res Lett 14:094001
van Vuuren DP, van Vliet J, Stehfest E (2009) Future bioenergy potentials under various natural constraints. Energy Policy 37(11):4220–4230
Venema HD, Rehman IH (2007) Decentralized renewable energy and the climate change mitigation-adaptation nexus. Mitig Adapt Strateg Glob Chang 12:875–900
Vollmer F, Zorrilla-Miras P, Baumert S, Luz AC, Woollen E, Grundy I et al (2017) Charcoal income as a means to a valuable end: scope and limitations of income from rural charcoal production to alleviate acute multidimensional poverty in Mabalane district, southern Mozambique. World Dev Perspect 7:43–60
von Maltitz G et al (2014) The rise, decline and future resilience benefits of jatropha in southern Africa. Sustainability 6:3615–3643
von Maltitz G et al (2016) Jatropha cultivation in Malawi and Mozambique: impact on ecosystem services, local human wellbeing and poverty alleviation. Ecol Soc 21(3):3
von Maltitz GP, Henley G, Ogg M, Samboko PC, Gasparatos A, Ahmed A, Read M, Engelbrecht F (2019) Institutional arrangements of outgrower sugarcane production in southern Africa. Dev South Afr 36:175–197
Wanjiru H, Nyambane A, Omedo G (2016) How Kenya can transform the charcoal sector and create new opportunities for low-carbon rural development. Stockholm Environ Inst, Nairobi, Kenya
WGBU (2009) Future bioenergy and sustainable land use. German Advisory Council on Global Change (WGBU). Earthscan, London
Wicke B, Brinkman MLJ, Gerssen-Gondelach S, van der Laan C, Faaij APC (2015) ILUC prevention strategies for sustainable biofuels: synthesis report from the ILUC prevention project. Utrecht University, Utrecht, Netherlands. http://www.geo.uu.nl/iluc

Wiggins S., Henley G, Keats S (2015) Competitive or complementary?—Industrial crops and food security in sub-Saharan Africa. Overseas Development Institute Report, 41pp

World Bank (2009) Environmental crisis or sustainable development opportunity?: transforming the charcoal sector in Tanzania. Policy note. World Bank, Washington, DC

Wright CK, Wimberly MC (2013) Recent land use change in the Western Corn Belt threatens grasslands and wetlands. Proc Natl Acad Sci 110(10):4134

Zulu LC (2010) The forbidden fuel: charcoal, urban woodfuel demand and supply dynamics, community forest management and woodfuel policy in Malawi. Energy Policy 38 (7):3717–3730

Zulu LC, Richardson RB (2013) Charcoal, livelihoods, and poverty reduction: evidence from sub-Saharan Africa. Energy Sustain Dev 17(2):127–137

Jarzebski, M. P., Ahmed, A., Boafo, Y. A., Balde, B. S., Chinangwa, L., Saito, O., ... & Gasparatos, A. (2020) Food security impacts of industrial crop production in sub-Saharan Africa: a systematic review of the impact mechanisms. Food Security, 12(1), 105-135.

Pugesgaard, S., Schelde, K., Larsen, S. U., Lærke, P. E., & Jørgensen, U. (2015) Comparing annual and perennial crops for bioenergy production–influence on nitrate leaching and energy balance. GCB Bioenergy, 7(5), 1136-1149.

Chapter 3
Linking Industrial Crop Production and Food Security in Sub-Saharan Africa: Local, National and Continental Perspectives

Marcin Pawel Jarzebski, Abubakari Ahmed, Alice Karanja, Yaw Agyeman Boafo, Boubacar Siddighi Balde, Linda Chinangwa, Sileshi Degefa, Eric Brako Dompreh, Osamu Saito, and Alexandros Gasparatos

3.1 Introduction

Food security is one of the major sustainability challenges and development priorities facing sub-Saharan Africa (SSA) (UNECA 2015; FAO 2018; Sasson 2012) (Chap. 1 Vol. 1). Currently about a tenth of the global population is undernourished, but this rate is even higher in SSA standing at about 20% (FAO 2015) (Chap. 1 Vol. 1; Chap. 1 Vol. 2). Despite some progress, most SSA countries fall at the lowest ranks globally for many indicators related to hunger, child stunting and calorie deficit, among several others (FAO, IFAD, UNICEF, WFP and WHO 2019). Despite the constant increase in food supply across all sub-regions during the past decades (Fig. 3.1), there has been little progress in many countries to halt food inadequacy despite large economic growth in terms of gross domestic product (GDP) (Fig. 3.2).

M. P. Jarzebski (✉) · B. S. Balde · E. B. Dompreh
University of Tokyo, Tokyo, Japan

A. Ahmed
Department of Planning, University for Development Studies, Wa, Ghana

A. Karanja · A. Gasparatos
Institute for Future Initiatives (IFI), The University of Tokyo, Tokyo, Japan
e-mail: gasparatos@ifi.u-tokyo.ac.jp

Y. A. Boafo
University of Ghana, Accra, Ghana

L. Chinangwa · S. Degefa
United Nations University, Tokyo, Japan

O. Saito
Institute for Global Environmental Studies (IGES), Hayama, Japan

Institute for Future Initiatives (IFI), The University of Tokyo, Tokyo, Japan
e-mail: saito@unu.edu

A. Gasparatos et al. (eds.), *Sustainability Challenges in Sub-Saharan Africa I*,
Science for Sustainable Societies, https://doi.org/10.1007/978-981-15-4458-3_3

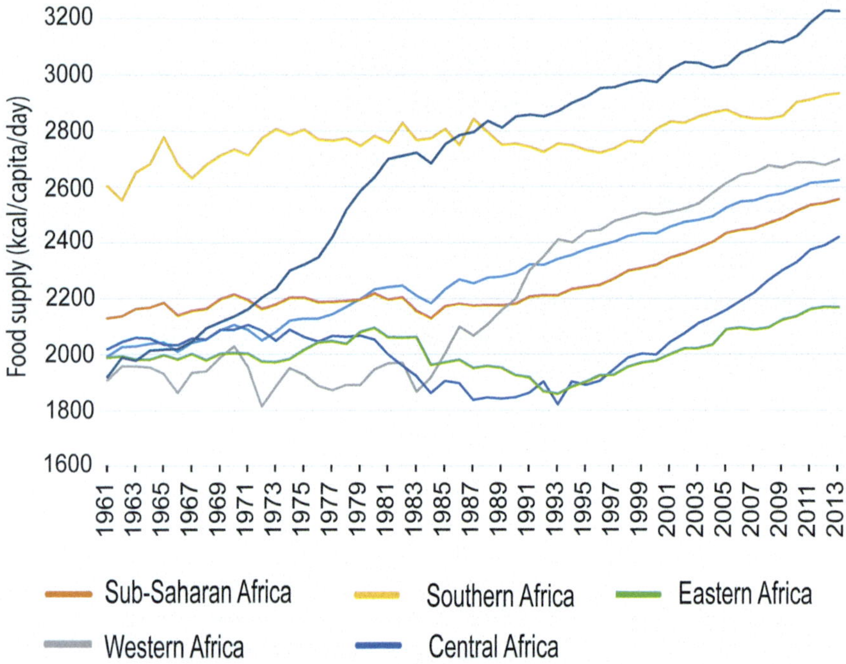

Fig. 3.1 Food supply in sub-Saharan Africa and its sub-regions. Source: FAOSTAT (2019)

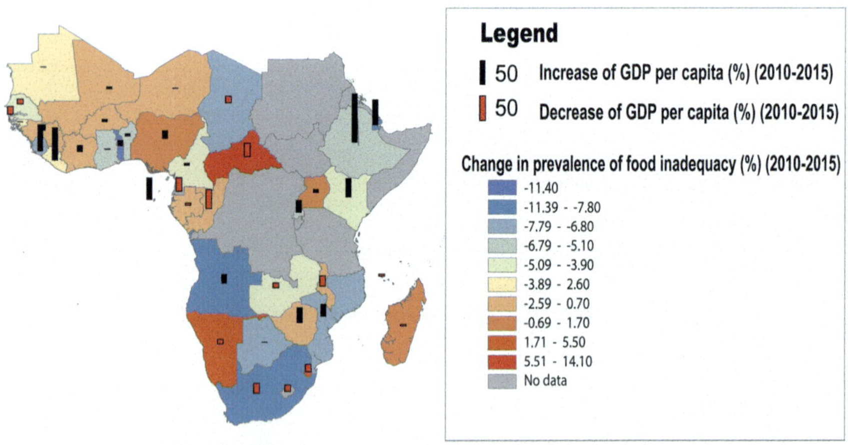

Fig. 3.2 GDP per capita and prevalence of food inadequacy in sub-Saharan Africa. Source: FAOSTAT (2019)

Thus ending hunger and ensuring food security have been major development targets at the national and international level. For example, many local, national and international policies, programmes and interventions have aimed at reducing hunger and achieving food security at different levels across the continent (FAO and ECA 2018; FAO and ECDPM 2018) (Chap. 1 Vol. 2). Ending hunger was a central part of the Millennium Development Goals (MDG1: "eradicate extreme poverty and hunger"). However, despite some success stories, many SSA countries failed to meet the goal by a wide margin (UNDP 2015).

Ending hunger and ensuring food security has become a strong element of the follow-up sustainable development goals (SDGs), with SDG2 (Zero Hunger) aiming to "end hunger, achieve food security and improved nutrition and promote sustainable agriculture" (Chap. 1 Vol. 1). However, the path to meeting SDG2 is particularly challenging as it is linked with many other SDGs including SDG1 (no poverty), SDG4 (quality education), SDG5 (gender equality), SDG6 (clean water), SDG8 (decent work and economic growth), SDG13 (climate action), SDG14 (life below water) and SDG15 (life on land), among others. In fact SDG2 has many interlinkages with several other SDGs, suggesting the centrality, but at the same time difficulty and importance in meeting the goal (ICSU 2017).

Indeed there seems to be no silver-bullet approach towards achieving food security in SSA, as multiple socio-economic and environmental factors combine to affect it (FAO and ECA 2018; FAO, IFAD, UNICEF, WFP and WHO 2019). Often the actual mechanisms are very context-specific, suggesting that what might enhance food security in one context might have a different effect elsewhere (Palm et al. 2010). This is largely because food security is a multi-dimensional concept that combines several different elements and has different interpretations and definitions across international and national contexts (Gibson 2012; Jones et al. 2013). The most prevailing definition includes the four dimensions/pillars proposed by the UN Food and Agriculture Organization (FAO), namely availability, stability, access and utilization[1] (FAO 2006).

Enhancing food crop production has become a central element of reducing hunger and improving food security in SSA, especially related to food availability (FAO and ECA 2018). This is because a large fraction of the SSA's population is either involved in subsistence agriculture or occupied in the agricultural sector (Gollin 2010; IFAD 2013; Ricciardi et al. 2018; FAO, IFAD, UNICEF, WFP and WHO 2019) (Chap. 1 Vol. 1). However, despite the very low past performance there has been evidence of a modest agricultural production growth in past two decades (Wiggins et al. 2015; FAO, IFAD, UNICEF, WFP and WHO 2019). Indeed agricultural output (and its contribution to national economic growth) have been steadily increasing, but with important discrepancies between sub-regions (Fig. 3.3).

[1]Food availability relates to the general supply of food, and addresses issues of food production, stocks and trade. Access to food (both economic and physical) reflects the sufficient supply of food at the national and international levels. Aspects related to income, expenditure, markets and prices are critical in enabling food access. Food utilization relates to nutrient utilization within the human body, as the basis of the good nutritional status of individuals. Food stability reflects the stability of the three aforementioned dimensions over time, which is a necessary pre-condition for sustaining a stable food intake and overcoming periodic risks and food insecurity.

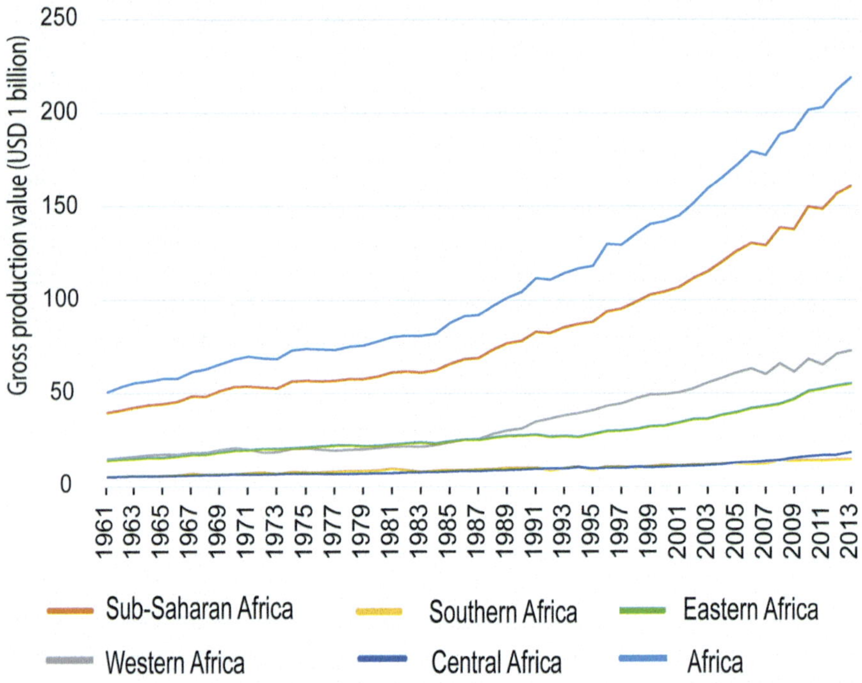

Fig. 3.3 Agricultural gross production value in sub-Saharan Africa and its sub-regions. Source: FAOSTAT (2019)

Apart from food crops, many SSA countries produce crops that have non-food uses for fibre, bioenergy and other industrial products (Singh 2010). These industrial crops[2] cannot be usually used for food consumption (e.g. cotton, tobacco, jatropha) or that have non-food uses but are also integral components of the food industry without being staple crops (e.g. oil palm, sugarcane) (Wiggins et al. 2015; Singh 2010; Jarzebski et al. 2019). Many SSA governments perceive industrial crops as opportunities for modernizing and diversifying the agricultural sector, and thus as potent engines of economic growth and rural development (Schoneveld et al. 2011; Gasparatos et al. 2015). Thus, many SSA countries have opened up their rural frontiers for foreign direct investments (FDIs) related to industrial crop production (Giovannetti and Ticci 2016) (Chap. 4 Vol. 1). This was especially pronounced

[2]Some industrial crops can be mono-functional, used, for example, only for fibre (e.g. cotton), recreation (e.g. tobacco) or energy (e.g. jatropha). Others such as sugarcane can have multiple functions including such as food supplements (i.e. sugar), fuel (i.e. ethanol), and industrial products (i.e. bioplastics). Major staple food crops in SSA such as maize and cassava also have non-food uses (e.g. alcohol production), so they can also be considered to be industrial crops. However, these uses are miniscule in across most of SSA, as these crops are overwhelmingly the main staples across most of the continent.

during the recent land rush that was partly spearheaded by bioenergy crops such as jatropha and sugarcane (Schoneveld et al. 2011).

However, industrial crop production can have various socioeconomic and environmental impacts depending on factors as diverse as the crop, the mode of production, the context within which production takes place and the institutions that govern industrial crop production, use and trade (e.g. Gasparatos et al. 2015) (Chap. 2 Vol. 1; Chap. 5 Vol. 2). Many recent studies have explored a series of impacts related to industrial crop production in SSA including economic growth, poverty alleviation, loss of livelihoods and environmental degradation, among many others (Gasparatos et al. 2015; Arndt et al. 2010; Hess et al. 2016; Strona et al. 2018).

Food security has emerged as one of the most extensively studied impacts (Jarzebski et al. 2019), largely due to the rapid expansion of biofuel projects during the recent land rush, and the popularity of negative narratives such as land grabbing, land dispossession and "food vs. fuel" (Kuchler and Linnér 2012; Tenenbaum 2008; Tomei and Helliwell 2016). Many studies have explored the food security outcomes of industrial crop production in SSA across different scales, (i.e. from the household level to the national and international level) (e.g. Zeller and Sharma 2000; Komarek 2010; Wood et al. 2012; Negash and Swinnen 2013). The perceptions and outcomes of industrial crops articulated in these studies are very polarized, ranging from "industrial crops as major risks to food security" (e.g. Molony and Smith 2010; Matondi et al. 2011; HLPE 2013) to "important agents of economic growth and rural transformation" whose expansion could have positive food security outcomes (e.g. Arndt et al. 2009; Arndt et al. 2012; Hartley et al. 2019; HLPE 2013). However, it is extremely complicated to unravel the food security outcomes of industrial crop production as they are mediated by many different mechanisms (Wiggins et al. 2015; Jarzebski et al. 2019).

The above suggest that the interface of industrial crop production and food security presents a major sustainability challenge for many SSA countries. On the one hand industrial crops can drive the modernization and diversification of SSA agriculture, offering important income and employment opportunities locally and economic growth nationally. On the other hand, they can have profound environmental and socioeconomic impacts, which are mediated through multiple different mechanisms. Collectively these effects can have important ramifications for food security, which can be positive or negative depending on many context-specific factors.

The aim of this chapter is to provide a broad overview of how industrial crop production intersects with food security in SSA. We provide insights about these intersections at different levels (i.e. local, national, continental) through a literature review, secondary data analysis and policy analysis. Section 3.2 outlines the methodological approach. Section 3.3 highlights the history and drivers of industrial crop production in SSA (Sect. 3.3.1) and the mechanisms through which industrial crop production intersects with food security (Sect. 3.3.2). Section 3.4 discusses some of the main literature patterns (Sect. 3.4.1), and policy implications in the context of the SDGs (Sect. 3.4.2).

3.2 Methodology

Initially we conduct a secondary data analysis and policy analysis to understand the drivers, production patterns and policies supporting industrial crop production in different parts of SSA (Sect. 3.3.1). We extract agricultural production data from the statistical database of the FAO (FAOSTAT 2019), and compare them with basic food security indicators such as the "Depth of Food Deficit" and "Global Hunger Index" for selected countries (FAOSTAT 2019; von Grebmer et al. 2018). We focus on countries with different experiences and trajectories of industrial crop production, namely Burkina Faso, Ethiopia, Ghana, Kenya, Malawi and Swaziland.

Subsequently, we conduct a systematic review to synthesize the evidence about the impacts of industrial crop production on food security (Sect. 3.3.2). In particular, we identify the mechanisms through which industrial crop production intersects with food security, and the underlying processes mediating these mechanisms. As a starting point, we use the common food security mechanisms identified in the literature across the pillars of food availability, access, utilization and stability (Wiggins et al. 2015; Jarzebski et al. 2019) (Table 3.1). We use various criteria and variables to categorize the different studies and find patterns within the literature, including study characteristics (e.g. author affiliation, crops, study scale), methods, mechanisms and food security outcomes (Table 3.2). The food security outcomes are categorized based on the presence of mechanisms and whether the mechanism has the predicted effect on food security, the opposite effect or a bi-directional effect.

For the systematic review, we mainly focus on peer-reviewed literature (i.e. journal papers, relevant book chapters), complementing it with limited relevant grey literature. The search keywords were combinations of industrial crops and the names of SSA countries that are the major producers of these crops as identified from the database of the Food and Agricultural Organization of the United Nations (FAOSTAT 2019) (e.g. "cocoa" + "Ghana", "coffee" + "Ethiopia"). The main studied crops include cocoa, coffee, cotton, jatropha, rubber, sugarcane, shea, tea, tobacco, oil palm and vanilla. We identify the reviewed literature though Elsevier Scopus and Web of Science (for peer-reviewed studies), and Google Scholar (technical reports) (Gentles et al. 2016). Other relevant literature was also identified through snowballing (e.g. through the references of reviewed studies). Literature selection was performed in October 2017.

After locating the relevant documents, we screen the abstracts to establish their relevance to the objective of the review (i.e. food security). If the document is relevant, we proceed with the full review. If the abstract does not allow us to conclusively determine relevance, we undertake a full review to avoiding missing any relevant literature. Overall, we fully review 118 papers, *of which 90 are peer-reviewed studies*, with the full list included in the four tables of Sect. 3.3.2.

Table 3.1 Impact mechanisms and predicted outcomes for each pillar of food security

Impact mechanism	Food security outcome	
Pillar A: Food availability		
A1. Food crop area	Industrial crop expansion reduces areas that provide food crops, possibly reducing the production of food crops (land use change: Food crop farms converted for industrial crop production)	↓
A2. Wild food area	Industrial crops expansion reduces areas that provide wild food such as plants, mushroom and game, possibly reducing wild food availability for household use (land use chance: Forest area converted for industrial crop production)	↓
A3. Livestock grazing area	Industrial crops expansion reduces grazing areas for livestock, possibly reducing livestock production (land use change: Grazing area converted into areas for industrial crop area)	↓
A4. Labour/capital diversion	Involvement in industrial crop production (e.g. as plantation workers, industrial crop smallholders) diverts labour and capital from activities related to food production (e.g. food crop cultivation, livestock rearing), reducing thus the production of food crops and livestock	↓
A5. Farming inputs	Industrial crop production increases access to farming inputs, such as seeds, fertilizers, pesticides and herbicides, which can be used/diverted for food crop production increasing food crop yields	↑
A6. Technology	Industrial crop production (and related investments) can introduce new farming technologies (e.g. irrigation, mechanization), which can be used/diverted for food crop production increasing food crop yields	↑
A7. Other	Other mechanism linking industrial crop production to food availability	↑ or ↓
Pillar B: Access to food		
B1. Infrastructure	Industrial crop production can catalyse infrastructure development especially in plantation settings (e.g. opening/enhancing roads and local/public transportation for large-scale production), increasing access to markets to purchase food	↑
B2. Market linkages	Industrial crop production facilitates market linkages (e.g. enhances access of local food/industrial crop producers to buyers), which gives incentive to local farmers to expand food crop production for sale	↑
B3. Food prices	Industrial crops expansion can increase food prices, making it less affordable (and thus accessible) particularly to urban poor and rural landless households	↓
B4. Income generation	Income generated directly through the involvement in industrial crop production (e.g. as plantation workers or smallholders) can be used to purchase food	↑
B5. Job creation	Employment generation by industrial crop estates, processing plants, and other downstream activities can become a dependable and stable source of livelihoods, ensuring a dependable and stable access to food	↑
B6. Land compensation	Lack of compensation for land loss due to industrial crop production does not compensate for the loss of income opportunities and food crop production, causing short- or long-term loss of access to food	↓

(continued)

Table 3.1 (continued)

Impact mechanism	Food security outcome	
B7. Other	Other mechanism linking industrial crop production to access to food	↑ or ↓
Pillar C: Food utilization		
C1. Time diversion	Female engagement in industrial crop activities (e.g. plantation employment, smallholder production) diverts their time from child care, nutrition and unpaid care work, taking a toll on food preparation and especially child nutrition	↓
C2. Energy security	Engagement in industrial crop production value chains can affect household energy choices and the adoption of clean and improved energy options through multiple pathways, improving food preparation practices	↑
C3. Other	Other mechanism linking industrial crop production to food utilization	↑ or ↓
Pillar D: Food stability		
D1. Natural disasters	Engagement in industrial crop activities acts as a buffer against food security risks posted by natural disasters, especially by offering a stable source of livelihoods that can help households cope during such events	↓
D2. Market stability	Industrial crops increases risks of food price fluctuation (i.e. unpredictable increases and decreases) thus reducing food stability	↓
D3. Food imports	Industrial crop production and export generates foreign exchange that can enable food import, enhancing thus the stability of food within the country	↑
D4. Food assistance	Industrial crop plantations/companies can provide food assistance, reducing incidence of seasonal hunger during periods of high food insecurity	↑
D5. Political stability	Industrial crop production can contribute manifold to the national economy and prosperity, catalysing political stability that can have a stabilizing effects for multiple other mechanisms and the wider economy	↑
D6. Women empowerment	Involvement in industrial crop value chains can increase women's access to land, income, and training/education opportunities, enabling them to provide better for their families.	↑
D7. Environmental impacts	Industrial crop introduction can affect local environmental conditions through land use and cover change, soil quality degradation, water quality degradation, and depletion of water resources depletion, among others. Collectively such effects reduce the capacity of local agro-ecosystems to produce food in a stable manner	↓
D8. Other	Other mechanism linking industrial crop production to food stability	

Note: (↑) denotes an expected positive effect on food security, and (↓) a predicted negative effect.
Source: Wiggins et al. (2015), Jarzebski et al. (2019)

Table 3.2 Variables and values used in the systematic review

Variable	Range of values
Peer-reviewed	1 = Yes 2 = No
Crop	1 = Oil pam 2 = Rubber 3 = Sugarcane 4 = Coffee 5 = Jatropha 6 = Shea 7 = Tobacco 8 = Tea 9 = Cotton 10 = Cocoa 11 = Vanilla
Author affiliation	1 = Local (affiliated with institution in the study SSA country) 2 = Regional/African (affiliated with institution in another SSA country) 3 = Outside Africa (affiliated with an institution outside SSA) 4 = Collaboration of African and non-African institutions 5 = Not specified/not known
Research funding	1 = Local (offered by an institution within the study SSA country) 2 = Regional/African (offered by an institution from another SSA country) 3 = Outside Africa (offered by an institution outside SSA) 4 = Co-funded by African and non-African institutions 5 = Not specified/not known
Spatial scale of study	1 = Local 2 = Sub-national 3 = National 4 = Regional 5 = Continental 6 = Global 7 = Not specified
Methodology	1 = Qualitative 2 = Quantitative 3 = Mixed method
Use of direct measure of food security	1 = Yes 2 = No 3 = Both 4 = Not applicable
Type of outcome	1 = Historical/baseline (current state) 2 = Predictive (future state) 3 = Both 4 = Not specified

3.3 Results

3.3.1 Drivers and Production Patterns

3.3.1.1 Continental Perspectives

Some industrial crops such as cotton have been produced for a long period in some parts of SSA, sometimes predating the colonial period (e.g. Kriger 2005). During the colonial era, the production of cotton, cocoa, coffee and rubber expanded rapidly across many parts of SSA to meet the large demand in Europe (Brun 1991). In the post-colonial era, industrial crop production became a major component in the economies of several SSA countries. For example, tobacco and sugarcane are the major export commodities in Malawi (Chinangwa et al. 2017), while cocoa has contributed substantially to the national economies of Ghana and the Ivory Coast (Breisinger et al. 2008). Sugarcane and cotton have dominated the economies of countries in Southern and Western Africa, such as Swaziland and Mauritius (Terry and Ogg 2017; Kwong 2005) and Burkina Faso and Mali respectively (Tschirley et al. 2009; Vitale 2018).

Cocoa is the most widely produced industrial crop in SSA in terms of area (11.37 Mha in 2012), with area under cocoa increased sharply from the mid-1980s onwards, surpassing coffee by a wide margin (Fig. 3.4). Sugarcane is the leading industrial crop in terms of tonnage (94.6 Mt. in 2012) (Fig. 3.5). Export revenues from the major industrial crops have also increased rapidly in the past decades (USD 19.6 billion in 2011) (Fig. 3.6), with industrial crops being the main export commodities across many of the sub-regions (Table 3.3).

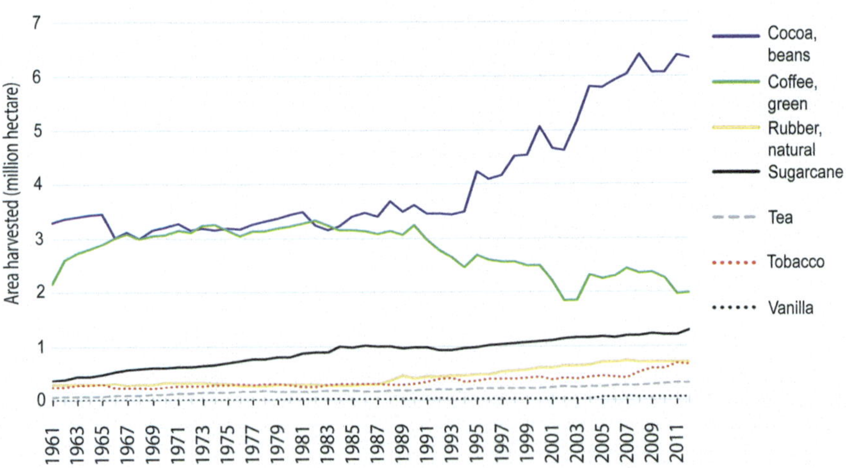

Fig. 3.4 Harvested area for major industrial crops in sub-Saharan Africa. Source (FAOSTAT 2019)

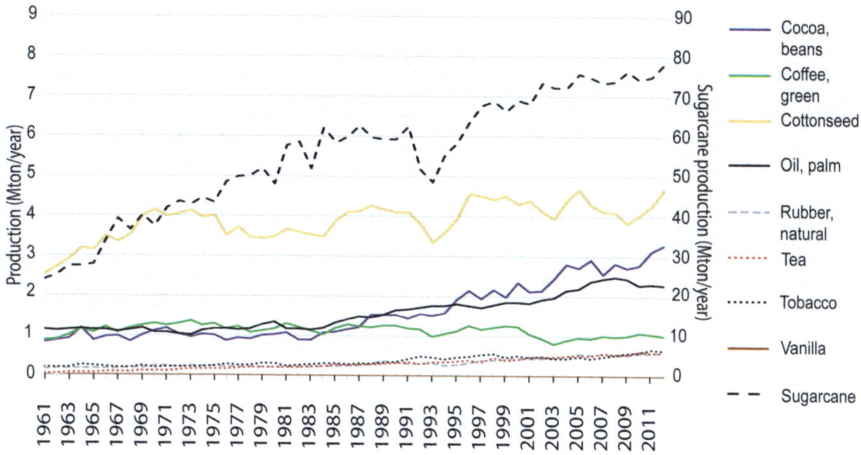

Fig. 3.5 Production output of major industrial crops in sub-Saharan Africa. Source (FAOSTAT 2019). Note: The secondary y-axis in the right denotes sugarcane production.

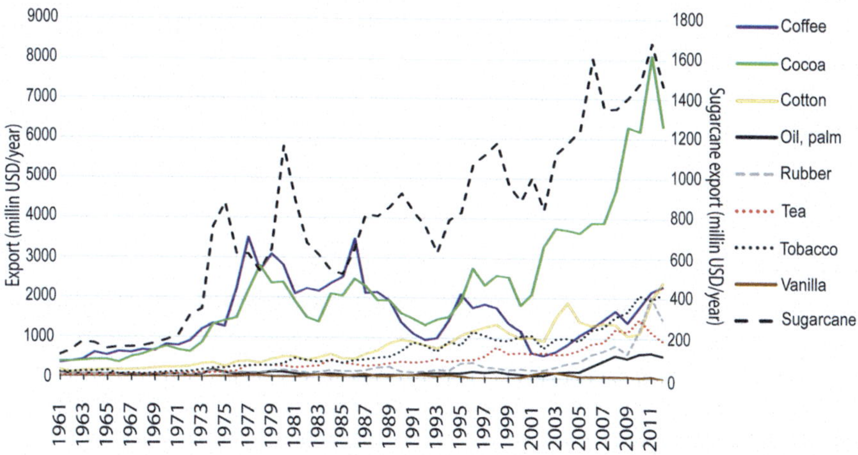

Fig. 3.6 Export value of major industrial crops in sub-Saharan Africa. (Source: FAOSAT 2019). Note: Estimates include raw materials, refined materials and by-products of cocoa, coffee, cotton, oil palm, rubber, tea, tobacco, vanilla and sugarcane

In the past couple of decades, bioenergy crops such as jatropha and sugarcane accounted for most industrial crop expansion, often through foreign-led large-scale land acquisitions (Schoneveld 2014) (Chap. 4 Vol. 1). Estimates suggest that jatropha investments accounted *"for 31.1% of the total area acquired; with the largest areas acquired in Madagascar (979,610 ha), Zambia (707,476 ha), and Ghana (671,951 ha)"* (Schoneveld 2014: 39). However, it is unclear how much land was eventually allocated and converted before the widespread collapse of the sector (Gasparatos et al. 2015; von Maltitz et al. 2014), with evidence suggesting lower

Table 3.3 Prevalence of undernourishment in sub-Saharan Africa and its sub-regions (in %) between 2014 and 2016

Region	Prevalence of food inadequacy (%)	Undernourishment (%) (average between 2014–2016)	Main exports ordered by value (as of 2013)
Middle Africa	47.6	41.3	1. Cocoa 2. Bananas 3. Cotton 4. Rubber natural dry 5. Coffee (green)
Eastern Africa	40.3	31.5	1. Tobacco (unmanufactured) 2. Coffee (green) 3. Tea 4. Crude materials 5. Sugar (raw and refined)
Southern Africa	9.1	5.2	1. Wine 2. Maize, 3. Oranges 4. Sugar raw centrifugal 5. Grapes
Western Africa	13.8	9.0	1. Cocoa 2. Cotton 3. Rubber natural dry 4. Cashew nuts 5. Oil, palm
Sub-Saharan Africa	29.4	23.0	
Africa	25.5	19.8	

Source: FAOSTAT (2019)

land conversion in most countries (Locke and Henley 2013). Recently, palm oil production has increased sharply in Central and Western Africa to meet an the burgeoning regional and global demand (Ordway et al. 2017; Carrere 2013).

The main modes of industrial crop production include (a) smallholder-based schemes, (b) large plantations and (c) hybrids systems that combine large-scale and smallholder-based production (Fig. 3.7). These production systems are integrated in radically different land uses, usually representing a mosaic of agricultural land and natural ecosystems (Gasparatos et al. 2018).

Smallholder-based production is most typically integrated into current farming practices (e.g. intercropping, hedge growing) or totally displaces prior cropping practices (e.g. small block plantations within smallholder farms), with the latter being common in sugarcane production (von Maltitz et al. 2018) (Box 3.1). Smallholder-based systems are often small in size (usually below 5 ha) using labour from inside the household (Lowder et al. 2016). In some cases smallholders have access to agricultural inputs, irrigation and credit, usually through rural development efforts of the respective governments, organization of smallholders in growers' groups and/or extension efforts of government and/or industrial crop companies (von Maltitz et al. 2019; Burney et al. 2013).

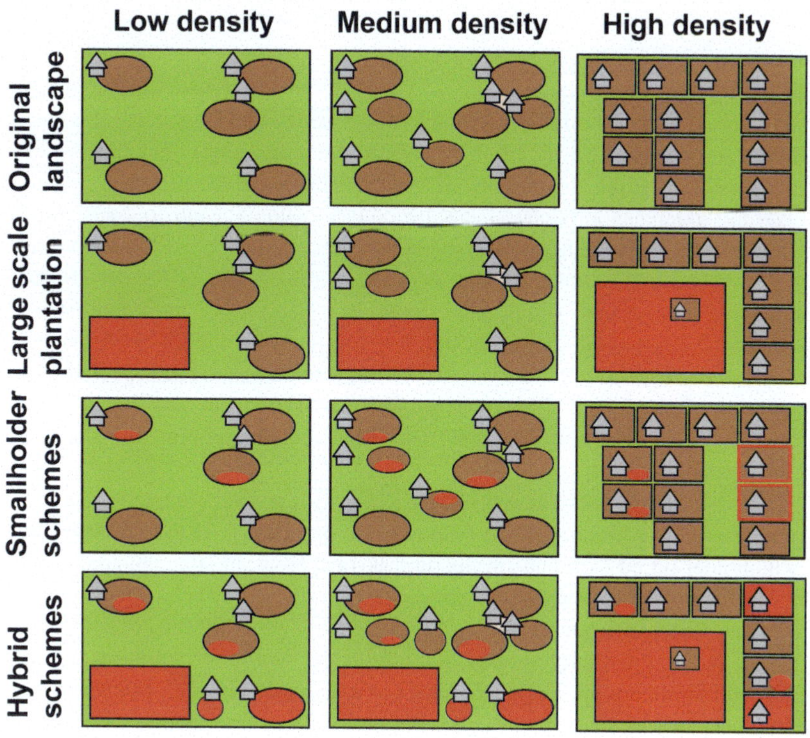

Fig. 3.7 Main modes of industrial crop production in sub-Saharan Africa. Source: Gasparatos et al. (2018). Note: Green denotes natural ecosystems (e.g. woodland, grassland), brown denotes agricultural land, and red industrial crop production

Box 3.1: Food Security Impact Mechanisms in Smallholder-Based Industrial Crop Production Systems

The Komati Downstream Development Project (KDDP) is a smallholder community development programme promoted by a para-statal agency in Swaziland (SWADE) in the late 1990s. Its main aim was to develop capacity for irrigated sugarcane production among smallholders (Terry and Ogg 2017). Smallholders were incentivized to pool their cropland to accommodate sugarcane production, forming 28 associations spanning an estimated 5500 ha (Gasparatos et al. 2018). These associations operate as independent cooperatives or companies, with the involved smallholders becoming equal partners and receiving annual dividends through sugarcane sales to the two mills operated by the Royal Swazi Sugar Company (RSSC) in Mhulme and Shimunye.

(continued)

Box 3.1 (continued)

KDDP was instrumental in providing irrigation and knowledge for sugarcane production to thousands of smallholders (Terry and Ogg 2017). Furthermore, it generated local employment in the community plantations for households that did not join sugarcane production (Mudombi et al. 2018). Overall, sugarcane production has become a major avenue to boost rural development and livelihood diversification (Terry and Ogg 2017), succeeding to reduce poverty among smallholders (Mudombi et al. 2018), despite the extensive conversion of dryland agriculture and degraded ecosystems (Romeu-Dalmau et al. 2018). In this respect, many of the different mechanisms of food security impacts intersect in the broader area, and especially A1 "Food crop area"; A3 "Livestock grazing area"; A5 "Farming inputs"; A6 "Technology"; B1 "Infrastructure"; B2 "Market linkages" and B4 "Income generation."

The BioEnergy Resources Ltd. (BERL) was a private company that promoted jatropha production in Malawi. In particular, BERL incentivized smallholder-based jatropha production in hedges along the boundaries of small family farms, buying in return the produced jatropha. This model assumes that farm boundaries are underutilized (von Maltitz et al. 2014, 2016), and that planting jatropha in these hedges would have minimal trade-offs with food crop production. However, the income generated from jatropha production was relatively low, having little effect on poverty alleviation (Mudombi et al. 2018). Even though BERL targeted around 100,000 farmers across Malawi, only a fraction of these farmers took up and maintained jatropha production. In this respect only a few of the different mechanisms of food security impacts intersect in the broader area, and especially A4 "Labour/capital diversion"; B2 "Market linkages" and B4 "Income generation."

Large industrial crop plantations produce industrial crops in large blocks that can extend from a few tens of hectares to several thousand hectares depending on the crop and the area (e.g. Hall et al. 2017; Smalley 2013). Industrial crop production usually follows intensive monocultural practices and requires extensive land consolidation processes, often displacing rural communities and converting/degrading natural ecosystems (Hall et al. 2017). However, large plantations often generate employment and income, and develop infrastructure (e.g. roads) in poor rural areas that lack such options (Smalley 2013). Several studies have found that plantation employment has very different characteristics depending on the context, ranging from insecure, precarious and lowly paid (Ahmed et al. 2019a), to highly beneficial and appreciated by some local communities (Hall et al. 2017). In any cases plantation employment can divert substantial amount of labour from other local activities (e.g. food crop production), often under questionable working practices and wages. Plantation owners and investors can be private companies, state agencies, parastatal bodies or joint partnerships.

Hybrid production systems usually comprise of a core large plantation surrounded by hundreds or thousands family farmers (Brüntrup et al. 2018) (Box 3.2). These farmers can be contractually linked to a single buyer that is usually the core plantation (i.e. outgrowers), or sell to multiple buyers depending on markets signals (i.e. independent producers) (von Maltitz et al. 2019). Hybrid systems are more common for crops such as sugarcane and oil palm that are perishable and whose production benefits from economies of scale (von Maltitz ct al. 2019; Carrere 2013).

Box 3.2: Food Security Impact Mechanisms in Hybrid Industrial Crop Production Systems

The Ghana Oil Palm Plantation Development Company (GOPDC) has adopted a hybrid production system that combines large-scale oil palm production undertaken by GOPDC in a large plantation surrounded by individual outgrowers and independent smallholders (Ahmed et al. 2019a). The core plantation is located in the Kwae community and spans 8200 ha and is surrounded by approximately 10,000 ha cultivated by outgrowers and independent smallholders. This hybrid production system has converted extensive areas of natural vegetation and other agricultural land (mainly under food crops and cocoa).

The GOPDC currently employs approximately 4500 staff members, over 70% of which are occupied in plantation-related activities such as harvesting, fertilizer application and weeding. Workers are engaged in both permanent and seasonal employment, with salaries and wages paid on a monthly basis. The GOPDC also directly supports about 7200 outgrowers that have contractual relationships with the company, which in turn co-finances the establishment of small oil palm farms. The company serves as the direct market for these outgrowers, as well as a possible market option for independent growers willing to sell fresh palm fruits to GOPDC. Usually independent growers decide whether to sell to the GOPDC mills or other independent processors depending on local prices. Employment and access to this mature market secure income flows to workers, outgrowers and independent growers, improving the livelihoods of many types of households (Ahmed et al. 2019a).

The GOPDC also maintains a road network spanning approximately 500 km within the Kwaebibirem district that facilitates the transportation of oil palm from outgrowers and independent growers to the GOPDC mills. The company also assists local communities in the rehabilitation and construction of community markets. Such infrastructure facilitates the transport and sale of oil palm, as well as the distribution, sale and marketing of food crops within the district.

In this respect many of the different mechanisms of food security impacts intersect in the broader area, and especially A1 "Food crop area"; A2 "Wild food area"; A4 "Labour/capital diversion"; A5 "Farming inputs"; B1 "Infrastructure"; B2 "Market linkages"; B4 "Income generation"; B5 "Job creation"; C1 "Time diversion"; D7 "Environmental impacts."

3.3.1.2 National Perspectives

General Patterns

Different countries across SSA have promoted a wide variety of industrial crops (FAOSTAT 2019), through different modes of production and for different policy goals (Gasparatos et al. 2015). Economic growth, rural development, energy security and conducive policy environments are some of the most common drivers of industrial crop expansion in SSA (Table 3.4). While rural development and economic growth have been the main driver of industrial crop production in many of SSA countries, energy security and conducive policy environments have played a major role in some countries (Table 3.4).

Similarly, the dynamic interface of industrial crop expansion and national food security has been rather different across SSA countries. Figure 3.8 highlights patterns of industrial crop production (in terms of allocated land) and food security (in terms of depth of food deficit and hunger) for some of the major industrial crop producers in SSA. In some of these countries, the long-term expansion of industrial crop production has been relatively moderate (even reaching a plateau in some countries such as Ghana), and has coincided with a constant reduction of hunger and food deficit (Fig. 3.8). In some of these countries the reduction of hunger and food deficit has been rapid (e.g. Ghana), while in others more gradual (e.g. Ethiopia, Malawi). However, such food security outcomes might reflect broader national socio-economic processes, with industrial crop expansion being one of the activities that might influence national food security (de Graaff et al. 2011).

Conversely, the expansion of single industrial crops has been very aggressive in countries such as Burkina Faso and Swaziland, eventually dominating the respective national economies (Vitale 2018; Terry and Ogg 2017) (Fig. 3.8). In these countries, the periodical rapid industrial crop expansion or contraction cycles have also tended to coincide with rapid changes in hunger and food deficit (Fig. 3.8). This possibly suggests a closer linkage between industrial crop production and national food security in countries whose economies are dominated by industrial crop production, and especially of a single crop. However, this most certainly also reflects many other related factors including reforms in the industrial crop sector (Kaminski et al. 2009) and international commodity prices and trading regimes (Terry and Ryder 2007).

It is important to note that industrial crop production is rarely uniform within individual countries. In fact, the production of many industrial crops is concentrated in specific areas. This is especially the case for crops such as sugarcane and oil palm

Table 3.4 Drivers of industrial crop expansion in selected SSA countries

Driver	Kenya	Ethiopia	Malawi	Burkina Faso	Ghana	Swaziland
Economic growth	–	✓	✓	✓	✓[a]	✓
Energy security	✓	✓	–	–	✓	–
Rural development	✓	✓	✓	✓	✓	✓

Note: [a]Excluding cotton

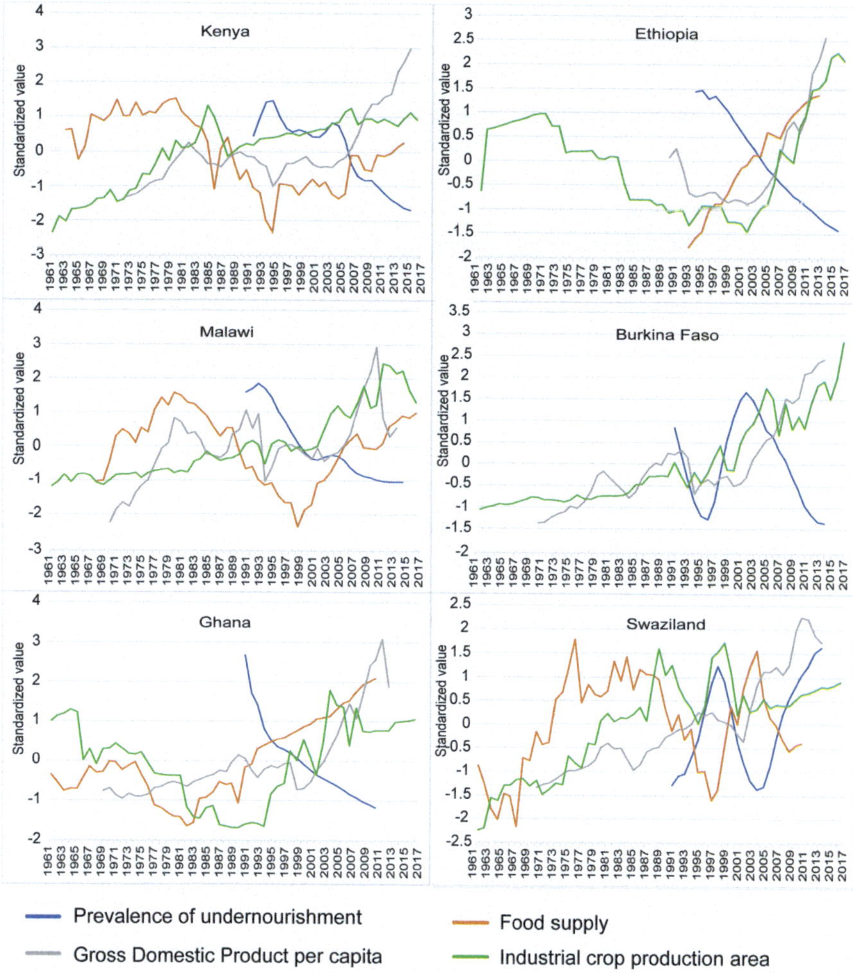

Fig. 3.8 Patterns of food security and industrial crop production in selected SSA countries. Note: Data points represent indicator levels for each year compared to the base indicator value in the year 2016. Hence data points do not report absolute indicator values, but reflect increases or decreases in indicator level compared to the 2016 baseline. Data for "Industrial crop production area" report aggregate production area figures for all industrial crops produced in each SSA country and considered in this study (Sect. 3.2), with the exception of jatropha (FAOSTAT 2019). Hence, land under industrial crop production is likely to be underestimated for countries that have experienced large jatropha expansion such as Ghana (Schoneveld 2014). Data for "Depth of Food Deficit" and "Global Hunger Index" are collected from (FAOSTAT 2019; von Grebmer et al. 2018). The right-hand y-axis in Fig. 3.8a (Ghana) reports values for the "Depth of Food Deficit"

that require large investments to achieve economies of scale. For example, sugarcane production in Malawi and Swaziland is concentrated in just two relatively small areas in each country; Nchalo and Dwangwa in Malawi (Chinangwa et al. 2017), and Big Bend and Northern Lowveld in Swaziland (Terry and Ogg 2017). Conversely, industrial crops geared mainly towards smallholders (e.g. coffee, cocoa, cotton, tobacco) tend to be produced in wider areas that offer conducive agro-ecological conditions. For example, tobacco in Malawi is mainly produced in the Central, Northern and Southern regions (Chinangwa et al. 2017). In Ethiopia, coffee is produced in many parts of the country, and mainly in the southern regions of Sidamo, Harrar, Ghimbi and Limu (Moat et al. 2017). Cotton is grown in most regions of Burkina Faso (except for the arid north regions), with the eastern cotton zone of Bobo-Dioulasso accounting for most of cotton output (Boafo et al. 2018). Cocoa production spans large parts of Ghana and especially southern Ghana. This concentration of industrial crop production can have more pronounced and easily tracked outcomes on local food security (rather than national food security), as discussed throughout this paper (see list of mechanisms in Sect. 3.3.2).

Below we unpack industrial crop production patterns, drivers and policies for some of the main producing countries in SSA, namely Kenya (Sect. 3.1.2.2), Ethiopia (Sect. 3.1.2.3), Malawi (Sect. 3.1.2.4), Burkina Faso (Sect. 3.1.2.5), Ghana (Sect. 3.1.2.6) and Swaziland (Sect. 3.1.2.7).

Kenya

Sugarcane is the major industrial crop produced in Kenya in terms of output followed by tea and coffee (FAOSTAT 2019). The sugar industry supports an estimated two million Kenyans and contributes about USD 540 million to the national GDP (FAOSTAT 2019), with smallholder farmers supplying more than 92% of the sugarcane processed by the domestic sugar mills (KSI 2009; KSB 2010). Sugarcane output increased rapidly between the early 1960s to the early 1980s (from 570,000 tons to 4.5 million tons) (FAOSTAT 2019). For the next 15 years, the sugarcane output oscillated around this level, but then experienced a rapid increase from 4.7 million tons in 2004 to 7.2 million tonnes in 2016 (FAOSTAT 2019). However, the sugarcane output experienced a rapid decline falling to 4.8 million tonnes in 2017 (33.3% decline) (FAOSTAT 2019). This drastic production reduction causes the sharp decline in the value of marketed sugarcane, from USD 234 million in 2016 to USD 195 million in 2017 (17% decline) (GoK 2018). This decline sugarcane output has been linked to multiple interconnected challenges related to low yields such as the (a) widespread use of low-quality sugarcane varieties; (b) shortage of irrigation water; (c) poor agricultural and land management practices and (d) delayed harvesting of mature sugarcane due to weather variability and/or logistical constraints (Mulwa et al. 2005; Mulianga et al. 2015; Onyango et al. 2012; Lindell and Kroon 2011; Hess et al. 2016). The Kenyan sugar industry has also faced critical challenges related to the (a) trade liberalization under the COMESA and

WTO protocols; (b) high production costs compared to other sugar-producing countries in the region (e.g. Tanzania); (c) poor governance and management; (d) insufficient funding and (e) inadequate research and extension services (KSI 2009; GoK 2007a).

The coffee sector plays an important role in the national economy in terms of income/employment generation, foreign exchange earnings and tax revenue generation. Coffee production increased rapidly in the first two decades following the 1963 independence, with the total coffee output from large estates and smaller cooperatives increasing from 43,778 tons in 1963, to 130,000 tons in 1988. Since then, however, the coffee industry has been on a downward trend, with the total output reaching only 53,400 metric tonnes in 2007 (GoK 2007b). Coffee output further declined in the following decade, reaching about 41,000 tons in 2017 (FAOSTAT 2019). As a result, the contribution of the coffee sector to the national economy has declined appreciably (Thuku et al. 2013). The sector faces many challenges including monopolistic practices, cooperative mismanagement, repeated droughts, decreasing international prices and weak infrastructure (Condliffe et al. 2008).

Tea cultivation and production have expanded from 18,000 tonnes and 21,488 ha (Nyagito 2001) (in 1963), to 328,500 tonnes and 141,300 ha, respectively, in 2005 (CBS 2005). The tea output and cultivation area increased further to 293,670 tonnes and 218,538 ha, respectively, in 2017 (FAOSTAT 2019). In 2002, Kenya was second only to Sri Lanka in exports of black tea (Bassett 2010). The success of tea sector has been attributed to the (a) supportive government policies following independence that have integrated successfully the small-scale growers into the sector, (b) adoption of high-yielding varieties mainly developed nationally by the Tea Research Foundation of Kenya (TRFK) and (c) selective application of herbicides and improved planting and cultivation methods (Kagira et al. 2012; Onduru et al. 2012). On the other side, obstacles facing tea smallholders include the prolonged droughts, lack of credit facilities and poor road infrastructure to transport the produced tea (Gesimba et al. 2005).

Since 2001, several policy initiatives have sought to support the agricultural sector, both targeting the main industrial crops discussed above and the broader sector. The Poverty Reduction Strategy Paper (2001–2004) developed different initiatives aiming to achieve sustainable growth in the agricultural sector through improved extension services, provision of credit to smallholders, improvement of rural infrastructure, development of stronger marketing links and capacity-building for institutions implementing these initiatives (GoK 2001). The Strategy for Revitalizing Agriculture (2004–2014) aimed at reversing the declining performance of the agricultural sector by introducing new management approaches including drastic changes in the operation ministries overseeing the sector and their interaction with other key stakeholders. The strategy emphasizes on the role of private–public partnerships as a means of facilitating competition, enhancing market performance and raising resource utilization efficiency (GoK 2004). The Kenya Rural Development Strategy (2002–2017) has been a longer term framework emphasizing on food security as the first step towards poverty alleviation and equitable growth and rural development (GoK 2002). Finally, the Strategic Plan (2008–2012) was a 5-year strategic management plan promoted by the Ministry of Agriculture that catalysed institutional, policy and civil service reforms. Particular emphasis was paid to governance bottlenecks, food insecurity and volatile trade and financial regimes that have large influence on national agricultural production (GoK 2008).

Ethiopia

Coffee is the main industrial crop and export revenue stream in Ethiopia, generating approximately 25–30% of the national total export earnings (Moat et al. 2017). Ethiopia is the leading coffee producer in SSA and the fifth largest producer globally. In 2018, coffee production reached 7.1 million 60-kg bags (~426,000 metric tons), with exports forecasts at 3.98 million bags (~239,000 metric tons) (USDA 2018). Coffee production is predominantly characterized by traditional farm management systems, limited use of fertilizers/pesticides and manual cultivation systems and drying methods (Tefera and Tefera 2014). The estimated area used for coffee production (525,000 ha) (Tefera 2015) shrinks occasionally largely due to increasing population, land use conflicts, extensive deforestation, expansion of other industrial crops and other agricultural practices (Minten et al. 2017; Sisay 2018).

Sugarcane is another major industrial crop produced in the country. The sugarcane sector has undergone extensive transformation in the country, which has been mostly driven by the government. Even though smallholder-based sugarcane production has been prevalent for centuries, its large-scale cultivation started in the early 1950s with the establishment of the Wonji Sugar Factory (Wendimu et al. 2016). Subsequently the government pushed for the development of additional sugar plants to meet the increasing domestic sugar demand. The Sugar Corporation of Ethiopia currently administers six sugar factories (e.g. Wonji-Shoa, Metahara, Finchaa, Tendaho, Arjo-dedessa, Kessem), and nine sugar development projects at Kuraz, Tana Beles and Welkayit (Gashaw et al. 2018). Annual production has currently reached approximately 100,000 ha of sugarcane, 400,000 tons of sugar and 25,000 m^3 of ethanol, with the new sugar factories expected to expand significantly the production of sugar and other energy co-products through ethanol distilleries and bagasse cogeneration facilities (Gashaw et al. 2018).

Economic growth and rural development have been the two major drivers fuelling coffee expansion in the country. Conversely, energy security and economic growth have been the main drivers for sugarcane production (Tefera 2015; Hailemariam et al. 2019). Sugarcane ethanol production and bagasse co-generation are seen as possible avenues to increase domestic energy security considering Ethiopia's high reliance on imported fossil fuels (Berhanu et al. 2017). Indeed, the Ethiopian government has identified the sugar sector as a focal point in its efforts to become a middle-income country by 2025.

The Ethiopian government has fully supported the production of industrial crops through multiple relevant initiatives. In 1995, the Agricultural Development-Led Industrialization programme (ADLI 1995–2005) sought to boost the performance of the agricultural sector by transforming smallholder-based agriculture into a vehicle to catalyse the shift towards an industrial economy. The "Plan for Accelerated and Sustainable Development to End Poverty" (PASDEP) (2005–2010) emphasized on economic growth through agricultural commercialization. Essentially the national government and other key development partners such as the World Bank perceived large-scale commercial agriculture as essential for increasing food production, enhancing economic growth, increasing foreign exchange earnings, generating employment, enabling technology transfer to smallholders, modernizing agriculture and developing infrastructure and basic services to local communities (Rehmato 2011). Around that period, the government started to actively promote large-scale commercial agriculture (especially of sugarcane) by allocating land for large-scale agricultural investments to domestic and international investors. Other relevant government policies seeking to boost the performance of the agricultural sector partly through industrial crops include the (a) Sustainable Development and Poverty Reduction Program (2002–2005), (b) Plan for Accelerated and Sustainable Development to End Poverty (2005–2010), (c) the first Growth and Transformation Plan (2010–2015) and (d) the second Growth and Transformation Plan (2016–2020).

Malawi

Tobacco, sugarcane, tea coffee and cotton are some of the main industrial crops produced in Malawi. Of these tobacco and sugarcane have traditionally contributed significantly to the national economy and rural livelihoods. They collectively account for approximately 79% of the national foreign exchange earnings and 22% of the gross domestic product (GDP) (Chirwa 2011).

Tobacco production started in the early twentieth century, but only reached high production levels in the 1960s. Total production increased from about 15,000 tonnes in the mid-1960s to about 160,000 tonnes in 1997, before declining sharply (FAOSTAT 2019). Production bounced back and reached an all-time high of about 175,000 tonnes in 2011, before declining again (FAOSTAT 2019). Many different reasons have contributed to tobacco expansion in Malawi such as (a) shifts in tobacco demand from developed to developing countries; (b) declining political support for tobacco production in developed countries; (c) cost competitiveness and the relatively high profitability of the crop; (d) technical and financial support and (e) large investments from international companies (Jaffee 2003). Currently, tobacco is produced overwhelming by smallholders that grow it either on their own land or as tenant farmers (Kulik et al. 2017). The declining tobacco output and prices in recent years have been possibly due various interconnected reasons such as exceeding farmer quotas, poor quality product, international competition, global demand decline and certain corporate strategies (Kulik et al. 2017).

Conversely, sugarcane production has experienced a large and constant expansion since the late 1960s. The two main plantations in Malawi started operating in 1968 (Nchalo) and 1978 (Dwangwa) (Chisanga and Zulu-Mbata 2017). The sugar industry was privatized in 1998, with the South African company Illovo taking over the two production estates at Nchalo and Dwangwa. This meant that Illovo practically accounted for all sugar milling and refining capacity in the country. Overall, sugarcane output increased from about 170,000 tonnes in 1968, to over 2.9 million tonnes in 2017 (FAOSTAT 2019). Even though the two large estates in Nchalo and Dwangwa account for most sugarcane production, smallholder-based production has been promoted in both areas through various policies and incentives put in place by the national government (Chisinga et al. 2017). Furthermore, there has been continuous sugarcane ethanol production since the early 1980s, as a response to the energy crises (Gasparatos et al. 2015) (Chap. 2 Vol. 1; Chap. 5 Vol. 2). Malawi has been constantly blending high proportions of ethanol in gasoline, and is thus considered one of the pioneer countries globally in the transition to alternative transport fuels (Johnson and Silveira 2014).

Tobacco production seems to be overwhelmingly driven by national government efforts to boost economic growth and rural development. As mentioned above, tobacco is by far the main export in the country and is almost entirely produced nowadays by smallholders. On the other hand, sugarcane production has been mostly driven for economic growth. Even though Illovo undertakes most sugarcane production, a relatively large number of irrigated and rain-fed smallholders have been involved in the sector. Interestingly, despite sugarcane ethanol being blended in high proportions with gasoline, energy security does not seem to have been a major driver of sugarcane production. However, this might be gradually changing considering the current ongoing efforts to diversify ethanol feedstock production to boost the national ethanol output.

Many national policies and initiatives have been put in place to assist industrial crop production in Malawi. One of the first such regulations was the 1970 Tobacco Act that regulated tobacco production (Wiggins et al. 2015). Since 1981 structural adjustment programmes funded by the World Bank and the International Monetary Fund (IMF) were implemented, precipitating many institutional and policy changes in the agricultural sector in general, and food marketing in particular (Chilowa and Chirwa 1997). The liberalization of marketing and production inputs was implemented in 1990 under the Agriculture Sector Adjustment Credit initiative, which allowed the entry of private traders (of varying sizes in terms of scale of operations) in the marketing of farm inputs. Changes in the Special Crops Act of 1994 allowed for smallholder farmers participation in burley tobacco production, which was formerly undertaken only by large-scale estates. The smallholder tobacco sector benefited from the targeted Farm Input Subsidy Programme (FISP) (2006–2010), which offered poor farmers access to seeds and fertilizers. Similarly, the Malawi government has been supporting smallholder sugarcane growers with loans and grants mainly through two outgrower management companies, namely the Dwangwa Cane Growers Ltd. (DCGL) in Dwangwa and the Kasinthula Cane Growers Ltd. in Nchalo (Chisanga and Zulu-Mbata 2017). The National Export Strategy of 2012 was established to maintain the stable production of sugarcane for sugar and ethanol production (Wiggins et al. 2015).

Apart from these policies there is a constellation of national policies and institutions that affect industrial crop production. Some of the most prominent include:

(a) Malawi Growth Development strategy I (2006–2011) and II (2011–2016), which have been the overarching national development framework; (b) National Agriculture Policy Framework (NAPF) 2010–2016; (c) Agricultural Sector Wide Approach (ASWAP) 2010–2014; (d) Green Belt Initiative 2011–present; (e) National environmental Policy 2004 and (e) Malawi Energy policy, 2003.

Burkina Faso

Cotton was produced in Burkina Faso during the pre-colonial times as a secondary crop by family farmers. During the colonial period, cotton was produced in a top–down technocratic system that treated poorly these cotton smallholders (Vitale 2018). However like many other countries Western Africa, Burkina Faso re-oriented its economy towards cotton production in the post-colonial period. Despite some variability, cotton output increased sharply, from a few thousand tonnes in the early 1960s, to more than 370,000 tonnes in the late 1990s (FAOSTAT 2019). In the early 2000s, sudden cotton price drops precipitated a large decline across the sector, leading the country to insolvency. However, despite price volatility and concerns amidst stakeholders about the sector's viability (Vitale 2018), cotton output reached record levels in the early 2010s at more than 400,000 tonnes (FAOSTAT 2019). Up to the early 1980s, output gains materialized largely through the huge yield improvements catalysed by the introduction of animal traction and agricultural inputs (Vitale 2018). However, as yields leveled off in the mid-1980s, it has been cotton land expansion that caused the increases in output (World Bank 2013).

Economic growth has been perhaps the major driver of cotton production in Burkina Faso. Indeed cotton is the main agricultural commodity produced in the country and has historically constituted a substantial fraction of the GDP. On average cotton export revenues have accounted about 2.5% of GDP over the past decade, offering a stable source of foreign exchange that has catalysed economic development in other sectors (Vitale 2018).

Since independence and up until the early 2000s, SOFITEX (a government parastatal company) and CFDT (a privately owned French company) had complete control over cotton processing and marketing. In this "one-stop" cotton farming system SOFITEX provided on credit all of the production inputs to growers and maintained exclusive rights to purchase the seed cotton from farmers (Vitale 2018). A series of economic failures influenced major donors to push for reforms in the sector including among others: (a) changes in the laws for establishing farmer groups (1994, 1996–1999), (b) the establishment of the national cotton union (UNPCB) (1996–2001), (c) the partial privatization of SOFITEX (1999), (d) the delegation of major responsibilities from government and SOFITEX to UNPBC (2000–2006); (e) the introduction of new players in the sector such as private input providers, new regional private cotton monopsonies (SOCOMA, FASOCOTON) and private transport companies (2002–2006) and (f) changes in price-setting mechanisms to reflect better the prevailing global circumstances (2006–2008) (Kaminski 2009). Currently, three cotton companies operate in the major production areas, namely Burkina Faso–SOFITEX, Faso Coton and SOCOMA. These companies purchase cotton at the same price and follow a common pricing scheme. However, despite maintaining the "one-stop" approach, cotton prices are now negotiated among the principal stakeholders within the sector (Vitale 2018). Despite the increase of cotton output in past decades, there are mixed perspectives about the success of these reforms in increasing the sustainability of the sector (Vitale 2018; Kaminski 2009; Boafo et al. 2018).

Ghana

The main industrial crops produced across the different agro-ecological zones of Ghana include cocoa, cotton, rubber, coffee, shea, sugarcane, tobacco, oil palm and citronella. Below we examine the drivers and production patterns for three rather different industrial crops, namely cocoa, cotton and jatropha. These crops have experienced radically different trajectories influenced by their different agronomic characteristics, modes of production, national policy priorities they cater for, and international circumstances.

Cocoa has been the most important export crop of Ghana since its introduction in the mid-nineteenth century. Ghana is currently the world's second largest producer of cocoa behind the Ivory Coast (FAOSTAT 2019). As one of the main export crops, cocoa has been central to Ghana's development, economic growth and poverty alleviation efforts, as it supports about 800,000 households (Vigneri and Kolavalli 2018; GSS 2014). The evolution of the Ghanaian cocoa sector spans four distinct phases: (a) introduction and exponential growth (1888–1937); (b) stagnation followed by a brief but rapid growth following independence (1938–1964); (c) near collapse (1965–1982) and (d) recovery and expansion following the introduction of the Economic Recovery Program (ERP) (1983–present) (Quarmine et al. 2014; Williams 2009). It is worth mentioning that during the collapse phase cocoa production fell from about 580,000 tonnes (in 1964) to less than 170,000 tonnes (in 1984) (FASOTAT 2019). Since then, the cocoa output has been increasing constantly reaching record output in 2017, standing at more than 880,000 tonnes (FAOSTAT 2019).

Similar to Burkina Faso, the first coordinated efforts to promote cotton production in Ghana in the northern semi-arid regions of the country started following independence. Cotton output increased significantly until the late-1990s, albeit not to the same levels as other West African countries (i.e. 26,000 tonnes in 1998) (FAOSTAT 2019). The international cotton price decline of the 1990s affected the sector to the extent that it is now virtually at the point of collapse (Howard et al. 2012), with the total output currently amounting to less than 1% of cotton output in Western Africa, despite excellent growing conditions in the northern part of the country. Overall, cotton production has been driven by various factors during different periods including, rural development, exports for foreign exchange earnings and the production of raw material for local textile industry (Boafo et al. 2018). Similarly, the reasons behind the sector's underperformance are multi-faceted (Boafo et al. 2018), including the lack of institutional and regulatory frameworks that could promote effectively cotton to meet its rural development potential (MOFA 2013).

Jatropha was the most recent widely promoted industrial crop in Ghana. Jatropha received substantial attention in mid-2000s as a biofuel feedstock crop, with the National Jatropha Project Planning Committee pushing for the development of 1 Mha of jatropha plantations within a period of 5–6 years (by 2015) (Ahmed et al. 2017). However due to a set national circumstances such as the discovery of offshore oil deposits and the death of the early proponent of jatropha promotion, the impetus for jatropha expansion passed from the government to the private sector, and especially foreign interests (Ahmed et al. 2017). Many FDIs between the late 2000s to mid-2010s targeted large expanses of land to produce jatropha for export, and especially in the EU (Ahmed et al. 2017) (Chaps. 2 and 4 Vol. 1). This extensive allocation of land essentially made jatropha production the main driver of the land rush in Ghana (Schoneveld 2014). However, the sector has practically collapsed by the mid-2010s due to a combination of (a) low jatropha productivity, (b) weak business planning, (c) community conflicts, (d) institutional barriers and (e) civil society opposition (Ahmed et al. 2017, 2019b).

The promotion of these different industrial crops in Ghana was driven by slightly different factors. For example, cocoa production was largely driven by the availability of land in rural agro-ecosystems, and the need to generate foreign exchange to boost national economic development and poverty alleviation (Kolavalli and Vigneri 2011). Cotton, on the other hand, was promoted for rural economic development (especially targeting the arid and semi-arid regions of the country), as well as to boost exports of merchandize exports as part of the export substitution development strategy (Hussein et al. 2005; Baffes 2005). The drivers of jatropha production also varied over time. Initially, national energy security was the main driver behind the early promotion of jatropha, as implied by the strong focus on transport mandates of the early biofuels projects (Energy Commission 2006; Ahmed et al. 2017). However, by 2008, rural development through FDIs became the major driver of jatropha expansion in Ghana, given its prospects to create jobs in plantations and income for rural communities (Ahmed et al. 2017; Iddrisu and Bhattacharyya 2015; Boamah 2014).

The cocoa sector was shaped by the early establishment of the Ghana Cocoa Marketing Company in 1947, which formed into the Ghana Cocoa Board that seeks to promote the production, processing and marketing of high quality cocoa. Following the expansion and near collapse of the sector in the 1980s due to diseases, the government initiated two rehabilitation programmes (Cocoa Rehabilitation Programme I and II), which were unsuccessful (Amoah 1995; Kolavalli and Vigneri 2017). More importantly this period saw the implementation of the Economic Recovery Programme, which promoted, among others, the liberalization of the cocoa sector, the dissolution of all para-statal agencies and the adoption of a rehabilitation drive. Currently, the cocoa sector is partially liberalized with the Ghana Cocoa Board being responsible for cocoa exports and price-setting (Kolavalli and Vigneri 2017). To further boost cocoa production, the Cocoa Sector Development Strategy I (CSDS I) was implemented in 1999. Among others CSDS I permitted Licensed Buying Companies to export cocoa, which was however ignored by the government at that time (Kolavalli and Vigneri 2017). Policy revisions led to the CSDS II, which deals with emerging issues affecting cocoa production such as child labour, certification and climate change. CSDS II ensures consistency with the national and international development agenda. Currently, multiple government-led initiatives through the Productivity Enhancement Programmes (PEPs) aim to double yields, from the current average of 450 kg/ha to 1000 kg/ha by 2027. The PEPs encapsulate a series of programmes such as the Farm Rehabilitation Programme, the Diseases and Pest Control Programme, the Soil Fertility Management through High Technology Programme, the Irrigation of Cocoa Farms Programme and the Artificial Pollination Programme among others (MoFA 2018). In addition, there are multiple efforts to promote value addition and youth employment in the cocoa sector within the broader framework of the government-led initiative "Investing for Jobs" (MoFA 2018).

The cotton sector has undergone major policy changes throughout the past decades. The first major action was the establishment of the Cotton Development Board (CDB) in 1968, a para-statal agency tasked to oversee the cotton sector. By the mid-1980s, the national government faced strong pressure from the World Bank to reform the sector, mostly as part of the Structural Adjustment Programmes of the 1980s (Peltzer and Röttger 2013). In the mid-1980s, the cotton sector was deregulated, and the CDB was transformed into the Ghana Cotton Company Limited (GCCL), with the government retaining 30% of the shares. Over the next decade, there was a proliferation of private companies participating in the cotton industry, and in 1995 the government sold its 30% share of the GCCL. At around that period free agricultural input supply to cotton smallholders changed into input credit, and a price-setting mechanism was put in place. In 1997–1998, the Cotton Development Project 1 (CDP-1) was launched, but there were unprecedented malpractices such as poaching of farmers and adulteration of product. In the early 2000s, the financing of cotton companies ceased due to their high accumulated debts, with several companies exiting the cotton sector. In 2000–2002, MOFA introduced a zoning policy to address malpractices in the sector, and in 2004 the African Development Bank took control of the GCCL through a debt-equity swap. Shortly after that point the GCCL was liquidated and ceased its operation, but in 2010–2011 there was an effort to revive the sector through the "White Gold" campaign. The last major policy initiative was the formal inauguration of the board of Cotton Development Authority in 2016 to carefully examine the challenges of the sector.

The jatropha sector received a relatively short, but intense policy interest in Ghana that coincided with the expansion of the bioenergy sector. The starting point was the 2005 draft biofuel policy and the 2006 Strategic National Energy Plan, which were anchored on energy security concerns and established biofuel blending mandates for the transport sector. Subsequently, rural development priorities started dominating the sector, with the 2007–2008 Biofuel Implementation Group and the National Jatropha Planning Committee seeking to establish 1 Mha of jatropha plantations in 53 districts. Subsequent relevant policies have included the 2010 bioenergy policy (which included revised blending mandates) and the 2011

Renewable Energy Act (which set incentives for large-scale land acquisitions and formal licensing/permitting processes). The 2012 and 2016 guidelines for large-scale land acquisitions sought to regulate such processes, especially in the bioenergy sector that had already experienced extensive land allocation (Ahmed et al. 2017).

Swaziland

Sugarcane and cotton are the major industrial crops produced in Swaziland. Sugarcane production increased almost ten-fold since the early 1960s, from 61.7000 tonnes in 1961 to 573.7000 tonnes in 2017 (FAOSTAT 2019). Practically all sugarcane production is concentrated in two areas, with 77% produced in large-scale plantations operated by the Royal Swaziland Sugar Corporation (RSSC) and Illovo (Terry and Ogg 2017). Conversely cotton production has been largely based on smallholders Hinderink and Strenkenburg 1987, experiencing very divergent production trends, namely a large increase in production (1961–1990) followed by a large decline (1991–2017). The all-time high cotton output was achieved in 1990 (26.0,000 tonnes), dwindling to 1.1000 tonnes in 2017 (FAOSTAT 2019).

Both sugarcane and cotton production have been promoted as avenues of economic growth and economic development (Table 3.4). In terms of policies, the Sugar Act of 1967 mainly regulates sugarcane production in the country. This act essentially stipulates the main rules governing sugarcane growing, processing and marketing, with the sector regulated by the Swaziland Sugar Association and the Quota Board (UNCTAD 2000). Large-scale irrigation development projects such as KDDP (Box 3.1) were promoted through para-statal agencies to develop smallholder-based sugarcane production capacity as a means of rural development and poverty alleviation (Terry 2012). Sugarcane production further flourished through the funds and technical assistance provided by the EU, as a consequence of the reforms in its sugar regime in 2006 (Richardson 2012). In particular, Swaziland received about EUR 134 million through an "Aid for Trade" programme called the Accompanying Measures for Sugar Protocol countries (AMSP) to enhance the competitiveness of its sugar industry, diversify economic activity in sugarcane areas and address social and environmental impacts (Richardson 2012).

Cotton on the other hand received much less policy attention. Cotton production was promoted through the government in some rural development areas (RDAs) primarily in the Lowveld and secondarily in the Lubombo plateau, with production in other parts of the country relying on the efforts of individual farmers (Hinderink and Sterkenburg 1987). The cotton sector is vertically integrated and state-controlled, with the Swaziland Cotton Board being the main institutional body coordinating the cotton sector, including overseeing cotton research, production and marketing and providing a secure market for cotton producers (ABC 2017). The Board also administers the Credit Scheme, which finances agricultural inputs and other cotton activities, as a means of facilitating smallholder engagement in the sector (ACB 2017).

3.3.2 *Impacts of Industrial Crops on Food Security*

3.3.2.1 Main Literature Patterns

Figure 3.9 outlines the number of identified studies for each of the targeted industrial crops. Industrial crops such as coffee and cocoa that have a long history of production in SSA, account for many of the studies. Surprisingly there are very few studies for crops with an equally long history of production such as coffee, tobacco and cocoa. Even more surprising is that jatropha studies account for a very large portion of the overall studies, considering the very brief history of the crop in SSA (since mid-2000s) and its widespread collapse across the continent only after a few years. This might reflect the increasing interest that industrial crops have received in the academic literature in the 2010s, which coincides with the recent land rush and the rising prominence of the "food vs. fuel debate" (Sects. 3.1 and 3.4.1).

More than half of the studies report work solely from institutions outside SSA (55%), 18% are collaborative research between SSA and non-SSA institutions, and only 15% solely from SSA institutions (Fig. 3.10a). The majority of these studies focuses on local level effects (52%), with also a significant fraction focusing at the sub-national (17%) and national (19%) levels (Fig. 3.10b). Qualitative, quantitative and mix-method research approaches are represented relatively evenly between studies (Fig. 3.10c). Most studies focus on historic and baseline trends, with only a minority using predictive approaches to understand the possible outcomes of future industrial crop expansion (Fig. 3.10d).

"Access to food" and "food availability" are by far the most widely studied pillars of food security, with 80 and 93 studies, respectively (Fig. 3.11b). "Food stability" is also relatively well-represented (58 studies), whereas "food utilization" is the least studied (20 studies) (Fig. 3.11b). Relatively numerous jatropha and cotton studies

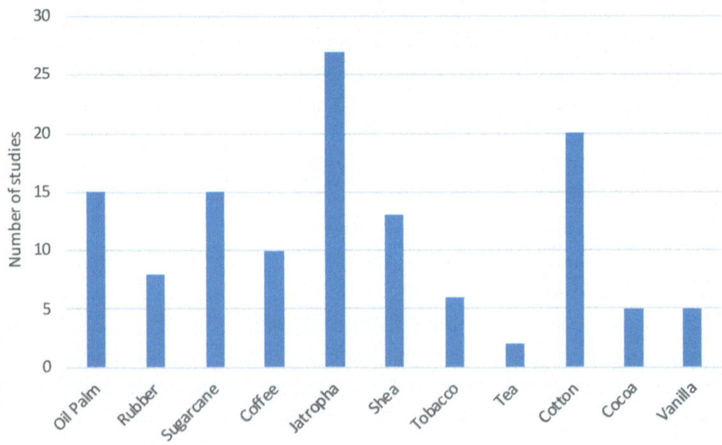

Fig. 3.9 Number of studies for each crop

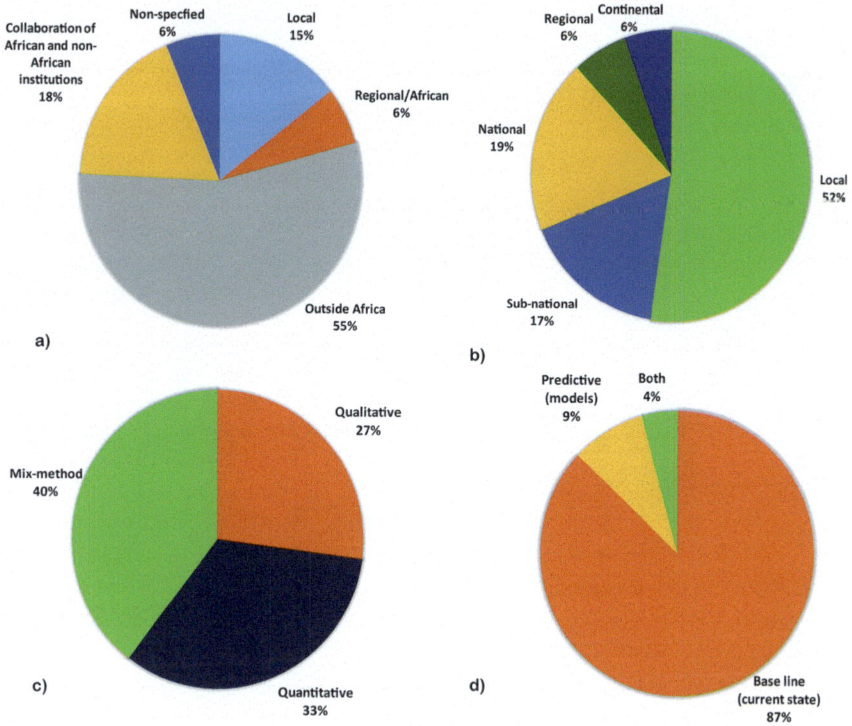

Fig. 3.10 Key characteristics of the reviewed studies: (**a**) Institutional affiliation, (**b**) spatial scale, (**c**) methodology and (**d**) temporal scale

span the four pillars of food security. Most of the tobacco, sugarcane and cocoa studies focus on mechanisms related to "food availability" and "access to food", whereas most rubber and oil palm studies focus on the "food availability" and "food stability" pillars (Fig. 3.11a, b).

The most frequently studied are (in descending order): "Income generation" (B4), "Food crop area" (A1), "Job creation" (B5), "Labour/capital diversion" (A4), "Market linkages" (B2), "Farming inputs" (A5), "Environmental impacts" (D7), "Market stability" (D2), "Women empowerment" (D6) and "Technology" (A6) (Fig. 3.12). Of these mechanisms A1, A4, B2, A5, D2 and D7 tend to have a negative outcome on food security, while B4, B5 and A5 tend to have a positive outcome (Fig. 3.13). When reading deeper the different studies it is possible to identify the underlying processes the mediate these effects on food security. Sections 3.3.2.2–3.3.2.5 discuss in more depth these underlying processes for most of the main mechanisms across the four pillars of food security.

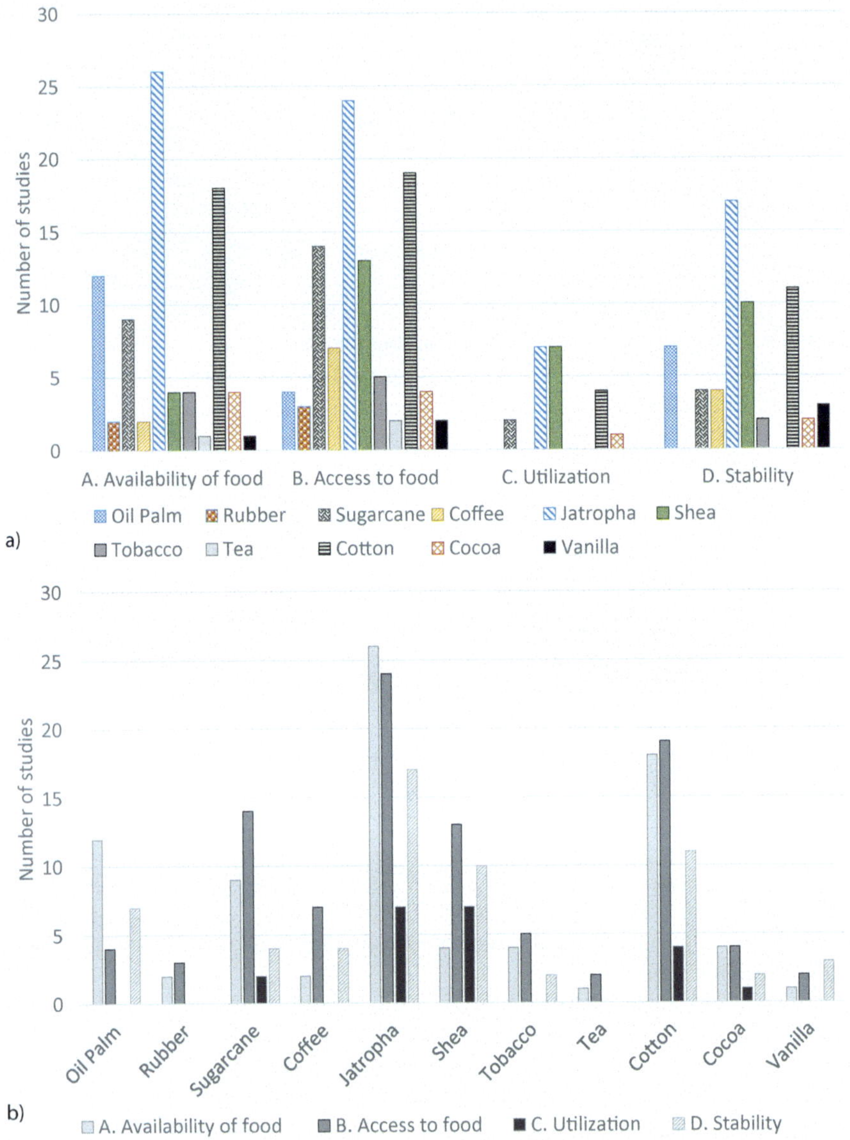

Fig. 3.11 Number of studies for each food security pillar by industrial crop (**a**) and for each industrial crop by food security pillar (**b**)

3.3.2.2 Food Availability

Most of the "food availability" studies focus on the "food crop area" mechanism (A1) (Table 3.5). Several of these studies highlight the negative food security outcomes of land competition with food crop production, whether through large-

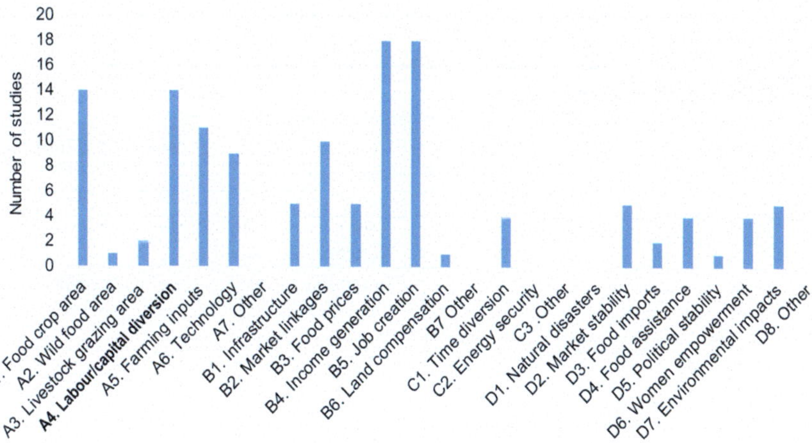

Fig. 3.12 Number of studies for each impact mechanism across all spatial scales. Note: Studies are double counted if they consider more than one food security pillar

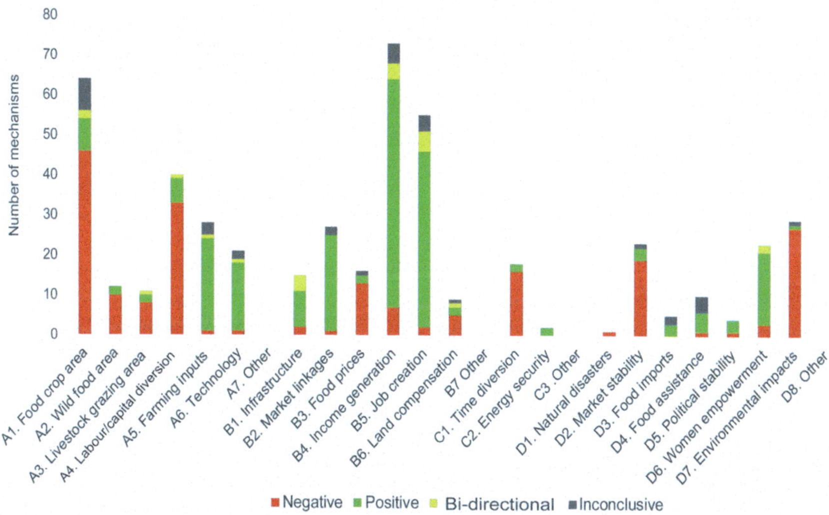

Fig. 3.13 Effect of different mechanisms on food security for all studies. Note: The total number of studies in this figure is lower to Fig. 3.11, as some studies focus on multiple crops and/or impact mechanisms

scale land acquisitions for plantation development or household-level decisions for engagement in smallholder production (e.g. Onoji et al. 2016; Terry 2007; Waswa et al. 2012; Arndt et al. 2011; Romijn et al. 2014; Timko et al. 2014; van Eijck et al. 2014; von Maltitz et al. 2016). However, many studies also point to the complementarity of industrial crops and food crops in smallholder settings that have positive food security outcomes (e.g. Wiggins et al. 2015), especially through

Table 3.5 Underlying processes related to each food availability impact mechanism

Mechanisms and underlying processes	Studies
A1. Food crop area	
Positive	
Intercropping of food crops and industrial crops allows for their joint production in smallholder farms	Uckert et al. (2015); Duvenage et al. (2012); Dyer et al. (2012); Favretto et al. (2014); German et al. (2011); Grimsby et al. (2012); Romijn et al. (2014); Leathers (1999); Komarek (2010)
Engagement in industrial crop production increases the availability of and access to agricultural inputs, which encourages the expansion of food cropland	Brambilla and Portoy (2011); Govereha and Jayne (2003); Laris et al. (2015); Theriault and Tschirley (2014)
Positive spillover effects from industrial crop production e.g. roads developed by plantations, encourages the expansion of food cropland	Laris and Foltz (2014); Ripoche et al. (2015); Vitale et al. (2011)
Engagement in industrial crop production has favorable socioeconomic outcomes e.g. education and income gains – See below, which influences smallholders to maintain better their family farms	Mponela et al. (2011); Romijn et al. (2014)
Negative	
Smallholders switch to industrial crop production from food crop production	Onoji et al. (2016); Terry (2007); Waswa et al. (2009a); Waswa et al. (2009b, 2012); Beghin (2016); Negash and Swinnen (2013); Arndt et al. (2011); Romijn et al. (2014); Timko et al. (2014); van Eijck et al. (2012); van Eijck et al. (2014); Von Maltitz et al. (2016); Theriault and Tschirley (2014); Anderman et al. (2014); Oluyole et al. (2009)
Loss of food cropland through its acquisition for industrial crop plantations e.g. international and state-owned industrial crop companies	Matondi et al. (2011); Carrere (2013); Delarue (2007); Greenpeace International (2012); Greenpeace International (2016); World Rainforest Movement (2012); World Rainforest Movement (2013); World Rainforest Movement (2015); World Rainforest Movement (2011); Acheampong and Campion (2014); Schoneveld et al. (2011)
A2. Wild food areas	
Negative	
Conversion of forest for industrial crop production causes the loss of access of forest-dependent communities to forestland, and the communal benefits it provides e.g. wild food	Gerber (2008); World Rainforest Movement (2011); Acheampong and Campion (2014); Kalinda et al. (2015); Schoneveld et al. (2011); Timko et al. (2014); Laube (2015)
A3. Livestock grazing area	
Positive	
Industrial crop growers can afford to keep more livestock	Matenga (2016); Terry (2012); Bosch and Zeller (2013)

(continued)

Table 3.5 (continued)

Mechanisms and underlying processes	Studies
Negative	
Conversion of grazing land for industrial crop production reduces livestock production	Waswa et al. (2009a, b); Von Maltitz et al. (2016); Williamson et al. (2005)
Conversion of cropland used for animal feed for industrial crop production reduces livestock production	Timko et al. (2014)
A4. Labour or capital	
Negative	
Smallholder-based industrial crop production requires intensive labour, thus reducing or diverting labour from food crop production on family farms	Negash and Swinnen (2013); Arndt et al. (2011); Bosch and Zeller (2013); German et al. (2011); Grimsby et al. (2012); Kalinda et al. (2015); van Eijck et al. (2012); Abudulai (2016); Naughton et al. (2017); Fortucci (2002); Laris et al. (2015); Leathers (1999); Moseley (2001); Anderman et al. (2014); Kiewisch (2015); Oluyole et al. (2009); Wiredu et al. (2011)
Employment in industrial crop plantations reduces or diverts the labour from food crop production on family farms	Romijn et al. (2014); Von Maltitz et al. (2016); Brambilla and Portoy (2011); Kaminski et al. (2011); Laris and Foltz (2014); Sodjinou et al. (2015); Vitale et al. (2011); Ismael et al. (2002)
High labour demand in industrial crop plantations increases the local agricultural labour costs	Borman et al. (2013)
Production of organic industrial crops increases labour intensiveness, further reducing or diverting labour from food crop production on family farms	Williamson et al. (2005)
Adoption of fair-trade and sustainable certification standards and practices for industrial crop production prohibits children from contributing to food crop production on family farms	van Eijck et al. (2014); Bassett (2010)
A5. Farming inputs	
Positive	
Engagement in smallholder-based industrial crop production improves access to fertilizers, which enhances food crop yields	Matenga (2016); Terry (2007); Brambilla and Portoy (2011); Mohammed et al. (2013); Delpeuch and Leblois (2014); Govereha and Jayne (2003); Kaminski et al. (2011); Laris and Foltz (2014); Laris et al. (2015); Moseley (2001); Ripoche et al. (2015); Sodjinou et al. (2015); Theriault and Tschirley (2014); Wiredu et al. (2011)
Engagement in smallholder-based industrial crop production improves access to other non-fertilizer agricultural assets and inputs e.g. seeds, extension, training, credit, which improves food crop production and yields	Bussolo et al. (2007); Negash and Swinnen (2013); Bosch and Zeller (2013); Mohammed et al. (2013)

(continued)

Table 3.5 (continued)

Mechanisms and underlying processes	Studies
Engagement in smallholder-based industrial crop production especially fairtrade/certified increases access to environmentally-friendly and reduces the use of harmful agricultural inputs, which improves food crop production and yields	Bassett (2010);Vitale et al. (2011)
Negative	
Engagement in smallholder-based industrial crop production does not improve access to fertilizers or knowledge how to use them effectively	Duvenage et al. (2012)
Industrial crop production increases local fertilizer prices, reducing their local accessibility to poorer households	Borman et al. (2013); van Eijck et al. (2012); Kiewisch (2015)
Industrial crop production increases the use of and access to harmful pesticides	Williamson et al. (2005)
A6. Technology	
Positive	
Industrial crop companies and grower associations provide technical support, knowledge and training for food crop production	German et al. (2011); Kalinda et al. (2015); Bello-Bravo et al. (2015); Hatskevich et al. (2011); Ismael et al. (2002); Jacques et al. (2009); Laris and Foltz (2014); Vitale et al. (2011); Wiredu et al. (2011)
Introduction of organic methods for industrial crop production improves food crops yield	Bassett (2010); Williamson et al. (2005)
Industrial crop production catalyses access to irrigation, which can also be used for food crop production	Terry (2007)
Negative	
Low technology transfer for industrial crop production brings no support for food crop growing, affecting low yield of food crops	Duvenage et al. (2012); Fortucci (2002)

Source: Jarzebski et al. (2019)

intercropping that minimizes land competition (e.g. Duvenage et al. 2012; Dyer et al. 2012; Favretto et al. 2014; Grimsby et al. 2012; Romijn et al. 2014; Leathers 1999).

Studies related to the "labour and capital" mechanism (A4) overwhelmingly focus on the high labour needs of smallholder-based industrial crop production often requires intensive labour, which often takes a toll on the availability of family labour for food crop production (e.g. Adams et al. 2016; Naughton et al. 2017; Fortucci 2002; Laris et al. 2015; Leathers 1999). Such employment diversion effects are also reported and for large-scale plantations (e.g. Ismael et al. 2002; Kaminski et al. 2011; Laris and Foltz 2014). Few studies have pointed that labour diversion from industrial crop production (especially in large plantation settings) can increase local labour costs, thus reducing affordable local labour options for food crop production (Borman et al. 2013).

Studies have linked some of the positive local food security outcomes of industrial crop production to the improved access of smallholders to farming inputs and technology transfer (A5–6). This includes improved access to various factors of production, including:

- Fertilizers (e.g. Brambilla and Portoy 2011; Delpeuch and Leblois 2014; Govereha and Jayne 2003; Kaminski et al. 2011; Herrmann et al. 2018; von Maltitz et al. 2019).
- Diverse varieties of food crops and seeds (Bussolo et al. 2007).
- Irrigation (Negash and Swinnen 2013; von Maltitz et al. 2019).
- Technical support (e.g. Kalinda et al. 2015; Bello-Bravo et al. 2015).
- Other environmentally friendly agricultural inputs (e.g. Bassett 2010).

Improved access to all of the above can enhance the yields of both industrial and food crops, reducing (or even reversing) the effects of land competition (mechanism A1, see above). However, some studies have also found that industrial crop production can increase local fertilizer demand, and thus its price, reducing its affordability to poorer households for food crop production (e.g. Borman et al. 2013; van Eijck et al. 2012).

3.3.2.3 Food Access

Most studies in the "food access" pillar by far focus on the positive local food security outcomes brought by income, employment and occupation generation (B4–5) (Table 3.6). For example, many studies have identified that involvement in industrial crop production can boost existing income (e.g. Banye 2015; Ferris et al. 2001; Suleman et al. 2014; Govereha and Jayne 2003), or create additional income sources (e.g. Onoji et al. 2016; Dyer et al. 2012; Favretto et al. 2014). Even though the generated income often quite low and depend on the crop type and mode of engagement (e.g. smallholder vs. plantation worker), this income can be constant and allow better planning within the household (German et al. 2011), or come during periods of low food security acting thus as a livelihood buffer (von Maltitz et al. 2016).

In terms of employment and occupation generation, many studies have reported the positive effect of industrial crop production on local employment opportunities, especially in poor rural areas with few formal employment options (Duvenage et al. 2012; Dyer et al. 2012; von Maltitz et al. 2016). Sometimes employment opportunities are seasonal (e.g. James and Woodhouse 2016; Matenga 2016), and can enhance self-employment (e.g. Kuntashula et al. 2014; Chivuraise 2011) or women employment (e.g. Bosch and Zeller 2013).

Table 3.6 Underlying processes related to each access to food impact mechanism

Mechanisms and underlying processes	Studies
B1. Infrastructure	
Positive	
Industrial crop companies develop infrastructure that improves water access and/or provides irrigation	Negash and Swinnen (2013); Romijn et al. (2014); Timko et al. (2014)
Industrial crop companies construct roads that improve access to food	Romijn et al. (2014)
B2. Market	
Positive	
Higher overall development in areas of industrial crop production results in the development of new food markets	Garba et al. (2015)
B3. Food price	
Positive	
Industrial crops smallholders associations also get involved in food crop production as a secondary activity, reducing local food prices	Terry (2007)
Negative	
Increased local demand for food coupled with reduced food production increases food prices	Arndt et al. (2011); Bosch and Zeller (2013); Timko et al. (2014); van Eijck et al. (2012); Kaminski et al. (2011); Laris and Foltz (2014) Anderman et al. (2014)
Expansion of industrial crop production causes the manipulation of market food prices	Kgathi et al. (2012)
B4. Income from industrial crops	
Positive	
Engagement in industrial crop production provides income incl. For women that is used to buy food, thus increasing access to food in these studies industrial crop income is the main household income stream	Onoji et al. (2016); Matenga (2016); Terry (2007); Terry (2012); Arndt et al. (2011); Dyer et al. (2012); Favretto et al. (2014); German et al. (2011); Grimsby et al. (2012); Kalinda et al. (2015); Romijn et al. (2014); Romijn et al. (2014); Timko et al. (2014); van Eijck et al. (2014); Adams et al. (2016); Annan (2013); Bello-Bravo et al. (2015); Hatskevich et al. (2011); Naughton et al. (2017); Ismael et al. (2002); Laris et al. (2015); Sodjinou et al. (2015)
Engagement in industrial crop production increases the total household income, thus increasing access to food in these studies it is not specified if the added income is the household main or it is additional	Greenpeace International (2012); World Growth (2010); FEWS Net Liberia (2016); Dam Lam et al. (2017); Kennedy (1989); Matenga (2016); Banye (2014); Ferris et al. (2001); Suleman et al. (2014); Bassett (2010); Fortucci (2002); Govereha and Jayne (2003); Kaminski et al. (2011); Theriault and Tschirley (2014); Oluyole et al. (2009)
Labour wages increase, providing additional income, thus increasing access to food	Borman et al. (2013); van Eijck et al. (2012)

(continued)

Table 3.6 (continued)

Mechanisms and underlying processes	Studies
Growing organic industrial crops increases income, which increases access to food	Williamson et al. (2005)
Modernizing management methods improves income, which increases access to food	Jacques et al. (2009)
B5. Job and occupation creation	
Positive	
Employment generation in plantations incl. For women	Acheampong and Campion (2014); Duvenage et al. (2012); Dyer et al. (2012); Favretto et al. (2014); German et al. (2011); Romijn et al. (2014); Romijn et al. (2014); Romijn et al. (2014); van Eijck et al. (2012); Von Maltitz et al. (2016); Banye (2014); Hatskevich et al. (2011); Hatskevich et al. (2014); Jamala et al. (2013); Laube (2015); Mohammed et al. (2013); Bassett (2010); Fortucci (2002); Vitale et al. (2011); Williamson et al. (2005); Komarek (2010); van Eijck et al. (2014); Von Maltitz et al. (2016); Adams et al. (2016); Sodjinou et al. (2015)
Women employment available	Arndt et al. (2011); Bosch and Zeller (2013); Bello-Bravo et al. (2015); Naughton et al. (2017); Sodjinou et al. (2015)
Self-employment as grower	Negash and Swinnen (2013); Kuntashula et al. (2014); van Eijck et al. (2014); Chivuraise (2011); Laris and Foltz (2014)
Seasonal employment available	James and Woodhouse (2016); Lazzarini (2016); Matenga (2016); Von Maltitz et al. (2016); Adams et al. (2016)
B6. Compensation for the land	
Positive	
Land compensated with arable land at small holder level	Laube (2015)
Compensation offered only after a conflict occurred	Romijn et al. (2014)
Negative	
No compensation received for taken land	Acheampong and Campion (2014); Romijn et al. (2014)

Source: Jarzebski et al. (2019)

3.3.2.4 Food Utilization

Studies related to the "food utilization" pillar are relatively scarce, and focus mostly on the negative effects of female engagement in industrial crop production (and paid employment in plantations in particular) (Table 3.7). This is usually linked to time loss for household activities, and especially meal preparation, feeding children and general unpaid care time (e.g. Arndt et al. 2011; Bosch and Zeller 2013). Some

Table 3.7 Underlying processes related to each food utilization impact mechanism

Mechanisms and underlying processes	Studies
C1. Jobs for women	
Positive	
Exclusion of women from waged employment in industrial crop plantations and smallholder-based production allows them to spend more time for unpaid household care activities	Lazzarini (2016); Moseley (2001); Kiewisch (2015)
Negative	
Employment for women in industrial crop plantations diverts their time from unpaid household care activities	Arndt et al. (2011); Bosch and Zeller (2013); Romijn et al. (2014); Romijn et al. (2014); Romijn et al. (2014); Adams et al. (2016); Annan (2013); Banye (2014); Bello-Bravo et al. (2015); Garba et al. (2015); Jamala et al. (2013); Laube (2015)
Fairtrade-related schemes increase women engagement in smallholder-based industrial crop production, diverting their time from household care activities	Kiewisch (2015); Waswa et al. (2009b); Acheampong and Campion (2014)
C2. Energy security	
Positive	
By-products of industrial crops can be used as cooking fuel, reducing thus the time spent for fuelwood collection	Von Maltitz et al. (2016)

Source: Jarzebski et al. (2019))

studies have also identified that men dominate employment opportunities along industrial crop value chains, maintaining thus, traditional societal structures and household roles for women (Lazzarini 2016; Waswa et al. 2009a, 2009b; Moseley 2001). Even though such outcomes might not be socially desirable (Sect. 3.4.2), they still seem to have possible positive effects related to food utilization.

3.3.2.5 Food Stability

Most of the studies related to the "food stability" pillar focus on two specific mechanisms, namely women empowerment (D6) and environmental stability (D7) (Table 3.8). Engagement in industrial crop value chains is occasionally associated with women empowerment. For example, some studies have identified that females involved in industrial crop value chains often assume greater control in intra-household income allocation decisions (e.g. Adams et al. 2016; Banye 2015). Furthermore, involvement in industrial crop value chains can enhance training opportunities for women (Williamson et al. 2005; Suleman et al. 2014) and the development of women groups that can negotiate crop prices from a stronger position (e.g. Favretto et al. 2014; Annan 2013).

Table 3.8 Underlying processes related to each food stability impact mechanism

Mechanisms and underlying processes	Studies
D1. Natural disasters	
Positive	
Jatropha provides a natural "fence" that protects crops from wind or floods	Favretto et al. (2014)
D2. Market stability	
Positive	
Smallholder women cooperatives increase market stability	Banye (2014)
D4. Hunger reduction	
Positive	
Industrial crop producers are able to purchase food during periods of food shortages	Bosch and Zeller (2013); Favretto et al. (2014)
Some industrial crops (e.g. shea) can be used for food purposes during hunger period	Hatskevich et al. (2011)
D6. Women empowerment	
Positive	
Engagement in industrial crop value chains enhances income (and income opportunities) for women	Carrere (2013); Arndt et al. (2011); Romijn et al. (2014); Romijn et al. (2014); Timko et al. (2014); Adams et al. (2016); Banye (2014); Bello-Bravo et al. (2015); Laube (2015); Suleman et al. (2014); Sodjinou et al. (2015)
Engagement in industrial crop value chains catalyses the formation of women groups that are in a better position to negotiate prices	Favretto et al. (2014); Annan (2013); Garba et al. (2015)
Industrial crop schemes provide training opportunities for women	Williamson et al. (2005)
Negative	
Women do not control the income generated through engagement in industrial crop value chain, and are excluded from related decision-making	Zommers et al. (2012); Anderman et al. (2014); Kiewisch (2015)
Women lose access to education and paid employment due to their engagement in industrial crop value chains	Lazzarini (2016); Moseley (2001)
Women are paid less for their engagement in industrial crop value chains	Matenga (2016)
D7. Environmental stability	
Positive	
Industrial crop production reduces deforestation, which affects one of the other three pillars of food security (food availability)	Uckert et al. (2015)
Negative	
Industrial crop production causes deforestation, which has a negative effect to one of the other three pillars of food security (food availability)	Carrere (2013); Greenpeace International (2012); Greenpeace International (2016); Moser (2008); Grimsby et al. (2012); Naughton et al. (2017); Chivuraise (2011); Moser (2008); Patel (2007); Randriamalala and Liu (2010)

(continued)

Table 3.8 (continued)

Mechanisms and underlying processes	Studies
Industrial crop production causes biodiversity loss, which has a negative effect to one of the other three pillars of food security (food availability)	Carrere (2013); Gerber (2008); Oyono (2013); Zommers et al. (2012); Beyene et al. (2012); Senbeta and Denich (2006); German et al. (2011); van Eijck et al. (2014); Laube (2015)
Industrial crop production causes water depletion, which has a negative effect to one of the other three pillars of food security (food availability)	Gerber (2008); Beyene et al. (2012); Von Maltitz et al. (2016)
Industrial crop production causes soil degradation, which has a negative effect to one of the other three pillars of food security (food availability)	Carrere (2013); World Rainforest Movement (2011); Duvenage et al. (2012)
Industrial crop production causes water quality degradation, which has a negative effect to one of the other three pillars of food security (food availability)	Oyono (2013); Favretto et al. (2014)

Source: Jarzebski et al. (2019)

However, industrial crop production often causes negative environmental impacts (e.g. deforestation) that threaten the stability of food production (e.g. Grimsby et al. 2012; Naughton et al. 2017; Chivuraise 2011). Other negative environmental impacts linked to negative food security outcomes include biodiversity loss (e.g. Senbeta and Denich 2006; German et al. 2011), water depletion (e.g. Gerber 2008; Von Maltitz et al. 2016) and soil degradation (e.g. Duvenage et al. 2012).

3.4 Discussion

3.4.1 Knowledge Synthesis

Overall, most of the reviewed studies at the interface of food security and industrial crop production have been published in the 2010s (Sect. 3.3.2.1). This implies that the recent global land rush (Schoneveld et al. 2011) has possibly catalysed and shaped more than anything else the literature on industrial crops and food security. Indeed narratives related to land grabbing, land competition and "food vs. fuel" have been key underlying themes in the reviewed literature (see also Borras and Franco 2013; Kaag and Zoomers 2014; Nalepa et al. 2017; Zoomers 2010; Kuchler and Linnér 2012; Shortall 2013; Tenenbaum 2008) (Chap. 4 Vol. 1). In fact a substantial fraction of the reviewed studies comes from countries that were targeted extensively for large-scale commercial agriculture investments during the recent land rush such as Zambia, Mozambique, Ghana and Ethiopia (Schoneveld 2014). Additionally, many of these studies have focused on jatropha, a relatively new and untested crop that received very sudden attention in SSA before its eventual collapse (Sect. 3.3.2.1) (von Maltitz et al. 2014; Ahmed et al. 2019b).

Most studies target a sub-set of mechanisms related to the "access to food" and "food availability" pillars of food security. Furthermore, the number and type of mechanisms captured are highly variable between crops. For example, cotton, jatropha, shea and sugarcane are the best-studied crops in terms of the number of captured mechanisms (Sect. 3.3.2.1). Conversely, important crops such as oil palm, rubber, cocoa and coffee feature in comparatively fewer studies, despite their extensive history and ongoing expansion across the continent (Sect. 3.1).

Industrial crop production facilitates local access to food by generating income and employment (Sect. 3.3.2.3). At the same time, engagement in industrial crop production (whether as plantation workers or smallholders) can divert family labour from food crop production, thus reducing local food availability (Sect. 3.3.2.2). As a result, there is a great need to ensure the generation of secure employment, and reliable and sustainable income, while minimizing the negative effects of time and labour diversion from food crop production in family farms (see also Sect. 3.4.2). Sustaining income and employment benefits would render the engagement in industrial crop value chains a worthwhile and risk-free endeavour to plantation workers and smallholders. This is very important lesson learnt from the almost total collapse of the jatropha sector and lack of materialization of the expected benefits in many rural contexts of SSA (Ahmed et al. 2019b; von Maltitz et al. 2015).

There is also strong evidence to suggest that industrial crop production often causes direct land competition with food crop production (Sect. 3.3.2.2). However, the actual land use change effects depend on the mode of production. However, there can also be indirect land use change effects that are nevertheless difficult to estimate accurately and in a non-controversial manner (Khanna and Crago 2012; Finkbeiner 2014). For example, industrial crop smallholders can either choose to expand (or not) food crop production to other areas to compensate for the land allocated to industrial crops. Similarly large-scale plantations might displace farmers, who might in turn clear land elsewhere to establish farms. In any case such direct and indirect land competition can affect local food availability either through the loss of food cropland or the loss of communal pasture/forest.

In smallholder settings increasing the use of agricultural inputs use (e.g. fertilizers, pesticides) and/or adopting improved production practices (e.g. irrigation, intercropping) could minimize the possible negative effects of land loss on food availability by increasing crop yields (Sect. 3.3.2.2). However, such agricultural intensification might lead to negative environmental impacts related to freshwater depletion, water pollution and soil degradation, all of which have been shown to have a negative effect to food stability (Sect. 3.3.2.5). The above suggest hard trade-offs between the food availability and food stability pillars.

Many of the reviewed studies pointed how gender issues mediate the positive or negative impacts of industrial crop production on food security. This happens especially through some mechanisms such as income and employment generation (B4–5), time diversion from unpaid care work (C1) and women empowerment (D6).

However, even though women are a major part of the labour force in SSA (Bryceson 2018) (see Chap. 1 Vol. 1), they are often barred from formal employment in industrial crop plantations due to lack of skillset, cultural reasons and the fact that formal waged jobs are often seen as a male domain (Sect. 3.3.2.4). Female entry in smallholder-based industrial crop production is also not straightforward, as they often lack land titles and decision-making power in family farms (Bryceson 2018).

Finally, as already discussed in Sect. 3.3.1 and above, the mode of industrial crop production can have important effects on food security through multiple impact mechanisms. However, deciding the fitness of particular modes of production is not straightforward and can depend on various factors. Some of the most important such factors include (a) crop characteristics/agronomy (b) production characteristics (e.g. labour intensity, returns to investment); (c) marketing characteristics (e.g. potential buyers, processing needs); (d) exogenous economic and political factors (e.g. land scarcity, population density) and (e) endogenous economic and political factors (e.g. input markets, low producer capacity, weak property rights enforcement) (Benfica et al. 2002).

3.4.2 Policy and Practice Implications and Recommendations

As outlined in Sect. 3.3, there are multiple trade-offs at the interface of industrial crop production and food security in SSA. When critically viewing the evidence outlined throughout this chapter it is possible to identify some priority policy and practice domains that can be targeted to enhance the positive (or reduce the negative) food security outcomes of industrial crop production. Below we outline three priority domains that are at the intersection of multiple SDGs:

- Safeguard the long-term economic and employment benefits accruing from engagement in industrial crop production,
- Enhance farm output for both industrial crops and food crops, while avoiding negative environmental impacts,
- Enhance female participation in industrial crop production, while reducing the negative effects of time diversion from unpaid household care work.

Regarding the first priority domain, income and employment generation are two of the main mechanisms through which industrial crop production has a positive effect on food security (Sects. 3.3.2 and 3.4.1). That said, government policies should support the economic viability and sustainability of industrial crop production systems. In particular, a key aim should be to achieve the better balance between competition and coordination, and provide appropriate incentives to better safeguard the interests of farmers when engaging in industrial crop value chains. This would entail different incentives to smallholders, plantation owners and other players across industrial crop value chains (e.g. millers, transporters). Such a coordinated approach could send strong signals about the long-term policy commitment in industrial crop production, which could alleviate investor uncertainty and help attract sustained investment (Chaps. 2 and 5 Vol. 1).

There have been such coordinated national efforts in countries such as Burkina Faso, Malawi and Swaziland, where a strong policy commitment over decades has made cotton, tobacco and sugarcane the cornerstones of their respective economies (Terry and Ogg 2017; Johnson and Silveira 2014; Boafo et al. 2018; Tschirley et al. 2009) (Sect. 3.3.1). This would entail strengthening industrial crop markets and streamlining all stages of the value chain, from the land acquisition, to the economic aspects of crop production, refinement, valourization of waste and final product use. As the actual interventions and long-term strategies might be crop- and area-specific, it would be necessary to factor national and local contexts. Some additional foci for large-scale plantations would be to ensure that (a) worker salaries are sufficient to buy food locally, (b) seasonal/part-time employment is based on a standard salary rate (and not on a picking rate), (c) flexible employment is possible during important periods of crop calendar year (e.g. during food crop sowing, harvesting). All of these actions could create very strong linkages among various SDGs namely SDG8 (descent work and economic growth), SDG9 (industry, innovation and infrastructure), SDG1 (no poverty) and of course SDG2 (zero hunger).

Regarding the second priority domain, it would be necessary to understand the possible environmental and food production trade-offs from industrial crop expansion. This could be achieved through robust baseline studies undertaken prior to the development of industrial crop projects that should seek to understand the decision-making processes of local farmers in terms of land allocation, and the adoption of crops and farming practices. This information can help build a strong evidence-base to guide the development and implementation of context-specific interventions that seek to minimize the negative trade-offs of industrial crop production. However, these aspects are rarely considered during project design, even in commonly used instruments such as environmental impact assessments (EIAs) (e.g. German et al. 2013). Thus it would be necessary to "enforce" the provision of this type of information (and identify possible mitigation option in the face of important trade-offs) prior to the approval of large-scale industrial crop projects.

Furthermore, it is important to rationalize the use of agrochemicals and irrigation water both in large plantations and smallholder settings. The former could help reduce the negative environmental impacts of industrial crop production, and the latter could ensure water availability to other water users, especially during periods of water scarcity. However, EIAs have little power to deal with such effects in smallholder settings as individual farmers can unilaterally decide to start and halt industrial crop production based on market signals. Even though engagement in industrial crop production improves access to agricultural inputs (and sometimes irrigation) (Sect. 3.3.2.2), smallholders often lack the capacity to utilize them in environmentally and socially responsible manners (Morris et al. 2007). Capacity-building efforts and the promotion of responsible production practices through the support packages offered to industrial crop smallholders could go a long way towards achieving this. However, increasing the environmental and social responsibility of smallholders through such actions would require coordinated efforts between industrial crop companies (e.g. buyers from smallholders), other private sector players (e.g. certification schemes), government agencies and civil society.

Enhancing farm output while avoiding negative environmental impacts as discussed above can have positive effects to multiple SDGs, such as SDG12 (responsible consumption and production), SDG15 (life on land), SDG6 (clean water and sanitation) and of course SDG2 (zero hunger).

Regarding the third priority domain, clearly it is not justifiable to prevent females from engaging in industrial crop value chains on grounds of safeguarding their crucial role in unpaid care work. However, we should also be aware of the almost complete lack of social services in rural SSA (ILO 2018) (Chap. 1 Vol. 1). Thus, it would be important to identify avenues to enhance female participation in industrial crop value chains as a means of accessing income/employment opportunities and achieving broader female empowerment benefits, while at the same time reducing the negative effects of time diversion from unpaid household care work.

Large-scale plantations can develop (or contribute to the development of) infrastructure that compensates for this loss of unpaid care work (e.g. schools with feeding programmes), where female employees can have priority access. Furthermore, flexible working arrangements and standardized salaries for seasonal and part-time employees can further ensure both that female workers are not discriminated and that food utilization trade-offs are minimized. A very interesting approach specific to bioenergy crops would be to provide (or support the acquisition of) improved stoves and fuel such as ethanol stoves and fuel (Chap. 2 Vol. 1). This could reduce time diversion to fuelwood collection and cooking, which is substantial in most parts of rural SSA and takes a toll on females and girls (Karanja et al. 2020; Köhlin et al. 2011).

However, it might not be as straightforward to enable the aforementioned gender inclusion interventions in smallholder settings. Possible avenues could be to valourize further female participation in industrial crop production by offering higher premiums from certification schemes (Parvathi 2017). Supporting the development of female grower associations or increasing the decision-making power of female growers in mixed-grower associations could further improve female negotiating and decision-making power. This could help maximize the gender empowerment benefits of engagement in smallholder-based industrial crop production. Furthermore, government agencies, civil society organizations and the private sector should increase the number (and improve the quality) of training activities geared towards female industrial crop growers. Apart from offering important knowledge for enhancing industrial crop production (see above), such training should further educate women about the possible negative trade-offs of their involvement in industrial crop value chains. However, even though such interventions could possibly enhance the benefits that women receive from engagement in industrial crop production, they can have a less direct effect to food security.

3.5 Conclusions

This sought to unravel the food security outcomes of industrial crop production in SSA. We reviewed the main production patterns, drivers and underlying policies in some of the most important producing countries in the region and reviewed systematically the existing literature for 11 industrial crops and 25 impact mechanisms across the four pillars of food security. Thc quantity of the current evidence varies considerably between crops, with jatropha, cotton sugarcane and shea being the most studied crops. Food access and availability are the only pillars of food security with substantial evidence and consensus about the direction and magnitude of the impact mechanisms. Much less literature exists for mechanisms related to food stability and utilization.

Overall, the current literature landscape is fragmented with most studies considering a sub-set of crops, modes of production regions and/or impact mechanisms. This is a major barrier for balanced policy and practice inferences at the interface of industrial crops and food security. Future research should better conceptualize the possible pathways through which industrial crop production can affect food security. Empirical studies should, to the extent possible, use compatible methods to allow for the better understanding of these mechanisms across SSA.

Policy and practice priorities should include to (a) safeguard the long-term economic and employment benefits accruing from engagement in industrial crop production; (b) enhance farm output (for both industrial and food crops) while avoiding negative environmental impacts; (c) enhance female participation in industrial crop production, while reducing the negative effects of time diversion from unpaid household care work.

Acknowledgements Authors acknowledge the financial support of the Japan Science and Technology Agency (JST) for the Belmont Forum project FICESSA. Abubakari Ahmed and Alice Karanja were supported by Monbukagakusho scholarships offered by the Japanese Ministry of Education, Culture, Sports, Science, and Technology (MEXT) through the Graduate Program in Sustainability Science – Global Leadership Initiative (GPSS-GLI), at the University of Tokyo. Linda Chinangwa acknowledges the support of the Japanese Society for the Promotion of Sciences (JSPS), through JSPS KAKENHI Grant number JP15F15765.

References

ACB (2017) GM cotton push in Swaziland: next target for failed BT cotton. African Centre for Biodiversity (ACB), Johannesburg

Acheampong E, Campion BB (2014) The effects of biofuel feedstock production on farmers' livelihoods in Ghana: the case of Jatropha curcas. Sustain 6:4587–4607. https://doi.org/10.3390/su6074587

Adams A-M, Abudulai I, Bashiru M (2016) The Shea industry and rural livelihoods among women in the WA municipality. J Soc Sci Stud 3:40. https://doi.org/10.5296/jsss.v3i2.8706

Ahmed A, Jarzebski MP, Gasparatos A (2017) Using the ecosystem service approach to determine whether jatropha projects were located in marginal lands in Ghana: implications for site selection. Biomass Bioenergy 114:112–124. https://doi.org/10.1016/j.biombioe.2017.07.020
Ahmed A, Dompreh E, Gasparatos A (2019a) Human wellbeing outcomes of involvement in industrial crop production in Ghana: evidence from sugarcane, oil palm and jatropha sites. PLoS One 14(4):e0215433
Ahmed A, Campion B, Gasparatos A (2019b) Towards a classification of the drivers of jatropha collapse in Ghana elicited from the perceptions of multiple stakeholders. Sustain Sci 14:315–339
Anderman TL, Remans R, Wood SA, DeRosa K, DeFries RS (2014) Synergies and tradeoffs between cash crop production and food security: a case study in rural Ghana. Food Secur 6:541–554. https://doi.org/10.1007/s12571-014-0360-6
Annan AT (2013) Transforming the lives of poor rural women in the Shea butter industry through entrepreneurship; A case of Sagnarigu Shea butter and soap centre. International Institute of Social Studies, Hague
Arndt C, Benfica R, Tarp F, Thurlow J, Uaiene R (2009) Biofuels, poverty, and growth: a computable general equilibrium analysis of Mozambique. Environ Dev Econ 15:81–105
Arndt C, Benfica R, Tarp F, Thurlow J, Uaiene R (2010) Biofuels, poverty, and growth: a computable general equilibrium analysis of Mozambique. Environ Dev Econ 15:81–105
Arndt C, Benfica R, Thurlow J (2011) Gender implications of biofuels expansion in Africa: the case of Mozambique. World Dev 39:1649–1662. https://doi.org/10.1016/j.worlddev.2011.02.012
Baffes J (2005) The "cotton problem". World Bank Res Obs 20(1):109–144. https://elibrary.worldbank.org/doi/abs/10.1093/wbro/lki004. Accessed 31 Oct 2019
Banye E (2014) Improving rural livelihoods through Shea butter businesses, a case of women in markets in northern Ghana 1–5. SNV Tanzania
Bassett TJ (2010) Slim pickings: Fairtrade cotton in West Africa. Geoforum 41:44–55. https://doi.org/10.1016/j.geoforum.2009.03.002
Beghin, J.C. (2016). The coffee-food security interface for subsistence households in Jimma zone Ethiopia
Bello-Bravo J, Lovett PN, Pittendrigh BR (2015) The evolution of Shea butter's "paradox of paradoxa" and the potential opportunity for information and communication technology (ICT) to improve quality, market access and women's livelihoods across Rural Africa. Sustain 7:5752–5772. https://doi.org/10.3390/su7055752
Benfica R, Tschirley D, Sambo L (2002) The impact of alternative AgroIndustrial investments on poverty reduction in rural Mozambique. MINAG/DE/DAP Research Report No 51E
Berhanu M, Jabasingh SA, Kifile Z (2017) Expanding sustenance in Ethiopia based on renewable energy resources–a comprehensive review. Renew Sust Energ Rev 75:1035–1045
Beyene A, Kassahun Y, Addis T, Assefa F, Amsalu A, Legesse W, Kloos H, Triest L (2012) The impact of traditional coffee processing on river water quality in Ethiopia and the urgency of adopting sound environmental practices. Environ Monit Assess 184:7053–7063. https://doi.org/10.1007/s10661-011-2479-7
Boafo YA, Balde B, Saito O, Gasparatos A, Lam RD, Ouedraogo N, Chamba E, Moussa ZP (2018) Stakeholder perceptions of the impact of reforms on the performance and sustainability of the cotton sector in Ghana and Burkina Faso: a tale of two countries. Cogent Food Agric 4:1477541
Boamah F (2014) Imageries of the contested concepts 'land grabbing' and 'land transactions': implications for biofuels investments in Ghana. Geoforum 54:324–334
Borman GD, Von Maltitz GP, Tiwari S, Scholes MC (2013) Modelling the economic returns to labour for Jatropha cultivation in southern Africa and India at different local fuel prices. Biomass Bioenergy 59:70–83. https://doi.org/10.1016/j.biombioe.2012.06.020
Borras SM, Franco JC (2013) Global Land Grabbing and Political Reactions 'From Below'. Third World Q 34:1723–1747. https://doi.org/10.1080/01436597.2013.843845
Bosch C, Zeller M (2013) The impacts of wage employment on a Jatropha plantation on income and food security of rural households in Madagascar - A panel data analysis. Q J Int Agric 52:119–140

Brambilla I, Portoy GG (2011) Market structure, outgrower contracts, and farm output. Evidence from cotton reforms in Zambia. Oxf Econ Pap 63:740–766. https://doi.org/10.1093/oep/gpr004

Breisinger C et al. (2008) Agriculture for development in Ghana new opportunities and challenges. Agriculture for Development. Agriculture 54, 1–6. The World Bank

Brun TA (1991) The nutrition and health impact of cash cropping in West Africa: a historical perspective. World Rev Nutr Diet 65:124–162

Brüntrup M, Schwarz F, Absmayr T, Dylla J, Eckhard F, Remke K, Sternisko K (2018) Nucleus-outgrower schemes as an alternative to traditional smallholder agriculture in Tanzania – strengths, weaknesses and policy requirements. Food Secur 10:807–826. https://doi.org/10.1007/s12571-018-0797-0

Bryceson DF (2018) Agrarian labour and resources in sub-Saharan Africa: gender and generational change within family farms, 1980–2015

Burney JA, Naylor RL, Postel SL (2013) The case for distributed irrigation as a development priority in sub-Saharan Africa. Proc Natl Acad Sci U S A 110:12513–12517. https://doi.org/10.1073/pnas.1203597110

Bussolo M, Godart O, Lay J, Thiele R (2007) The impact of coffee price changes on rural households in Uganda. Agric Econ 37(2–3):293–303. https://doi.org/10.1111/j.1574-0862.2007.00275.x

Carrere R (2013) Oil palm in Africa: past, present, and future scenarios. World Rainfor Mov 15:79

Chilowa W, Chirwa EW (1997) The impact of structural adjustment programmes on social and human development in Malawi. Bwalo I. Center for Social Research

Chinangwa L, Gasparatos A, Saito O (2017) Forest conservation and the private sector: stakeholder perceptions towards payment for ecosystem service schemes in the tobacco and sugarcane sectors in Malawi. Sustain Sci 12:727–746. https://doi.org/10.1007/s11625-017-0469-6

Chirwa E (2011) Analysis of the tobacco industry in Malawi, United Nations conference on trade and development. UNCTAD, New York and Geneva

Chisanga B, Zulu-Mbata B (2017) The changing food expenditure patterns and trends in Zambia: implications on agricultural policies. Working Paper 119. IAPRI, Lusaka

Chivuraise C (2011) Economics of smallholder tobacco production and its implication of deforastaion in Hurungwe District of Zimbabwe, MSc. Thesis, University of Zimbabwe, Harare, Zimbabwe

Condliffe K, Kebuchi W, Love C, Ruparell R (2008) Kenya coffee: a cluster analysis. Professor Michael Porter, Microeconomics of Competitiveness. Harvard Business School

Dam Lam R, Boafo YA, Degefa S, Gasparatos A, Saito O (2017) Assessing the food security outcomes of industrial crop expansion in smallholder settings: insights from cotton production in northern Ghana and sugarcane production in Central Ethiopia. Sustain Sci 12:677–693

de Graaff J, Kessler A, Nibbering JW (2011) Agriculture and food security in selected countries in sub-Saharan Africa: diversity in trends and opportunities. Food Secur. 3:195–213. https://doi.org/10.1007/s12571-011-0125-4

Delarue J (2007) Mise Au Point D'une Methode D'evaluation Systemique D'impact Des Projets De Developpement Agricole Sur Le Revenu Des Producteurs Etude De Cas En Region Kpele (Republique De Guinee) (in French)

Delpeuch C, Leblois A (2014) The elusive quest for supply response to cash-crop market reforms in sub-Saharan Africa: the case of cotton. World Dev 64:521–537. https://doi.org/10.1016/j.worlddev.2014.06.007

Duvenage I, Taplin R, Stringer L (2012) Bioenergy project appraisal in sub-Saharan Africa: sustainability barriers and opportunities in Zambia. Nat Resour Forum 36:167–180. https://doi.org/10.1111/j.1477-8947.2012.01453.x

Dyer JC, Stringer LC, Dougill AJ (2012) Jatropha curcas: sowing local seeds of success in Malawi?. In response to Achten et al. (2010). *J.* Arid Environ 79:107–110. https://doi.org/10.1016/j.jaridenv.2011.12.004

Energy Commission (2006) Strategic National Energy Plan 2006–2020. Energy Commission, Accra

FAO (2006) Food security, policy brief: 2. Food and Agriculture Organization, Rome
FAO (2015) The state of food insecurity in the world 2015. Food and Agriculture Organization, Rome
FAO (2018) Transforming food and agriculture to achieve the SDGs. Food and Agriculture Organization, Rome. http://www.fao.org/3/I9900EN/i9900en.pdf. Accessed on 31 Oct 2019
FAO and ECA (2018) Regional overview of food security and nutrition. Addressing the threat from climate variability and extremes for food security and nutrition. Food and Agriculture Organization, Accra. http://www.fao.org/3/CA2710EN/ca2710en.pdf. Accessed 1 Sept 2019
FAO and ECDPM (2018) Policy coherence for agricultural transformation in African least developed countries (LDCs): aligning agriculture and trade policymaking processes. Food and Agriculture Organization and European Centre for of Development Policy Management, Rome. http://www.fao.org/3/CA0025EN/ca0025en.pdf. Accessed 31 Oct 2019
FAO, IFAD, UNICEF, WFP and WHO (2019) The state of food security and nutrition in the world: safeguarding against economic slowdowns and downturns. Food and Agriculture Organization, Rome. http://www.fao.org/3/ca5162en/ca5162en.pdf. Accessed 31 Oct 2019
FAOSTAT (2019) Food and agriculture organization corporate statistical database. Food and Agriculture Organization, Rome
Favretto N, Stringer LC, Dougill AJ (2014) Unpacking livelihood challenges and opportunities in energy crop cultivation: perspectives on jatropha curcas projects in Mali. Geogr J 180:365–376. https://doi.org/10.1111/geoj.12053
Ferris RSB, Collinson C, Wanda K, Jagwe J, Wright P (2001) Evaluating the marketing opportunities for Shea nut and Shea nut processed products in Uganda. Technical Report. Natural Resources Institute, Chatham
FEWS Net Liberia (2016) Liberia food security outlook update. A delayed start to agricultural activities in the northwest and central areas
Finkbeiner M (2014) Indirect land use change – help beyond the hype? Biomass Bioenergy 62:218–221
Fortucci P (2002) The contribution of cotton to economy and food security in developing countries. Cott. Glob. Trade Negot
Garba ID, Sanni SA, Adebayo CO (2015) Analyzing the structure and performance of Shea butter market in Bosso and Borgu local government areas of Niger state, Nigeria. Int J Sci Technol 8:321–336. https://doi.org/10.14257/ijunesst.2015.8.2.31
Gashaw E, Mekbib F, Ayana A (2018) Sugarcane landraces of Ethiopia: Germplasm collection and analysis of regional diversity and distribution. Adv Agric 2018:7920724
Gasparatos A, von Maltitz G, Johnson FX, Lee L, Mathai M, Puppim de Oliveira J, Willis K (2015) Biofuels in Africa: drivers, impacts and priority policy areas. Renew Sust Energ Rev 45:879–901
Gasparatos A, Romeu-Dalmau C, von Maltitz G, Johnson FX, Shackleton C, Jarzebski MP, Jumbe C, Ochieng C, Mudombi S, Nyambane A, Willis KJ (2018) Mechanisms and indicators for assessing the impact of biofuel feedstock production on ecosystem services. Biomass Bioenergy 114:157–173
Gentles SJ, Charles C, Nicholas DB, Ploeg J, McKibbon KA (2016) Reviewing the research methods literature: principles and strategies illustrated by a systematic overview of sampling in qualitative research. Syst Rev 5:1–11. https://doi.org/10.1186/s13643-016-0343-0
Gerber J (2008) Résistances contre deux géants industriels en forêt tropicale: Populations locales versus plantations commerciales d'hévéas et de palmiers à huile dans le Sud-Cameroun
German L, Schoneveld GC, Gumbo D (2011) The local social and environmental impacts of smallholder-based biofuel investments in Zambia. Ecol Soc 16(4):12. https://doi.org/10.5751/ES-04280-160412
German L, Schoneveld G, Mwangi E (2013) Contemporary processes of large-scale land Acquisition in sub-Saharan Africa: legal deficiency or elite capture of the rule of law? World Dev 48:1–18. https://doi.org/10.1016/j.worlddev.2013.03.006

Gesimba RM, Langat MC, Liu G, Wolukau JN (2005) The tea industry in Kenya; the challenges and positive developments. J Appl Sci 5(2):334–336
Gibson M (2012) Food security—a commentary: what is it and why is it so complicated. Foods 1:18–27. https://doi.org/10.3390/foods1010018
Giovannetti G, Ticci E (2016) Determinants of biofuel-oriented land acquisitions in sub-Saharan Africa. Renew Sust Energ Rev 54:678–687. https://doi.org/10.1016/j.rser.2015.10.008
GoK (2001) The poverty reduction strategy paper (PRSP), 2001–2004. Government of Kenya (Gok): Ministry of Finance and Planning, Nairobi
GoK (2002) The Kenya Rural Development Strategy (KRDS), 2002-2017. Government of Kenya (GoK), Nairobi. Government printer
GoK (2004) Strategy for revitalizing agriculture 2004–2014, government of Kenya (GoK), Ministry of Agriculture and Ministry of Livestock and Fisheries Development, Nairobi
GoK (2007a) National Adaptation Strategy for the Sugar industry of Kenya. Government of Kenya (GoK), Nairobi. Government printer. http://www.kenyasugar.co.ke/
GoK (2007b) Structural implications of economic liberalization on agriculture and rural development in Kenya. Government of Kenya (GoK), Nairobi. Government printer
GoK (2008) Strategic plan (2008–2012). Ministry of Agriculture, Government of Kenya (GoK), Nairobi
GoK (2018) Economic Survey 2018. Ministry of planning, Kenya National Bureau of statistics. Government of Kenya (GoK), Nairobi
Gollin D (2010) Chapter 73 agricultural productivity and economic growth. Handbook of agricultural economics, vol 4, 1st edn. Elsevier B.V, Amsterdam. https://doi.org/10.1016/S1574-0072(09)04073-0
Govereha J, Jayne TS (2003) Cash cropping and food crop productivity. Agric Econ 28:39–50
Greenpeace International (2012) Palm oil's new frontier: How industrial expansion threantens Africa's rainforests
Greenpeace International (2016) Africa's forests under threat: Socfin's plantations in Cameroon and Liberia
Grimsby LK, Aune JB, Johnsen FH (2012) Human energy requirements in Jatropha oil production for rural electrification in Tanzania. Energy Sustain Dev 16:297–302. https://doi.org/10.1016/j.esd.2012.04.002
Hailemariam WG, Silalertruksa T, Gheewala SH, Jakrawatana N (2019) Water–energy–food nexus of sugarcane production in Ethiopia. Environ Eng Sci 36:798–807
Hall R, Scoones I, Tsikata D (2017) Plantations, outgrowers and commercial farming in Africa: agricultural commercialisation and implications for agrarian change. J Peasant Stud 44:515–537
Hartley F, van Seventer D, Tostão E, Arndt C (2019) Economic impacts of developing a biofuel industry in Mozambique. Dev South Afr 36:233–249. https://doi.org/10.1080/0376835X.2018.1548962
Hatskevich A, Jenicek V, Antwi Darkwah S (2011) Shea industry – a means of poverty reduction in northern Ghana. Agric Trop Subtrop 44:223–228
Hatskevich A, Karel S, Joseph EE (2014) Shea butter processing as an engine of poverty reduction in northern Ghana: case study of four communities in the Bolgatanga municipality. Afr J Agric Res 9:3185–3190. https://doi.org/10.5897/ajar2014.8907
Herrmann R, Jumbe C, Bruentrup M, Osabuohien E (2018) Competition between biofuel feedstock and food production: empirical evidence from sugarcane outgrower settings in Malawi. Biomass Bioenergy 114:100–111. https://doi.org/10.1016/J.BIOMBIOE.2017.09.002
Hess TM, Sumberg J, Biggs T, Georgescu M, Haro-Monteagudo D, Jewitt G, Ozdogan M, Marshall M, Thenkabail P, Daccache A, Marin F, Knox JW (2016) A sweet deal? Sugarcane, water and agricultural transformation in sub-Saharan Africa. Glob Environ Chang 39:181–194
Hinderink J, Sterkenburg JJ (1987) Agricultural commercialization and government policy in Africa. Routledge, Oxon
HLPE (2013) Investing in smallholder agriculture for food security. A report by the High Level Panel of Experts on Food Security and Nutrition of the Committee on World Food Security, Rome

Howard EK, Osei-Ntiri K, Osei-Poku P (2012) The non-performance of Ghana cotton industry: eco-friendly cotton production tech- nologies for sustainable development. Int J Fiber Text Res 2:30–38

Hussein K, Perret C, Hitimana L (2005) Economic and social importance of cotton in West Africa: role of cotton in regional development, trade and livelihoods. Sahel and West Africa Club Secretariat/OECD. http://hubrural.org/IMG/pdf/csao_etude_socio_eco_coton_eng.pdf. Accessed 31 Oct 2019

ICSU (2017) A guide to SDG interactions : from science to implementation. International Council for Science

Iddrisu I, Bhattacharyya SC (2015) Ghana's bioenergy policy: is 20% biofuel integration achievable by 2030? Renew Sust Energ Rev 43:32–39

IFAD (2013) Smallholders, food security, and the environment. International Fund for Agricultural Development (IFAD). https://www.ifad.org/documents/38714170/39135645/smallholders_report.pdf/133e8903-0204-4e7d-a780-bca847933f2e. Accessed 31 Oct 2019

Ismael Y, Bennett R, Morse S (2002) Benefits from Bt cotton use by smallholder farmers in South Africa. AgBioforum 5:1–5

Jacques W, Fagaye S, Jacques L (2009) Experiment-based prototyping to design and assess cotton management systems in West Africa. Agron Sustain Dev 29:545–556

Jaffee S (2003) Malawi's tobacco sector standing on one strong leg is better than on none. Africa Region Working Paper Series 55

Jamala G, Jada MY, Yidau JJ, Joel L (2013) Socio-economic contribution of Shea tree (Vitellaria paradoxa) in support of rural livelihood in Ganye, Southeastern Adamawa state, Nigeria IOSR J Environ Sci Toxicol. Food Technol 6:2319–2399

James P, Woodhouse P (2016) Crisis and differentiation among small-scale Sugar cane growers in Nkomazi, South Africa. J South Afr Stud 7070:1–15. https://doi.org/10.1080/03057070.2016.1197694

Jarzebski MP, Ahmed A, Baofo YA, Balde BS, Saito O, von Maltitz G, Gasparatos A (2019) Food security impacts of industrial crop production in sub-Saharan Africa: a systematic review of the impact mechanisms. Food Sec 12(1):105–135

Johnson FX, Silveira S (2014) Pioneer countries in the transition to alternative transport fuels: comparison of ethanol programmes and policies in Brazil, Malawi and Sweden. Environ Innov Soc Transitions 11:1–24. https://doi.org/10.1016/j.eist.2013.08.001

Jones MR, Singels A, Ruane A (2013) Simulated impacts of climate change on water use and yield of irrigated sugarcane in South Africa. Proc S Afr Sugar Technol Assoc 86:184–189

Kaag MMA, Zoomers EB (2014) The global land grab: beyond the hype. Zed Books, London

Kagira EK, Kimani SW, Githii KS (2012) Sustainable methods of addressing challenges facing small holder tea sector in Kenya: a supply chain management approach. J Mgmt Sustain 2:75

Kalinda C, Moses Z, Lackson C, Chisala LA, Donald Z, Darius P, Exildah CK (2015) Economic impact and challenges of Jatropha curcas L. projects in North-Western Province, Zambia: a case of Solwezi district. Sustain 7:9907–9923. https://doi.org/10.3390/su7089907

Kaminski J (2009) Cotton dependence in Burkina Faso: constraints and opportunities for balanced growth. World Bank. http://siteresources.worldbank.org/AFRICAEXT/Resources/258643-1271798012256/Burkina-cotton.pdf. Accessed 31 Oct 2019

Kaminski J, Headey D, Bernard T (2009) Navigating through reforms: cotton in Burkina Faso. In: In millions fed: proven successes in agricultural development. International Food Policy Research Institute, Washington, DC

Kaminski J, Headey D, Bernard T (2011) The Burkinabè cotton story 1992-2007: sustainable success or sub-saharan mirage? World Dev 39:1460–1475. https://doi.org/10.1016/j.worlddev.2010.12.003

Karanja A, Mburu F, Gasparatos A (2020) A multi-stakeholder perception analysis about the adoption, impacts and priority areas in the Kenyan clean cooking sector. Sustain Sci 15 (1):333–351. https://doi.org/10.1007/s11625-019-00742-4

Kennedy E (1989) The effects of sugarcane production on food security, health, and nutrition in Kenya - a longitudinal analysis. International Food Policy Research Institute, Washington, DC
Kenya Sugar Board (KSB) (2010) The Kenya Sugar industry value chain analysis: analysis of the production and marketing costs for sugarcane and Sugar related products. KSB, Nairobi, Kenya
Kgathi DL, Mfundisi KB, Mmopelwa G, Mosepele K (2012) Potential impacts of biofuel development on food security in Botswana: a contribution to energy policy. Energy Policy 43:70–79. https://doi.org/10.1016/j.enpol.2011.12.027
Khanna M, Crago CL (2012) Measuring indirect land use change with biofuels: implications for policy. Annu Rev Resour Econ 4:161–184
Kiewisch E (2015) Looking within the household: a study on gender, food security, and resilience in cocoa-growing communities. Gend Dev 23:497–513. https://doi.org/10.1080/13552074.2015.1095550
Köhlin G, Sills EO, Pattanayak SK, Wilfong C (2011) Energy, gender and development. What are the linkages? Where is the evidence? World Bank Policy Res 125:1–63. https://doi.org/10.1596/1813-9450-5800
Kolavalli S, Vigneri M (2011) Cocoa in Ghana: shaping the success of the economy. World Bank
Komarek AM (2010) Crop diversification decisions: the case of vanilla in Uganda. Q J Int Agric 49 (3):227–242
Kriger CE (2005) Mapping the history of cotton textile production in Precolonial West Africa. Afr Econ Hist 33:87–116
KSI (2009) Kenya Sugar Industry Strategic Plan 2010–2014. Nairobi, Kenya Sugar Industry, Kenya
Kuchler M, Linnér B-O (2012) Challenging the food vs. fuel dilemma: genealogical analysis of the biofuel discourse pursued by international organizations. Food Policy 37:581–588. https://doi.org/10.1016/j.foodpol.2012.06.005
Kulik MC, Bialous SA, Munthali S, Max W (2017) Tobacco growing and the sustainable development goals, Malawi Bull. World Health Organ 95:362–367
Kuntashula E, van der Horst D, Vermeylen S (2014) A pro-poor biofuel? Household wealth and farmer participation in Jatropha curcas seed production and exchange in eastern Zambia. Biomass Bioenergy 63:187–197. https://doi.org/10.1016/j.biombioe.2014.01.051
Kwong RNGK (2005) Status of sugar industry in Mauritius: constraints and future research strategies. Sugar Tech 7:5–10. https://doi.org/10.1007/bf02942411
Laris P, Foltz JD (2014) Cotton as catalyst? The role of shifting fertilizer in Mali's silent maize revolution. Hum Ecol 42:857–872. https://doi.org/10.1007/s10745-014-9694-0
Laris P, Foltz JD, Voorhees B (2015) Taking from cotton to grow maize: the shifting practices of small-holder farmers in the cotton belt of Mali. Agric Syst 133:1–13. https://doi.org/10.1016/j.agsy.2014.10.010
Laube W (2015) Global shea nut commodity chains and poverty eradication in northern Ghana: myth or reality? UDS Int J Dev 2(1):128–147
Lazzarini AH (2016) Gendered labour, migratory labour: reforming Sugar regimes in Xinavane, Mozambique. J South Afr Stud 7070:1–19. https://doi.org/10.1080/03057070.2016.1197695
Leathers H (1999) Transaction costs analysis of maize and cotton Marketing in Zambia and Tanzania, Sustainable Development Publication Series
Lindell I, Kroon GM (2011) Sugarcane and agroforestry farming in western Kenya: a comparative study of different farming systems in the Nyando district
Locke A, Henley G (2013) A review of the literature on biofuels and food security at a local level assessing the state of the evidence. ODI, London
Lowder SK, Skoet J, Raney T (2016) The number, size, and distribution of farms, smallholder farms, and family farms worldwide. World Dev 87:16–29
Matenga CR (2016) Outgrowers and livelihoods: the case of Magobbo smallholder block farming in Mazabuka District in Zambia. J South Afr Stud 7070:1–16. https://doi.org/10.1080/03057070.2016.1211402

Matondi PB, Havnevik K, Beyene A (2011) Biofuels, land grabbing and food security in Africa. The Nordic Africa Institute, Zed BooksLondon/New York
Minten B, Dereje M, Engida E, Kuma T (2017) Coffee value chains on the move: evidence in Ethiopia. Food Policy 83:370–383
Moat J, Williams J, Baena S, Wilkinson T, Demissew S, Challa ZK, Gole TW, Davis AP (2017) Coffee farming and climate change in Ethiopia: impacts, forecasts, resilience and opportunities. –summary. The strategic climate institutions Programme (SCIP). Royal Botanic Gardens, Kew, p 37
MOFA (2013) Study to Prepare Statutory Texts for Ghanaian Institutions for The Management and Regulation of The Cotton Sector. Final report by Ministry of Trade and Industry and Ministry of Food and Agriculture, Government of Ghana, Accra
Mohammed F, Boateng S, Al-Hassan S (2013) Effects of adoption of improved Sheabutter processing technology on women's livelihoods and their microenterprise growth. Am J Hum Soc Sci 1(4):244–250
Molony T, Smith J (2010) Briefing: biofuels, food security, and Africa. Afr Aff (Lond) 109:489–498. https://doi.org/10.1093/afraf/adq019
Morris M, Kelly VA, Kopicki RJ, Byerlee D (2007) Fertilizer use in African agriculture: lessons learned and good practices guide- line. World Bank, Washington, DC, p 144
Moseley WG (2001) Sahelian "White Gold" and Rural Poverty-Environment Interactions: the Political Ecology of Cotton Production, Environmental Change, and Household Food Economy in Mali. Univ. Georg. doi:https://doi.org/10.16953/deusbed.74839
Moser C (2008) An economic analysis of deforestation in Madagascar in the 1990s. Environ Sci 5:91–108. https://doi.org/10.1080/15693430801912170
Mponela P, Jumbe CBL, Mwase WF (2011) Determinants and extent of land allocation for Jatropha curcas L. cultivation among smallholder farmers in Malawi. Biomass Bioenergy 35:2499–2505. https://doi.org/10.1016/j.biombioe.2011.01.038
Mudombi S, Nyambane A, von Maltitz GP, Gasparatos A, Johnson FX, Chenene ML, Attanassov B (2018) User perceptions about the adoption and use of ethanol fuel and cookstoves in Maputo, Mozambique. Energy Sustain Dev 44:97–108. https://doi.org/10.1016/j.esd.2018.03.004
Mulianga B, Bégué A, Clouvel P, Todoroff P (2015) Mapping cropping practices of a sugarcane-based cropping system in Kenya using remote sensing. Remote Sens 7(11):14428–14444
Mulwa R, Nuppenau EA, Emrouznejad A (2005) Productivity growth in smallholder sugarcane farming in Kenya: a Malmquist TFP decomposition. Online access to Deutscher Tropentag, Deutscher Tropentag, 6
Nalepa RA, Short Gianotti AG, Bauer DM (2017) Marginal land and the global land rush: a spatial exploration of contested lands and state-directed development in contemporary Ethiopia. Geoforum 82:237–251. https://doi.org/10.1016/j.geoforum.2016.10.008
Naughton CC, Zhang Q, Mihelcic JR (2017) Modelling energy and environmental impacts of traditional and improved shea butter production in West Africa for food security. Sci Total Environ 576:284–291. https://doi.org/10.1016/j.scitotenv.2016.10.059
Negash M, Swinnen JFM (2013) Biofuels and food security: micro-evidence from Ethiopia. Energy Policy 61:963–976. https://doi.org/10.1016/j.enpol.2013.06.031
Nyagito HO (2001) Policy and legal framework for the coffee subsector and the impact of liberalization in Kenya. Kenya Institute of Public Policy and Analysis, Nairobi
Oluyole KA, Oni OA, Omonona BT, Adenegan KO (2009) Food security among cocoa farming households of Ondo state, Nigeria. ARPN J Agric Biol Sci 4(5):7–13
Onduru DD, De Jager A, Hiller S, Van den Bosch R (2012) Sustainability of smallholder tea production in developing countries: learning experiences from farmer field schools in Kenya. Int J Dev Sustain Eng 1(3):714–742
Onoji SE, Iyuke SE, Igbafe AI, Nkazi DB (2016) Rubber seed oil: a potential renewable source of biodiesel for sustainable development in sub-Saharan Africa. Energy Convers Manag 110:125–134. https://doi.org/10.1016/j.enconman.2015.12.002

Onyango JP, Waswa F, Mwamburi M (2012) Contract sugarcane farming and farmers' incomes in the Lake Victoria basin, Kenya
Ordway EM, Asner GP, Lambin EF (2017) Deforestation risk due to commodity crop expansion in sub-Saharan Africa. Environ Res Lett 12. https://doi.org/10.1088/1748-9326/aa6509
Oyono PR (2013) The narratives of capitalist land accumulation and recognition in coastal Cameroon. The Land Deal Politics Initiative (LDPI) Working Paper No. 29
Palm CA, Smukler SM, Sullivan CC, Mutuo PK, Nyadzi GI, Walsh MG (2010) Identifying potential synergies and trade-offs for meeting food security and climate change objectives in sub-Saharan Africa. Proc Natl Acad Sci U S A 107:19661–19666. https://doi.org/10.1073/pnas.0912248107
Parvathi P (2017) Economic divorce mixed crop-livestock farming to increase dietary diversity? Evidence from smallholders in Lao PDR. Paper presented at the Third GlobalFood Symposium, April 28–29, University of Goettingen, Germany
Patel ER (2007) Logging of rare rosewood and Palisandre (Dalbergia spp.) within Marojejy National Park, Madagascar. Madagascar Conserv Dev 2:11–16. https://doi.org/10.4314/mcd.v2i1.44124
Quarmine W, Haagsma R, van Huis A, Sakyi-Dawson O, Obeng-Ofori D, Asante FA (2014) Did the price-related reforms in Ghana's cocoa sector favour farmers? Int J Agric Sustain 12 (3):248–262
Randriamalala H, Liu Z (2010) Rosewood of Madagascar: between democracy and conservation. Madagascar Conserv Dev 5:11–22. https://doi.org/10.4314/mcd.v5i1.57336
Rehmato (2011) Land to investors: large-scale land transfers in Ethiopia. Forum for Social Studies, Addis Ababa
Ricciardi V, Ramankutty N, Mehrabi Z, Jarvis L, Chookolingo B (2018) How much of the world's food do smallholders produce? Glob Food Sec 17:64–72. https://doi.org/10.1016/j.gfs.2018.05.002
Richardson B (2012) Trade, aid and rural development: EU sugar policy and the experience of Swaziland. European Centre for Development Policy Management (ECDPM), Maastricht
Ripoche A, Crétenet M, Corbeels M, Affholder F, Naudin K, Sissoko F, Douzet JM, Tittonell P (2015) Cotton as an entry point for soil fertility maintenance and food crop productivity in savannah agroecosystems-evidence from a long-term experiment in southern Mali. Field Crops Res 177:37–48. https://doi.org/10.1016/j.fcr.2015.02.013
Romeu-Dalmau C et al (2018) Impacts of land use change due to biofuel crops on climate regulation services: five case studies in Malawi, Mozambique and Swaziland. Biomass Bioenergy 114:30–40
Romijn H, Heijnen S, Colthoff JR, de Jong B, van Eijck J (2014) Economic and social sustainability performance of jatropha projects: results from field surveys in Mozambique, Tanzania and Mali. Sustain 6:6203–6235. https://doi.org/10.3390/su6096203
Sasson A (2012) Food security for Africa: an urgent global challenge. Agric Food Secur 1:1–16. https://doi.org/10.1186/2048-7010-1-2
Schoneveld GC (2014) The geographic and sectoral patterns of large-scale farmland investments in sub-Saharan Africa. Food Policy 48:34–50. https://doi.org/10.1016/j.foodpol.2014.03.007
Schoneveld GC, German LA, Nukator E (2011) Land-based investments for rural development? A grounded analysis of the local impacts of biofuel feedstock plantations in Ghana. Ecol Soc 16 (4):10
Senbeta F, Denich M (2006) Effects of wild coffee management on species diversity in the Afromontane rainforests of Ethiopia. For Ecol Manag 232:68–74. https://doi.org/10.1016/j.foreco.2006.05.064
Shortall OK (2013) "Marginal land" for energy crops: exploring definitions and embedded assumptions. Energy Policy 62:19–27
Singh BP (2010) Overview of industrial crops. In: Singh BP (ed) Industrial crops and uses. CAB International, Wallingfrod/Cambridge, MA, pp 1–20

Sisay BT (2018) Coffee production and climate change in Ethiopia. In: Lichtfouse E. (eds). Sustain Agric Rev 33:99–113

Smalley R (2013) Plantations, Contract Farming and Commercial Farming Areas in Africa: A Comparative Review Rebecca Smalley Land and Agricultural Commercialisation in Africa (LACA) Project Working Paper Series

Sodjinou E, Glin LC, Nicolay G, Tovignan S, Hinvi J (2015) Socioeconomic determinants of organic cotton adoption in Benin, West Africa. Agric Food Econ 3(1):12. https://doi.org/10.1186/s40100-015-0030-9

Strona G, Stringer SD, Vieilledent G, Szantoi Z, Garcia-Ulloa J, Wich SA (2018) Small room for compromise between oil palm cultivation and primate conservation in Africa. Proc Natl Acad Sci U S A 115:8811–8816. https://doi.org/10.1073/pnas.1804775115

Suleman D, Collins AM, Ricky YN (2014) Urban poverty in northern Ghana: tracing the livelihood strategies of women in the Shea butter industry. J Hum Soc Sci Res 3(1):15–25

Tefera A (2015) Ethiopia: coffee annual report. GAIN report (number ET1514–26/5/2015). USDA Foreign Agricultural Service. p 6

Tefera A, Tefera T (2014) Ethiopia: Coffee Annual Report. Gain report ET-1402, USDA

Tenenbaum DJ (2008) Food vs. fuel: diversion of crops could cause more hunger. *Environ*. Health Perspect 116:A254–A257

Terry A (2007) The Komati downstream development. Tijdschr voor Econ en Soc Geogr 98:641–651

Terry A (2012) Evaluating the green revolution after a decade: a Swaziland case study. Int J Agric Sustain 10(2):135–149. https://doi.org/10.1080/14735903.2011.600828

Terry A, Ogg M (2017) Restructuring the Swazi sugar industry: the changing role and political significance of smallholders. J South Afr Stud 43(3):585–603. https://doi.org/10.1080/03057070.2016.1190520

Terry A, Ryder M (2007) Improving food security in Swaziland: the transition from subsistence to communally managed cash cropping. Nat Resour Forum 31:263–272. https://doi.org/10.1111/j.1477-8947.2007.00161.x

Theriault V, Tschirley DL (2014) How institutions mediate the impact of cash cropping on food crop intensification: an application to cotton in sub-Saharan Africa. World Dev 64:298–310. https://doi.org/10.1016/j.worlddev.2014.06.014

Thuku GK, Gachanja P, Almadi O (2013) Effects of reforms on productivity of coffee in Kenya. Int J Bus Soc Sci 4(15)

Timko JA, Amsalu A, Acheampong E, Teferi MK (2014) Local perceptions about the effects of Jatropha (Jatropha curcas) and castor (Ricinus communis) plantations on households in Ghana and Ethiopia. Sustain 6:7224–7241. https://doi.org/10.3390/su6107224

Tomei J, Helliwell R (2016) Food versus fuel? Going beyond biofuels. Land Use Policy 56:320–326. https://doi.org/10.1016/j.landusepol.2015.11.015

Tschirley D, Poulton C, Labaste P (2009) Organization and performance of cotton sectors in Africa learning from reform experience. World Bank, Washington, DC

Uckert G et al (2015) Increase without spatial extension: productivity in small-scale palm oil production in Africa—the case of Kigoma, Tanzania. Reg Environ Chang 15:1229–1241

UNCTAD (2000) Policies for small-scale Sugar growing in Swaziland. United Nations Conference on Trade and Development (UNCTAD), Geneva

UNDP (2015) Mdg Report 2015 Lessons learned in the implementing the MDGs: Assessing progress in Africa toward the Millennium Development Goals : https://www.undp.org/content/undp/en/home/librarypage/mdg/mdg-reports/africa-collection.html. Accessed 31 Oct 2019

UNECA (2015) Draft Africa Regional Report on the Sustainable Development Goals. United Nations Economic Commission for Africa, 2015 Addis Ababa, Ethiopia. https://www.uneca.org/sites/default/files/PublicationFiles/africa_regional_report_on_sdgs_fullreport_eng.pdf. Accessed 31 Oct 2019

Van Eijck J, Smeets E, Faaij A (2012) The economic performance of jatropha, cassava and eucalyptus production systems for energy in an east African smallholder setting. GCB Bioenergy 4:828–845. https://doi.org/10.1111/j.1757-1707.2012.01179.x

Van Eijck J, Romijn H, Smeets E, Bailis R, Rooijakkers M, Hooijkaas N, Verweij P, Faaij A (2014) Comparative analysis of key socio-economic and environmental impacts of smallholder and plantation based jatropha biofuel production systems in Tanzania. Biomass Bioenergy 61:25–45. https://doi.org/10.1016/j.biombioe.2013.10.005

Vitale J (2018) Economic importance of cotton in Burkina Faso. Food and Agriculture Organization, Rome

Vitale J, Ouattarra M, Vognan G (2011) Enhancing sustainability of cotton production systems in West Africa: a summary of empirical evidence from Burkina Faso. Sustainability 3:1136–1169. https://doi.org/10.3390/su3081136

von Grebmer et al (2018) Global hunger index forced migration and hunger. Welt Hunger Hilfe, Bonn

von Maltitz G, Gasparatos A, Fabricius C (2014) The rise, decline and future resilience benefits of jatropha in southern Africa. Sustainability 6:3615–3643

von Maltitz GP, Gasparatos A, Fabricius C, Morris A, Willis KJ (2016) Jatropha cultivation in Malawi and Mozambique: impact on ecosystem services, local human Well-being, and poverty alleviation. Ecol Soc 21. https://doi.org/10.5751/ES-08554-210303

von Maltitz GP, Henley G, Ogg M, Samboko PC, Gasparatos A, Read M, Engelbrecht F, Ahmed A (2018) Institutional arrangements of outgrower sugarcane production in southern Africa. Dev South Afr:1–23. https://doi.org/10.1080/0376835X.2018.1527215

von Maltitz GP, Henley G, Ogg M, Samboko PC, Gasparatos A, Read M, Engelbrecht F, Ahmed A (2019) Institutional arrangements of outgrower sugarcane production in southern Africa. Dev South Afr 36:175–197. https://doi.org/10.1080/0376835X.2018.1527215

Waswa F, Mcharo M, Netondo G et al (2009a) Enhancing household food and income security through crop diversification in the Nzoia and Mumias sugarbelts in Kenya. J Appl Biosci 23:1406–1415

Waswa F, Netondo G, Maina L, Naisiko T, Wangamati J (2009b) Potential of corporate social responsibility for poverty alleviation among contract sugarcane farmers in the Nzoia Sugarbelt, Western Kenya. J Agric Environ Ethics 22:463–475. https://doi.org/10.1007/s10806-009-9165-6

Waswa F, Gweyi-Onyango JP, Mcharo M (2012) Contract sugarcane farming and farmers ' incomes in the Lake Victoria basin, Kenya. J Appl Biosci 52:3685–3695

Wendimu MA, Henningsen A, Gibbon P (2016) Sugarcane outgrowers in Ethiopia:"forced" to remain poor? World Dev 83:84–97

Wiggins S, Henley G, Keats S (2015) Competitive or complementary? Industrial crops and food security in sub-Saharan Africa. Overseas Development Institute, London

Williams T (2009) An African success story: Ghana's cocoa marketing system, Institute of Development Studies Working Paper 318, 2009

Williamson S, Wilson J, Ton P, Ratter SG, Ferrigno S, Vodouhe DS (2005) Organic Cotton: A new development path for african smallholders? International Institute for Environment and Development, London

Wiredu AN, Mensah-Bonsu A, Andah EK, Fosu KY (2011) Hybrid cocoa and land productivity of cocoa farmers in Ashanti region of Ghana. World J Agric Sci 7:172–178

Wood B, Nelson CH, Kilic T, Murray S (2012) Up in smoke? Agricultural commercialization, rising food prices and stunting in Malawi. In 2012 Conference, August 18–24, 2012, Foz do Iguacu, Brazil, 2012 (Vol. 131261): International Association of Agricultural Economists

World Bank (2013) Burkina Faso: What is Driving Cotton Production, Stochastic Frontier Approach for Panel Data. Washington, DC. World Bank. https://openknowledge.worldbank.org/handle/10986/15989. License: CC BY 3.0 IGO

World Growth (2010) Palm oil and food security – the impediment of land supply. How environmentalists and "no conversion" are inflating food prices

World Rainforest Movement (2011) Oil palm in Benin: From small-scale production by women to large-scale corporate industry. In international day against monoculture tree plantations. Monthly Bulletin, Issue 170

World Rainforest Movement (2012) Cameroon: major impacts of oil palm agribusiness on food sovereignty and livelihoods. Monthly Bulletin, Issue 183 - October 2012

World Rainforest Movement (2013) Liberian women impacted by expansion of Sime Darby's oil palm plantations. Monthly Bulletin - Issue 187 - February 2013

World Rainforest Movement (2015) Wilmar International: Destroying lives for profit. Monthly Bulletin, Issue 213 – April 2015

Zeller M, Sharma M (2000) Many borrow, more save, and all insure: implications for food and micro-finance policy. Food Policy 25(2):143–167. https://doi.org/10.1016/S0306-9192(99)00065-2

Zommers ZA, Johnson PJ, Macdonald and D.W. (2012) Biofuels bonanza? Sugarcane production and poverty in villages surrounding Budongo Forest, Uganda. J East Afr Stud 6:177–195. https://doi.org/10.1080/17531055.2012.669569

Chapter 4
Large-Scale Land Acquisitions in Sub-Saharan Africa and Corporate Social Responsibility (CSR): Insights from Italian Investments

Marta Antonelli, Stefania Bracco, Margherita Emma Turvani, and Antonio Vicario

4.1 Introduction

Since the mid-2000s, there has been a large interest in agriculture-based private and public–private investments in many parts of Sub-Saharan Africa (SSA) (Schoneveld 2014). This trend has been associated with the acquisitions of large tracts of agricultural land through large-scale land acquisitions (LSLAs) such as (a) long-term leases that typically span between 55 and 99 years and/or (b) purchase agreements by private companies, state-owned companies, investment funds and public–private partnerships. Such land deals involve a broad range of different stakeholders during their distinct phases, including international investors, private interests in the host countries, national government agencies and eventually local communities (Onoja and Achike 2015). Investors involved in LSLAs either aim at producing agricultural commodities or obtaining the ownership of successive stages of crop value chains through vertical integration (Anseeuw et al. 2013).

Overall, since 2000, more than 26 Mha of agricultural land have been leased to foreign investors globally, with SSA being a particularly popular investment destination (FAO 2012; Nolte et al. 2016). Schoneveld (2014) estimates that 563 LSLAs

---

Marta Antonelli and Stefania Bracco contributed equally to this work

M. Antonelli (✉)
Euro-Mediterranean Centre on Climate Change (CMCC), Lecce, Italy

S. Bracco
University of Gastronomic Sciences, Pollenzo, Italy

M. E. Turvani · A. Vicario
IUAV University of Venice, Venice, Italy

A. Gasparatos et al. (eds.), *Sustainability Challenges in Sub-Saharan Africa I*,
Science for Sustainable Societies, https://doi.org/10.1007/978-981-15-4458-3_4

of over 2000 ha each have accounted for land transactions in the order of 22.7 Mha since the year 2000. This proliferation of LSLAs has been invariably referred to as the 'land rush' considering its magnitude and rapidity (Arezki et al. 2013). European investors emerged as important players in this land rush, as they have been involved in more than 300 deals, spanning almost 14 million ha of agricultural land, mostly in SSA (Antonelli et al. 2015; Bracco 2015).

Many studies have discussed the different drivers of LSLAs in SSA, including the (a) need to secure reliable and long-term food supplies (especially for land and/or water-scarce countries) (Allan et al. 2013); (b) increasing demand for liquid biofuels, especially in the EU (Antonelli et al. 2015; Bracco et al. 2015; Cotula et al. 2009) and (c) speculation on the future prices of agricultural land (De Schutter 2011a) (see Chaps. 2–3 Vol. 1, Chap. 5 Vol. 2). Actually, the production of agrofuels[1] and flexible crops (i.e. crops for food, fuel and industrial purposes) has been a key driver of both the realised and intended LSLAs pursued by many EU investors around SSA (Bracco et al. 2015; Bracco 2015; Schoneveld 2014).

Some of the key factors that have created this conducive environment for the surge in LSLAs in SSA include the high rates of return on capital for farming investments (up to 50–60% per year) (FAO in World Agronomy and VM Group 2011) and the combination of favourable conditions, such as cheap land/labour and guaranteed markets (Friends of the Earth Africa and Friends of the Earth Europe 2010). However, according to FAO (2011) of the land deals undertaken in SSA during the early stages of the land rush, only 20% resulted in productive investments. In fact by the mid-2010s, almost all of the biofuel investments related to jatropha that drove to a large extent the land rush (Schoneveld 2014) had collapsed across SSA (Ahmed et al. 2019).

Perspectives about the opportunities and risks associated with LSLAs in SSA have been quite polarised. For example, some organisations have pointed to the opportunities for economic and social development that may arise from the capital inflows in the target local and national economies (World Bank 2011; German Federal Ministry for Economic Cooperation and Development 2009). Other organisations have mostly delved on the negative negative effects of poorly operated and regulated LSLAs on local communities and the environment (Hufe and Heuermann 2017). Many scholars have highlighted the need to maximise the benefits of LSLAs while minimising their negative impacts (Allan et al. 2013; Cotula et al. 2009; Davis et al. 2014; Jägerskog et al. 2012; Deininger 2011; De Schutter 2009, 2011a; Matondi et al. 2011; FAO 2011) (see below).

LSLA can indeed have many different sustainability impacts in SSA, and thus affect progress towards meeting the Sustainable Development Goals (Dell'Angelo et al. 2017). On the one hand, LSLAs can boost economic growth, rural development

[1]The EU Renewable Energy Directive (EU RED. 2009/28/EC) has set by the year 2020 a mandatory national target of meeting 20% of a country's gross final energy consumption through renewable energy sources, as well as a 10% share of renewable energy in all forms of transport (EU 2009).

and national energy security if promoted carefully (see also Chap. 1 Vol. 1, Chap. 5 Vol. 2). For example, many scholars have identified how large-scale agricultural investments can boost rural employment and have wider economic and poverty alleviation benefits (Arndt 2010; Hall et al. 2017). Similarly, in some landlocked countries such as Malawi, LSLAs that produce biofuels can contribute to national energy security (e.g. Gasparatos et al. 2015) (see also Chaps. 1–3 Vol. 1; Chap. 5 Vol. 2).

However, LSLAs have also been associated with various negative impacts on ecosystems and local communities (see also Chaps. 1 and 3 Vol. 1). This includes many negative environmental impacts related to land use change, biodiversity loss and water depletion and pollution (Gasparatos et al. 2015; Hess et al. 2016). Similarly, LSLAs can have significant socioeconomic impacts mainly related to (a) food security and the loss of capacity to secure food supplies for locals (Matondi et al. 2011) (Chap. 3, Vol. 1); (b) the exclusion of local communities with customary access to land from the new agricultural development projects (Deininger and Byerlee 2011; Ahmed et al. 2018) and (c) increasing competition for land and water resources between local communities' needs and commercial uses (Jägerskog et al. 2012; Hess et al. 2016). In several cases, the positive effects of LSLAs on employment creation and infrastructure development may be offset by the negative impacts such as land degradation, conflicts and inadequate compensation (Hufe and Heuermann 2017).

There are multiple factors that affect the type and magnitude of these sustainability impacts, including the institutional setting of LSLAs. An expanding body of literature has approached this aspect through social and political lens (see among others, De Schutter 2011b, Messerli et al. 2013, Borras et al. 2012). For example, the scope of LSLAs, as well as the type of ownership, can vary considerably across countries and commodities. In Nigeria, for instance, land is held by the state government, which is therefore the only actor involved in land transactions and allocation (Onoja and Achike 2015). In countries such as Ethiopia, the national government played a very strong role in the promotion and operation of LSLAs (Keeley et al. 2014). In many other countries such as Ghana and Mozambique, LSLAs were driven largely by the private sector, with the government simply enacting (often with delay) rules about the promotion and operations of LSLAs (Ahmed et al. 2017; Schut et al. 2014). Lease terms and annual rental fees have also been very diverse across countries and regions. Furthermore, in most SSA countries, there is a strong traditional element in land governance, and it has been argued that it is imperative to take into account chiefs' involvement in LSLAs to avoid negative impacts (Ahmed et al. 2018; German et al. 2013).

Some of these negative impacts associated with LSLAs have emerged through the phenomenon of *land grabbing*. In a nutshell LSLAs can be defined as land grabs when they result in human rights violations, or are characterised by a lack of consent by host communities and an overal lack of transparency (ILC 2011). Several scholars have discussed the main mechanisms through which land-grabbing occurs in SSA (Hall 2011; White et al. 2012), as well as its main sustainability impacts especially related to food security (Matondi et al. 2011) (see also Chap. 3, Vol. 1).

It is in this context of multiple sustainability impacts and involved stakeholders that many organisations have raised the need to ensure that LSLAs are pursued in a 'responsible' manner in SSA (Deininger and Byerlee 2011; Global Witness 2012; HLPE 2011). Several initiatives have aimed to define what constitutes responsible agricultural investments, including the Principles for Responsible Investment in Agriculture and Food Systems (PRAI), the Voluntary Guidelines on the Responsible Governance of Tenure of Land, Fisheries and Forests and the UN Global Compact 'Food sustainability: A Guide for Private Sector Action'. CFS (2013) provides a summary of such international initiatives aiming to define and promote responsible agricultural investments. The EU Renewable Energy Directive (RED) also sets sustainability criteria that are not only taken into account when evaluating compliance with national targets but are also pre-conditions for eligibility for receiving financial support.

Avoiding or minimising the negative and maximising the positive impacts of LSLAs is a major sustainability challenge for many SSA countries. Achieving this can ensure that SSA countries targetted for LSLAs can benefit from their competitive advantages, without inadvertedly facing negative impacts on local communities and the environment. Indeed addressing the sustainability challenges posed by LSLAs is pivotal for attaining the sustainable development goals (SDGs) of the 2030 Agenda set by the United Nations in 2015. Engagement in responsible behaviour, including the adoption of Corporate Social Responsibility (CSR) practices, in the context of LSLAs is key for achieving different SDGs. This is especially true for those SDGs related to environmental sustainability (e.g. clean water and sanitation (SDG 6), affordable and clean energy (SDG 7), sustainable production and consumption (SDG 12)), but also to social sustainability and equality, e.g. reduced inequality (SDG 10) (Hopkins 2016).

Against this backdrop, the aim of this chapter is to explore the intersection of CSR and LSLAs in SSA. In particular, we outline whether corporate social responsibility (CSR) practices related to standard, certification and labelling (SCL) schemes can ensure that land investments are conducted responsibly. We particularly scrutinise LSLAs related to the production of biofuel crops as they can have significant impacts on food security in target countries as outlined above (see also Chap. 3 Vol. 1). We focus mainly on biofuel crops as they have dominated most of the current LSLA debate in SSA[2] (see above), and Italian investors given their growing role in the EU biofuel market. In particular, Italy is the second largest EU investor in SSA after the UK, and has mainly been involved in investments related to energy crops. Furthermore, Italy is committed to obtain 17% of its total energy use from renewable sources by the year 2020 (Directive 2009/28/EC). It is the third largest consumer of biofuels after Germany and France (EurObserv'er 2014) and imports almost all biofuel feedstock crops (Pignatelli and Clamentel 2006).

[2]Even though the actual extent of LSLAs arising related to the EU biofuel market is relatively moderate, their visibility in the current academic and policy debates is enormous. Similar concerns are also relevant and for other LSLAs are related to food and afforestation-related land deals (Ecofys 2013).

Section 4.2 outlines some of the key issues at the intersections of CSR, SCL and LSLAs in Sub-Saharan Africa. Section 4.3 outlines the methodology of this chapter, and Section 4.4 quantifies the extent and type of Italian LSLAs in SSA, as well as the adoption of different SCL schemes. Section 4.5 identifies some of the shortcomings of current practices for improving the sustainability of LSLAs and offers some related policy and practice recommendations.

4.2 Corporate Social Responsibility (CSR) and Large-Scale Land Acquisitions

CSR in the context of LSLAs can be intepreted as a corporate strategy that can be used to internalise the economic, social and environmental costs and benefits of land acquisitions (Bracco 2016). A key rationale of CSR is that when the private and social costs of an economic action (i.e. a land acquisition in this case), are not aligned, then markets are not sufficient to ensure societal well-being. In the presence of such externalities, non-market interventions are needed. At the same time, actors from the private sector (e.g. companies, investors) face pressures from legal systems, regulatory frameworks and, increasingly, the civil society, to manage such externalities in a way that is consistent with the social good (Heal 2008). Some of the factors that influence companies to consider the social and environmental ramifications of their operations (and adopt CSR strategies) include, environmental and social costs, fairness, reputation risks and shareholder and civil society pressure (Heal 2008; Weber 2008; Zezza 2013).

In the context of LSLAs in SSA, it is highly possible that land investments that have not taken into account their possible impacts on the environment and local communites, run the risk of escalating direct operation costs and delays in project implementation due to land disputes with local communities, NGOs and/or governments (Mirza et al. 2014; Bracco 2016). For example, apart from hindering company operations, such conflicts may contribute to the loss of financial support from public or financial institutions, therefore increasing the financial risks of investors, and tarnishing company and investor reputation (De Man 2013; Weber 2008). This is particularly true for big agro-food corporations with well-known brands that face high reputation costs if their name is associated with irresponsible behaviour (Bracco 2016; De Man 2013).

Foreign investors may avoid such risks by following Free, Prior and Informed Consent (FPIC) procedures and respecting the rights of local communities to decide about their land and resources (and be compensated appropriately if given away) (Global Witness 2012). Therefore, it may be in their best interest to identify all the stakeholders involved in the LSLA, and to inform, involve and eventually compensate them accordingly (Global Witness 2012).

A company may choose to join a standard, certification and labelling (SCL) scheme as a strategy to meet its corporate social and environmental responsibility requirements. We argue that, by joining such a SCL scheme, a private organisation essentially adopts a CSR strategy to ensure and prove the responsibility of its actions. Therefore, in this chapter, we consider that joining a SCL scheme is a manifestation of a CSR activity. Indeed, private companies have incentives to join such SCL schemes to prove their responsible behaviour to stakeholders, consumers, financiers and public legislators (Bracco 2016; Van Gelder and German 2011).

Participation in certification SCL schemes can also reduce the information gap between producers, consumers and regulators (Scarlat and Dallemand 2011; Bracco 2015; Zezza 2013), receive guidance and support to comply with environmental and social standards. In fact, SCL schemes aim at managing various stages of regulation such as agenda-setting, negotiations of standards, implementation and monitoring of practices and enforcement of principles (Roberts 1992). Process standards that track the environmental and social conditions under which agricultural commodities (e.g. food crops, biofuel feedstocks) are produced and/or traded are increasingly becoming popular (Zezza 2013). They have thus the potential to promote better production practices, quality and transparency along the entire supply chain of agricultural commodities (Van Dam et al. 2008, 2010).

However, individual companies and investors have different 'pressure points', which might be used to influence their behaviour in terms of responsibility and the decision to join voluntary standards and certification schemes (Cotula and Blackmore 2014). Energy policy can be such a leverage point to pressure investors to adopt responsible practices in order to avoid losing political and financial support and market access. In fact, the European Commission requires that the biofuels and feedstocks entering its market meet some environmental and social sustainability criteria in order to be eligible to receive governmental support and to count towards the Renewable Energy Directive targets (2009/28/EC, Article 17). For example, in order to comply with the EU RED requirements, biofuels should reduce greenhouse gas (GHG) emission by 35% (Art. 17.2) and be produced with feedstock originating from land with high biodiversity value and high carbon stocks (e.g. peatland) (Art. 17.3-4-5).

In order to verify compliance to these standards, the EU does not rely on a proper regulatory framework, but follows a principle of double delegation (Zezza 2013). Compliance with these sustainability criteria is verified through private voluntary schemes approved by the Commission (Table 4.1), national systems of compliance that each Member-State is required to develop, or by bilateral or multilateral agreement with third parties (2009/28/EC, Article 17).

Further to the voluntary schemes approved by the European Commission (Table 4.1), investors can adopt various other types of SCL schemes such as:

- commodity-specific initiatives (e.g. UTZ),
- schemes addressing forest-related investments (e.g. Forest Stewardship Council (FSC); Rainforest Alliance (RA)),
- initiatives related to the financial sector (e.g. the Equator Principles (EP)),
- other relevant governance instruments (e.g. the EU Eco-Management and Audit Scheme (EMAS), the United Nations Global Compact, standards from the International Organization for Standardization (ISO)),

Table 4.1 Voluntary schemes approved by the EU

Name	Commodity	Commodity Origin	Extent of supply chain covered
Abengoa RED Bioenergy Sustainability Assurance (expired)	Wide range of feedstocks	Global	Full supply chain
Biograce GHG calculation tool	Wide range of feedstocks	Global	Supply chain not covered
Biomass Biofuels voluntary scheme (2BSvs)	Wide range of feedstocks	Global	Full supply chain
Bonsucro EU RED	Sugarcane	Global	Full supply chain
Ensus voluntary scheme under RED for Ensus bioethanol production (expired)	Feed wheat	EU	From the first feedstock delivery point to the Ensus One bioethanol storage
Gafta Trade Assurance Scheme	Wide range of feedstocks	Global	From farm gate to first processor
Greenergy Brazilian Bioethanol verification programme (expired)	Sugar cane	Brazil	Full supply chain
HVO Renewable Diesel Scheme for Verification of Compliance with the RED sustainability criteria for biofuels	All feedstocks suitable for HVO-type biodiesel	Global	From the producer of HVO-type renewable diesel
International Sustainability and Carbon Certification (ISCC)	Wide range of feedstocks	Global	Full supply chain
KZR INIG System	Wide range of feedstocks	Europe	Full supply chain
NTA 8080 (expired 21 August 2017)	Wide range of feedstocks	Global	Full supply chain
Red Cert Decision	Wide range of feedstocks	Europe	Full supply chain
Red Tractor Farm Assurance Combinable Crops and Sugar Beet Scheme	Cereals, oilseeds, sugar beet	UK	Until the first feedstock delivery point
Round Table on Responsible Soy (RTRS) EU RED	Soybean	Global	Full supply chain
Roundtable on Sustainable Biofuels (RSB) EU RED	Wide range of feedstocks	Global	Full supply chain
Roundtable on Sustainable Palm Oil (RSPO) RED (Expired 14/12/2017)	Palm oil	Global	Full supply chain
Scottish Quality Farm Assured Combinable Crops (SQC)	Winter wheat, maize, oilseed rape	North Great Britain	Until the first feedstock delivery point
Trade Assurance Scheme for Combinable Crops	Combinable crops, such as cereals, oilseeds and sugar beet	UK	From farm gate to first processor
Universal Feed Assurance Scheme	Feed ingredients and compound feeds as well as combinable crops	UK	From farm gate to first processor

Source: Based on data collected from European Commission (2018) (accessed 22 February 2018)

The SCL schemes outlined above can be broadly categorised also according to their commitment and requirements in four main areas: product quality and economic, environmental and social standards. They normally require transparency in the supply chain and/or in the performance of the organisation (Bracco 2016). Economic standards usually cover management and business issues, while environmental standards generally address biodiversity conservation and the sustainable use of land and natural resources. Social requirements are often connected to human rights or labor rights based on the International Labour Organization (ILO) standards.

However, more often than not, the effective adoption and implementation of SCL schemes face important constraints. Some of the main challenges discussed below include (a) complexity of feedstock value chains, (b) legitimacy; (c) contradiction between policies in exporting and importing countries and (d) enforcement.

The voluntary schemes approved by the EU are usually set by firms, industrial/business groups, civil society (e.g. non-governmental organisations, civic participants, community actors), governmental agencies or hybrids of the above. In a way the EU 'delegates' to the voluntary initiatives the responsibility to ensure compliance to sustainability criteria. However, this is often not an easy task considering that biofuel supply chains are generally long and feedstocks originate from remote areas, making it difficult to assess and monitor effectively.

At the same time, many SCL schemes and specific criteria face legitimacy issues. For example, despite the strict targets on GHG emissions (see above), the EU RED encounters legitimacy problems when attempting to impose social and environmental standards as regulatory requirements due to World Trade Organization (WTO) regulation. Often, inside the WTO, social and environmental process standards are considered unfair trade barriers and are not accepted by developing countries (Van Stappen 2009; FAO 2013). For example, compliance with WTO/GATT regulation made the European Commission to require form Germany to modify its draft biofuel policies, which prohibited the import of palm oil and soy-based biodiesel until the compliance with sustainability criteria was verified (Van Stappen 2009; The Bioenergy Site News Desk 2009). However, various EU-approved voluntary schemes, such as the Roundtable on Sustainable Palm Oil (RSPO), the Roundtable on Responsible Soy (RTRS), the Roundtable on Sustainable Biomaterials (RSB), and Bonsucro, contain requirements relating to the protection and enhancement of biodiversity, ecosystems and conservation values (Chao et al. 2012; Scarlat and Dallemand 2011; De Man 2010). Moreover, the EU RED requires the European Commission to report every 2 years about the impact of biofuels on social sustainability, food prices and 'wider development issues' inside and outside the EU. However, the EU RED does not provide any specific social requirements to enter the EU market, such as FPIC.

Finally, problems may arise when the recognition of customary rights or the requirements for the production sites (e.g. avoiding feedstock from high biodiversity/carbon stock areas and peatlands) set by the EU RED or by other SCL schemes are not recognised under the national law of the target country (or even go against it). In this case, investors find themselves in a problematic situation, as they cannot in practice fulfil both the national regulation and the requirements set by the EU RED or SCL schemes. Similarly, sometimes the EU regulation has made it difficult for Member-States to apply sustainability criteria. More specifically, the certification through private voluntary initiatives requires the harmonisation of public and private regulation, and the trust between the actors involved at the different levels. In such contexts, programs with non-corporate governing bodies and/or involving NGOs can provide greater independence from business interests, increasing legitimacy and therefore trust (Zezza 2013).

It is worth mentioning that many of the SCL schemes provide only suggestions or require 'good practices', but only a few explicitly provide 'certification' to the adopting companies. For instance, compliance to ISO standards is voluntary, and certification is possible but not obligatory. In this sense, ISO does not perform certification per se, and therefore a company/organisation cannot be certified *by* ISO itself. Instead, non-mandatory external certification is performed by independent bodies (mostly private) to verify the compliance with ISO standards. An alternative to third-party certification is 'second-party' verification performed by the customers of the organisation or a self-declaration of conformity with the standard (ISO 2011). SCL shemes with mandatory requirements often require the adherence to relevant national or international regulations, such as the ILO standards. However, many of these SCL schemes are not always well aligned with key aspects related to LSLAs (Box 1).

Box 1 Land-related impacts in popular standards

The Food Safety System Certification (FSSC) 22000 addresses food safety and quality and was developed by the Foundation for Food Safety Certification. The FSSC 22000 scheme provides no requirements on the environmental and social impact of production activities. The OHSAS 18000 health and safety management standard has some requirements concerning the treatment of workers, but does not contain environmental or social criteria for feedstock production.

The ISO 14001 standard on environmental management is a stepping stone for the European Union Eco-Management and Audit Scheme (EMAS), a voluntary environmental management instrument created by the European Commission in 1993 (EMAS 2014). It aims at improving performance, credibility and transparency, helping the assessment, management and continuous improvement of the environmental performance of adopting organisations. The EMAS core indicators focus on performance in key environmental areas such as energy efficiency, material efficiency, water consumption, waste generation/management, biodiversity and GHG emissions. However, this environmental standard does not include criteria specific to land, with land use being only indirectly visible through biodiversity conservation (EMAS 2014). The ISO 9001 standard does not cover any aspect of environmental and

(continued)

Box 1 (continued)
social sustainability, but addresses issues concerning product quality management. The implementation of the ISO 55001 standard intends to reduce GHG emissions, energy costs and other related environmental impacts, through systematic management of energy (ISO 2011). However, similar to the above ISO standards, it does not contain any specific requirements on land use and land-use change.

4.3 Methodology

4.3.1 Establishing the Extent of Italian LSLAs in SSA

In order to identify foreign-led LSLAs in SSA, we use the Land Matrix Global Dataset (henceforth Land Matrix). The Land Matrix is an online public database that reports global land transactions,[3] which entail a transfer of rights to use, control or own land through sale, lease or concession. This implies the conversion of land from local land uses to large-scale commercial uses. The acquisitions reported in the Land Matrix are 200 ha or larger, and were concluded after the year 2000 (Anseeuw et al. 2013). The Land Matrix records cases of intended and realised land deals, involving foreign or domestic investors, and it reports the status of investments at different levels of implementation (i.e. under negotiation, start-up phase, in operation, failed).

For the purpose of this chapter, we extract directly from the Land Matrix data on the contract size of the different Italian land investments in SSA. When the contract size areas are not available, we extract production size data. Areas reported as 'intended size' are used only when neither contract nor production sizes are available. To investigate the possible competition between food and energy production, we also identify the final uses of the acquired land. Data are classified according to various levels of reliability and are crosschecked to the extent possible. We use basic descriptive statistics for data analysis. A more detailed explanation of the methodology used in this chapter can be found in Antonelli et al. (2015).

We should also mention that the Land Matrix has faced substantial criticism among scholars and practitioners. According to Pearce (2013), databases of land transactions might both over- and under-estimate the levels of land transactions. For example, the size of land acquisitions could be under-estimated, as conflict-ridden or fragile countries might provide incomplete information (Nolte et al. 2016; Anseeuw et al. 2013). Furthermore, many of the reported land transactions might not have materialised, or reached the operational stage, or even used less land than actually acquired (Verhoeven and Woertz 2012; Locke and Henley 2013). Due to these inherent limitations of large-scale global inventories on land deals, this study does

[3]For more information refer to: https://landmatrix.org/en/

not seek to quantify the exact amount of the land acquired by Italian investors, but rather to highlight the trends, drivers and patterns of investments in agricultural land. Yet, despite its limitations, the Land Matrix provides, up to date, the most comprehensive public database on LSLAs that is fit for the purpose of this study.

4.3.2 Adoption of CSR Practices by Italian Investors

To enhance the accuracy of the results (see above), we carefully investigated each deal, searching for additional secondary sources to complement and/or check the validity of the Land Matrix data. We used corporate sources (e.g. company statements, official declarations, company websites), academic studies and media reports to collect information about the commitment in (and stategies related to) CSR for all Italian investors involved in LSLAs in SSA. The analysis of the different SCL schemes was drawn through the consultation of official sources for each scheme (e.g. rules of participation, standards, requirements). Data and information about investor behaviour were first collected in 2014, and then double-checked in early 2017.

As explained in Sect. 4.2, we assume that by joining a SCL scheme, private companies essentially 'adopt' CSR strategies to ensure and prove the responsibility of their actions. For the purpose of this chapter, we distinguish among four types of CSR strategies:

- *CSR real:* the sustainability strategy is 'certified' by a SCL scheme.
- *CSR claimed*: in the absence of SCL adoption, investors reported being involved in charitable or philanthropic initiatives in the investment countries (e.g. participation in small education/development projects for communities).
- *No CSR*: sustainability and CSR are not declared as core business strategies.
- *No CSR info:* there is no available information about CSR activities (i.e. primary, secondary and tertiary sources were not available or reliable).

4.4 Results

4.4.1 Italian Large-Scale Land Acquisitions in Sub-Saharan Africa

Italian investors have been involved in 26 LSLAs, amounting to almost 1.7 Mha (as of January 2017) (Table 4.2). The negotiations over these LSLAs have been mainly concluded through written agreements (92%). Approximatelly 8% were intended for future projects and further expansion of the acquired land. Among investor countries, Italy ranks fourth in terms of the number of deals and seventh in terms of the land size of the deals. The investments reported in Liberia account for

Table 4.2 Large-scale land acquisitions by Italian investors in Sub-Saharan Africa (in ha)

	In operation	Concluded	Intended	Failed
Angola	–	–	12,000	–
Benin	–	–	–	200,000
Republic of Congo	–	–	70,000	–
Ethiopia	500	70,500	–	–
Ghana	1400	6699	–	–
Guinea	–	74,504	–	–
Kenya	–	–	–	50,000
Liberia	310,932	310,932	–	–
Madagascar	3500	25,558	–	–
Mozambique	15,800	35,300	–	–
Nigeria	11,292	11,292	–	–
Senegal	2574	63,250	–	–
Tanzania	–	500	–	–
TOTAL	345,998	598,535	82,000	250,000

Source: Own elaboration based on Land Matrix (2017)

Table 4.3 Socioeconomic characteristics of the countries targeted by Italian investors

	Human development index	Labour force in agriculture (%)	Population below poverty line (%)	Rate of population growth (%)	Corruption (0–100)
Angola	Low	85	41	2.7	23
Benin	Low	N/A	37.4	2.7	36
Republic of Congo	Medium	35	47	2	22
Ethiopia	Low	85	29.6	2.2	33
Ghana	Medium	44.7	24.2	2.2	46
Guinea	Low	76	47	2.6	24
Kenya	Low	75	43	1.8	27
Liberia	Low	70	64	2.4	38
Madagascar	Low	N/A	75	2.5	28
Mozambique	Low	81	52	2.5	30
Nigeria	Low	70	70	2.4	25
Senegal	Low	78	47	2.4	41
Tanzania	Low	80	68	2.8	33

For corruption, larger numbers indicate lower perceptions of corruption, whereas lower numbers indicate higher perceptions of corruption. Own elaboration based on CIA (2017), FAO (2014), Transparency International (2013)

about half of the total concluded Italian land deals, and entail two logging concessions in Timbo (57,262 ha), and Konobo/Gbeapo (253,670 ha).

The countries targeted by Italian investors exhibit generally high rates of population growth and employment in agriculture (Table 4.3). They also exhibit low socioeconomic development, high incidence of poverty, undernourishment and

Table 4.4 Land deals by sector, number of deals and size

Sector	Sub-sector	Number of deals	Intended size (ha)	Contract size (ha)
Energy	Conventional energy	4	141,000	25,300
	Renewable energy	13	1,184,250	257,303
Forestry	Logging	2	310,932	310,932
Other	Other (not specified)	3	225,500	201,000
Industrial	Industrial (food and non-food)	2	21,000	4000
	Industrial (non-food)	1	1000	0
	Industrial (food)	1	15,000	15,000

Source: Own elaboration based on Land Matrix (2017)

corruption (Table 4.3). The combined effects of the above make them vulnerable to the potential adverse impacts of LSLAs (see Sect. 4.1).

Investments by companies involved in the energy sector (including renewable energy) account for 65% of all Italian LSLAs and for most of the acquired land (Table 4.4). Industry- and forestry-related LSLAs account for 15% and 8% of the total deals, respectively, while there is no information available for 12% of the deals. Some LSLAs have been undertaken by small and medium-size companies operating in the industrial sector (e.g. food/non-food sub-sector). This includes companies involved in (a) the mechanical industry; (b) agribusiness, cosmetics and technology and (c) food processing (in particular sugar production). However, these LSLAs are quite marginal in terms of land size. In fact, most of the investments focused on the production of energy crops, mainly jatropha and oil palm.

4.4.2 Investor Engagement in CSR Strategies

As explained in Sect. 4.3.2, we use secondary information disclosed by investors and third parties to understand the extent to which Italian investors engage in CSR activities and adopt SCL schemes. Figure 4.1 shows the certified and uncertified LSLAs by Italian investors, as a percentage of total land deals. Fewer than half of the LSLAs were undertaken by investors that engaged in CSR activities (42%). This corresponds to 11 out of the 26 LSLAs, and only a small fraction of the total land transactions (Fig. 4.2). Only for one small acquisition CSR is 'claimed' but not proven by an external party ('CSR Claimed') (Fig. 4.2).

In terms of land size, most LSLAs fall within the 'No CSR' category, suggesting the widespread lack of investor engagement in CSR activities (Fig. 4.2). Interestingly there is a large discrepancy in this category between the intended and contracted land size, mainly due to an intended 700,000 ha jatropha investment in Guinea. The company that made this deal (and three other similar deals for jatropha production) went bankrupt in 2015, but as the company did not engage in any CSR activity as of 2014 (i.e. before bankrupcy), we classified these deals as 'No CSR'.

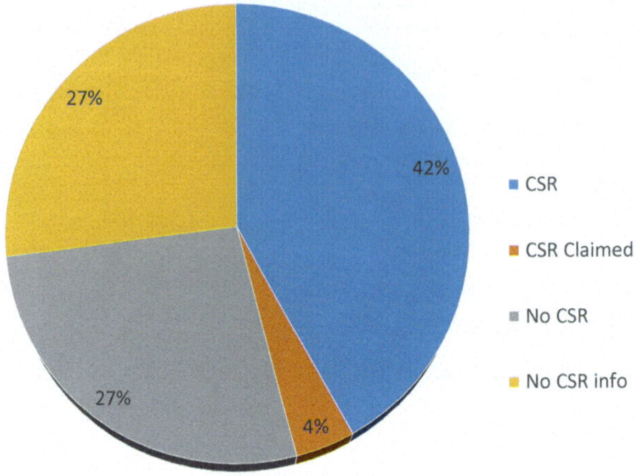

Fig. 4.1 CSR adoption as a percentage of total land deals. *Source:* Own elaboration based on Land Matrix (2017)

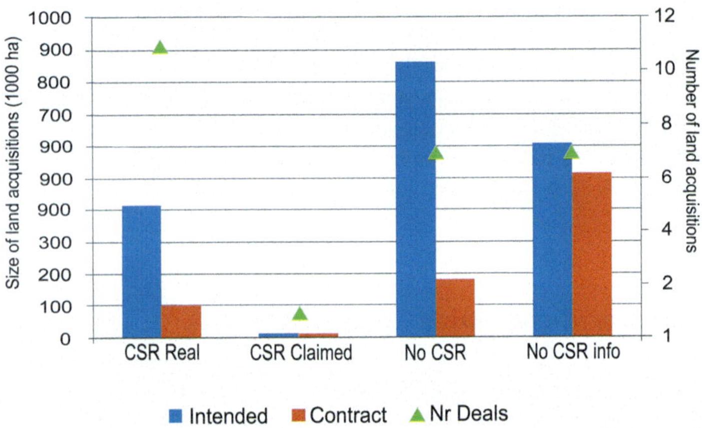

Fig. 4.2 CSR adoption in terms of the number and sizes of land deals. *Note: The* number of deals in each category is reported on the right-hand y-axis and indicated by the green triangles. The intended and the contract size of the LSLA are reported to the left-hand y-axis. Source: Own elaboration based on Land Matrix (2017)

There is no information regarding CSR adoption ('No CSR info') for seven deals, accounting for approximately 500,000 ha. It is worth mentioning that there has been little change over time. In fact, only one investor that claimed CSR in 2014 was actually certified through ISO 14001 in 2017. For all the other investors, there were no changes within this 3-year period (i.e. 2014–2017).

Table 4.5 shows the different schemes adopted by the Italian investors, the characteristics of each scheme, and the number of acquisitions covered by each scheme. Based on our analysis, no Italian investor involved in LSLAs in SSA has

Table 4.5 Adoption of schemes and coverage of sustainability criteria in Italian land deals

	SCL schemes	Standards						
		Quality	Economic	Environmental	Social	Requirements	Certification	Adoption
EU-approved voluntary schemes	Bonsucro	–	–	✓	✓	✓	✓	*
	ISCC	✓	✓	✓	✓	✓	✓	*
	RTRS	–	✓	✓	✓	✓	✓	*
Commodity-specific schemes	4C	✓	–	✓	✓	✓	✓	*
	UTZ	–	✓	✓	✓	✓	✓	*
Forest-related schemes	Rainforest Alliance	–	✓	✓	✓	✓	✓	*
Other standards	EMAS	–	✓	✓	–	✓	✓	6
	Fairtrade	–	✓	–	✓	✓	✓	*
	FSSC 22000	✓	–	–	–	–	✓	1
	ISO 9001	✓	✓	–	–	✓	–	4
	ISO 14001	✓	✓	✓	–	✓	–	8
	ISO 55001	✓	✓	✓	–	✓	–	2
	OHSAS 18001	–	✓	–	✓	✓	✓	5

The adoption of each scheme is measured through the number of Italian deals certified by it (column 'Adoption'); * refers to schemes adopted by the British Company ED&F, which is a partner investor of the Italian SFIR; Source: Own elaboration based on investors' disclosed information adopted any of the EU-approved voluntary schemes, national system or bilateral and multilateral agreements (see Sect. 4.2). In fact, only the British partner of the Italian corporation SFIR is certified by some of the voluntary schemes approved by the European Commission.

Conversely, Italian investors have mostly adopted the ISO 14001, EMAS, OHSAS 18001 and ISO 9001 standards (Table 4.5). Most of these standards do not specifically cover issues related to LSLAs, such as environmental impact and land rights (Box 1). None of the schemes deployed by the Italian investors refers to national or international legislation that covers policy domains such as indigenous rights, relevant for investments in land. Only one scheme, the OHSAS 18001, provides social requirements, but those address labour condition criteria, and are insufficient to guarantee, for instance, recognition of legal and customary rights by means of FPIC. On the other hand, the schemes that provide environmental requirements (i.e. EMAS; ISO 14001; ISO 55001) do not cover, for instance, environmental protection of areas holding high conservation values (HCV).

4.5 Discussion

4.5.1 Italian Investments and Compliance with EU RED Sustainability Criteria

Our analysis suggests that most of the LSLAs involving Italian investors in SSA relate to biofuel feedstock production (Sect. 4.4.1). Based on EU legislation, the biofuel feedstock imported into the EU needs to meet certain sustainability requirements. Up to date, the EU has approved 19 voluntary schemes for biofuel feedstocks as part of EU RED (Table 4.1). A few Member-States have also approved national systems to certify their biofuel feedstocks. However, no formal agreement has been signed between the EU (or its Member-States) and third parties (Ecofys 2013). To the best of our knowledge, only a few Member-States have set up a domestic system to certify biofuel feedstock production, with no evidence of bilateral or multilateral agreements to be found.

Italy is among the few EU countries which have transposed the EU RED sustainability criteria and set national certification schemes for biofuels. This was done through decrees adopted in January and March 2012 (Ecofys 2013). For example, the National System for Certification of the Sustainability of Biofuels and Bioliquids was established through an Inter-Ministerial Decree in January 2012 (Ministero dell'Ambiente 2012). This can be considered as a 'national system' as it allows for the information provided by relevant investors/companies to be verified by the appointed national authorities (Ecofys 2013). The Italian agency for the accreditation of organisations certifies compliance with the national system and is in charge of evaluating the SCL schemes (ACCREDIA 2012). However, none of the Italian investors engaged in LSLAs in SSA (and analysed in this study) refers to the Italian system of certification (Sect. 4.4.2).

Voluntary schemes are therefore the only instrument currently used to verify the compliance of Italian investors with sustainability criteria (Sect. 4.2). This creates a hybrid governance system that includes Member-States, NGOs/civil society, private companies and business groups (Zezza 2013). Unfortunately, as shown in our study, the biofuel investors and companies that engage in LSLAs in SSA rarely join such schemes (Sect. 4.4.2). This is particularly true for investments in their start-up phase or those that already failed. Indeed, many of the Italian investments studied in this chapter did not reach the phase of production, and some of them have failed. In these cases, the investors may not have joined an EU-approved scheme due to the simple fact that there is no produce to be certified, opting to adopt or join SCL schemes at a later stage. In fact, certification is only required for export to EU, and in this sense it does not necessarily need to occur until the exports are to start.

Moreover, as many SCL schemes are often geared towards large-scale agro-industry, the cost structure of certification may be out of reach for small- and medium-sized companies (FAO 2013). Multinational corporations are in fact far more likely to adopt SCL schemes, as they also tend to face a higher reputation risk (Bracco 2015). It is also interesting to note that the EU requirements are not applied for non-biofuel end uses. This means that when the acquisition is made for 'flexible' crops (i.e. feedstocks that can be used for both energy and non-energy purposes such as palm oil), the investor does not need to adopt the EU-approved certification schemes if the exports are for non-energy uses.

Furthermore, the delegation system set by the EU to ensure the compliance with the EU RED sustainability criteria (Sect. 4.2) requires a high level of trust for the processes adopted by Member-States and voluntary initiatives. This system expects, in a way, that the approved voluntary initiatives are able to strictly regulate, monitor and ensure the compliance of investors with the underlying sustainability criteria. However, it has often been reported that these voluntary schemes are sometimes unable to effectively monitor and ensure the compliance of their members (German and Schoneveld 2012). For instance, in several cases NGOs and civil society have indicated that the RSPO's mechanism for complaints is unable to guarantee the preservation of the rights of local communities (FPP 2011; Farm Land Grab 2014a, 2014b).

4.5.2 Policy Implications and Recommendations

Three main recommendations can be drawn from the analysis conducted in this chapter. First, the SCL schemes adopted by Italian investors operating in SSA are insufficient to ensure that the LSLAs are managed in a responsible and fair way. In particular, the schemes adopted by Italian investors fail in meaningfully covering a range of issues connected with LSLAs in SSA, such as land use and land tenure change, gender equity, food security, intergenerational equity and access to land and water resources, among others (Sect. 4.1; see also Chap. 2–3 Vol. 1, Chap. 5 Vol. 2). In order to guarantee better performance of LSLAs, investors should adopt voluntary initiatives that have specific social and environmental requirements including (but not limited to) the recognition of legal and customary land rights and the protection of areas of high conservation values (Bracco 2015; Chao et al. 2012; German and Schoneveld 2012). This would require both the updating of the many SCL schemes and also the broadening of the EU RED sustainability criteria to directly cover and address all the social and environmental aspects of LSLAs.

Secondly, in the context of LSLAs, it is necessary to acknowledge the right of local communities to FPIC, in order to assure their 'right to access information, right to participate in decision-making, and right to challenge such decisions' (ILC et al. 2012). This is important because granting and enforcing the right to access information can prevent to some extent the adverse effects of LSLAs, especially in highly corrupt political contexts. The recognition of the rights of local communities over land and natural resources should be a pre-requisite for LSLAs before they are signed, with contracts disclosed to ensure the full access to information about the LSLA, including its risks and impacts (ILC et al. 2012). Effective land governance policies can reconcile potentially conflicting interests in the context of LSLAs (Onoja and Achike 2015). In this sense, private operators also have some incentive to go beyond the weak property rights in many SSA countries to reduce their investment risk. A possible way to enhance the adoption of the FPIC principle would be to add it as an additional requirement for investors aiming to export feedstock for the EU biofuel market.

Thirdly, CSR schemes should ensure the operationalisation of conflict resolution mechanisms in the LSLA context. Such mechanisms must ensure that local communities have access to reliable and independent mechanisms for oversight and grievances. As suggested above, this is particularly relevant in SSA contexts with poor transparency and high levels of political corruption.

Finally, on a more practical note, existing land transactions databases should (a) validate the assessment methodologies and LSLA sizes, (b) cross-check with legal information the legality of the land deal and (c) monitor the LSLA to identify whether the intended LSLA actually materialised. Such monitoring and verification processes can provide updated information about the LSLA, as well as avoid labelling LSLAs as 'land grabs' when unfounded. The latter can possibly have negative impacts on foreign direct investment flows, which are important for rural development and economic growth in some SSA countries (Ahmed et al. 2017).

4.6 Conclusions

This chapter explored the interface of SCL schemes and LSLAs in countries of Sub-Saharan Africa, especially in the context of biofuel feedstock production. LSLAs for producing biofuel feedstock have emerged as an important policy topic over the past few years due to the scope, rapid evolution and lack of transparency generally associated with many of these investments. LSLAs for biofuel feedstocks have been highly controversial in many parts of SSA, raising many concerns over their environmental and social performance.

This chapter focused on Italian investors that have been involved in over 25 LSLAs across SSA, most of which were geared towards the production of biofuel crops to meet the targets set by the EU Renewable Energy Directive (2009/28/EC). Even though the EU RED has acknowledged certain sustainability criteria for feedstock production and identified various SCL schemes that can play a major role in addressing them, many important issues related to LSLAs have not been considered sufficiently.

The adoption of SCL schemes by Italian investors involved in SSA is quite poor. This raises important concerns about the sustainability of biofuel feedstock production geared towards the Italian market. It is not clear why the adoption of SCL schemes is so low, but it might be due to a combination of factors related to its voluntary nature and the maturity of the investments. Considering the above, the scope of SCL schemes must both expand and be enforced more strongly to ensure that the negative sustainability impacts of LSLAs in SSA are minimised.

References

ACCREDIA (2012) Prescrizioni per l'accreditamento degli Organismi che rilasciano certificati di conformità a fronte del Sistema Nazionale di Certificazione della sostenibilità dei biocarburanti e dei bioliquidi. Regolamenti Tecnici RT-31 rev.00. ACCREDIA Dipartimento Organismi di Certificazione ed Ispezione

Ahmed A, Campion B, Gasparatos A (2017) Biofuel development in Ghana: policies of expansion and drivers of failure in the jatropha sector. Renew Sust Energ Rev 70:133–149

Ahmed A, Campion B, Gasparatos A (2019) Towards a classification of the drivers of jatropha collapse in Ghana elicited from the perceptions of multiple stakeholders. Sustain Sci 14:315–339

Ahmed A, Kuusaana ED, Gasparatos A (2018) The role of chiefs in the political economy of large-scale land acquisitions in Ghana: insights from the jatropha sector. Land Use Policy., In Press. https://doi.org/10.1016/j.landusepol.2018.04.033

Allan JA, Keulertz M, Sojamo S, Warner J (2013) Handbook of Land and Water Grabs in Africa. Foreign direct investment and food and water security. IB Tauris, Abingdon

Anseeuw W, Lay J, Messerli P, Giger M, Taylor M (2013) Creating a public tool to assess and promote transparency in global land deals: the experience of the Land Matrix. J Peasant Stud 40 (3):521–530

Antonelli M, Siciliano G, Turvani ME, Rulli MC (2015) Global investments in agricultural land and the role of the EU: drivers, scope and potential impacts. Land Use Policy 47:98–111

Arezki R, Deininger K, Selod H (2013) What drives the global "land rush"? World Bank Econ Rev 29(2). 2015:207–233. https://doi.org/10.1093/wber/lht034

Arndt C (2010) Biofuels, poverty, and growth: a computable general equilibrium analysis of Mozambique. Environ Dev Econ 15(1):81–105

Borras S, Kay C, Gomez S, Wilkinson J (2012) Land grabbing and global capitalist accumulation: key features in Latin America. Can J Dev Stud Can d'études du développement 33:402–416

Bracco S (2015) The effectiveness of EU biofuels sustainability criteria in the context of land acquisitions in Africa. Renewable Sustainable Energy Rev Volume 50:130–143. https://doi.org/10.1016/j.rser.2015.05.006

Bracco S (2016) The economics of biofuels: the impact of EU bioenergy policy on agricultural markets and land grabbing in Africa. Routledge Explorations in Environmental Economics, UK. ISBN-13: 978-1138657854

Bracco S, Antonelli M, Turvani ME (2015) Foreign investment in land and corporate social responsibility: an investigation for Africa. Nairobi, Kenya: Acts Press, 2015. African Centre for Technology Studies (ACTS) Responsible Natural Resource Economy Programme Issue Paper 002/2015. ISBN 9966-41-171-2

Committee on World Food Security (CFS) (2013) Summary of international initiatives that provide guidance on responsible investment: key characteristics. Open Ended Working Group on principles for responsible agricultural investments which enhance food security and nutrition

Chao S, Colchester M, Jiwan N (2012) Securing rights through commodity roundtables? A comparative review. Forest Peoples Programme & Rights and Resources Initiative. https://www.forestpeoples.org/sites/fpp/files/publication/2012/11/securing-rights-through-commodity-roundtables-comparative-review.pdf

CIA (2017) The world factbook. https://www.cia.gov/library/publications/the-world-factbook/

Cotula L, Blackmore E (2014) Understanding agricultural investment chains: Lessons to improve governance. FAO and IIED, Rome and London

Cotula L, Vermeulen S, Leonard R, Keeley J (2009) Land grab or development opportunity? Agricultural investment and international land deals in Africa. IIED/FAO/IFAD, London/Rome

Davis KF, D'Odorico P, Rulli MC (2014) Land grabbing: a preliminary quantification of economic impacts on rural livelihoods. Popul Environ 36:180–192

De Man R (2010) Land issues in voluntary standards for investments in agriculture. Discussion paper submitted to The World Bank Annual Bank Conference on Land Policy and Administration. Washington, April 26 and 27, 2010

De Man R (2013) The business case for transparent land deals. Annual World Bank Conference on Land and Poverty 2013, Washington

De Schutter O (2009) Large-scale land acquisitions and leases: a set core of principles and measures to address the human rights challenge. http://www.oecd.org/site/swacmali2010/44031283.pdf

De Schutter O (2011a) How not to think of land-grabbing: three critiques of large-scale investments in farmland. J Peasant Stud 38:249–279

De Schutter O (2011b) The green rush: the global race for farmland and the rights of land users. Harv Int Law J 52(2):504–559

Deininger K (2011) Challenges posed by the new wave of farmland investment. J Peasant Stud 38 (2):217–247

Deininger K, Byerlee D (2011) Rising global interest in farmland: can it yield sustainable and equitable benefits? The World Bank, Washington DC

Dell'Angelo et al (2017) Threats to sustainable development posed by land and water grabbing. Curr Opin Environ Sustain 26–27:120–128

Ecofys (2013) Land grabs for biofuels driven by EU biofuels policies. By: Carlo Hamelinck. Ecofys 2013 by order of: ePURE

EMAS (2014) European Commission. Environment. http://ec.europa.eu/environment/emas/index_en.htm

EU [European Union] (2009) Directive 2009/28/EC of the European Parliament and of the Council of 23 April 2009 on the Promotion of the Use of Energy from Renewable Sources

EurObserv'er (2014) Online database. http://observer.cartajour-online.com/Interface_Standard/cart@jour.phtml?NOM_PROJET=barosig&NOM_USER=&Langue=Langue2&Login=OK&Pass=OK

European Commission (2018) Voluntary schemes. https://ec.europa.eu/energy/topics/renewable-energy/biofuels/voluntary-schemes_en?redir=1

Farm Land Grab (2014a) Liberia—RSPO's inability to address root causes of the conflict related to Sime Darby's operations. http://farmlandgrab.org/post/view/23479

Farm Land Grab (2014b) Nigeria—RSPO legal opinion ignores core of the complaint against Wilmar International subsidiary Biase Plantation Limited. http://farmlandgrab.org/post/view/23480

Food and Agriculture Organisation of the United Nations [FAO] (2011) Africa's changing landscape: securing land access for the rural poor. FAO regional office for Africa. www.fao.org/Africa

Food and Agriculture Organisation of the United Nations [FAO] (2012) Voluntary Guidelines on the Responsible Governance of Tenure of Land, Fisheries and Forests in the context of National Food Security, Rome. http://www.fao.org/docrep/016/i2801e/i2801e.pdf

Food and Agriculture Organisation of the United Nations [FAO] (2013) Biofuels and the sustainability challenge: A global assessment of sustainability issues, trends and policies for biofuels and related feedstocks. By: Aziz Elbehri, Anna Segerstedt and Pascal Liu. Trade and Markets Division Food and Agriculture Organization of the United Nations. Rome, 2013. ISBN 978-92-5-107414-5

Food and Agriculture Organisation of the United Nations [FAO] (2014) Online database. http://faostat3.fao.org/download/D/FS/E

FPP (2011) BioPalm plantation will lead to destruction of Bagyeli communities in Cameroon. http://www.forestpeoples.org/topics/palm-oil-rspo/news/2011/10/biopalm-plantation-will-lead-destruction-bagyeli-communities-camer

Friends of the Earth Africa and Friends of the Earth Europe (2010) Africa: up for grabs: The scale and impact of land grabbing for agrofuels

Gasparatos A, Von Maltitz G, Johnson F, Lee L, Mathai M, Puppim de Oliveira J, Willis K et al (2015) Biofuels in sub-Sahara Africa: drivers, impacts and priority policy areas. Renew Sust Energ Rev 45:879–901

German Federal Ministry for Economic Cooperation and Development (BMZ) (2009) Development policy stance on the topic of land grabbing – the purchase and leasing of large areas of land in developing countries. Discourse 015. BMZ, Bonn/Berlin

German L, Schoneveld G (2012) A review of social sustainability considerations among EU-approved voluntary schemes for biofuels, with implications for rural livelihoods. Energy Policy 51(2012):765–778

German LA, Schoneveld GC, Mwangi E (2013) Contemporary processes of large-scale land acquisition in Sub-Saharan Africa: legal failures, or elite capture of the rule of law? World Dev 48:1–18

Global Witness (2012) *Dealing with disclosure: improving transparency in decision-making over large-scale land acquisitions, allocations and investments*. Global Witness/Oakland Institute/International Land Coalition

Hall R (2011) Land grabbing in Southern Africa: the many faces of the investor rush. Rev Afr Polit Econ 38:193–214

Hall et al (2017) Plantations, outgrowers and commercial farming in Africa: agricultural commercialisation and implications for agrarian change. J Peasant Stud 44:3

Heal G (2008) When principles pay: corporate social responsibility and the bottom line. Columbia University Press

Hess et al (2016) A sweet deal? Sugarcane, water and agricultural transformation in Sub-Saharan Africa. Glob Environ Chang 39:181–194

High Level Panel of Experts [HLPE] (2011) Land tenure and international investments in agriculture, A report by the High Level Panel of Experts on Food Security and Nutrition of the Committee on World Food Security, Rome

Hopkins MJD (2016) Corporate social responsibility (CSR) and the United Nations sustainable development goals (SDG): the role of the private sector. https://www.csrfi.com/wp-content/uploads/2013/10/CSR-and-the-United-Nations-SDGs.pdf

Hufe P, Heuermann DF (2017) The local impacts of large-scale land acquisitions: a review of case study evidence from Sub-Saharan Africa. J Contemp Stud 35(2):168–189. https://doi.org/10.1080/02589001.2017.1307505

International Land Coalition (ILC) (2011) Tirana declaration. http://www.landcoalition.org/sites/default/files/aom11/Tirana_Declaration_ILC_2011_ENG.pdf

ISO (2011) Win the energy challenge with ISO 55000. ISO-06/3 000 ISBN 978-92-67-10552-9

Jägerskog A, Cascão A, Hårsmar M, Kim K (2012) Land acquisitions: how will they impact transboundary waters? Report Nr 30. SIWI, Stockholm

Keeley J, Seide WM, Eid A, Kidewa AL (2014) Large-scale land deals in Ethiopia: scale, trends, features and outcomes to date. IED, London

Land Matrix (2017) Online database. www.landmatrix.org

Locke A, and Henley G (2013) Scoping report on biofuels projects in five developing countries. ODI. https://www.odi.org/sites/odi.org.uk/files/odi-assets/publications-opinion-files/8394.pdf

Matondi PB, Havnevik K, Beyene A (eds) (2011) Biofuels, Land-Grabbing and Food Security in Africa. Zed Books, London

Messerli P, Heinimann A, Giger M, Breu T, Schönweger O (2013) From 'land grabbing' to sustainable investments in land: potential contributions by land change science. Curr Opin Environ Sustain 5:528–534

Ministero dell'Ambiente (2012) Normativa di riferimento sulla sostenibilità dei biocarburanti e bioliquidi.—http://www.minambiente.it/pagina/normativa-di-riferimento-sulla-sostenibilita-dei-biocarburanti-e-bioliquidi#sthash.aJAEKkBB.dpuf

Mirza H, Speller W, Dixie G and Goodman Z (2014) The practice of responsible investment principles in larger scale agricultural investments: implications for corporate performance and impact on local communities. Paper prepared for presentation at the 2014 World Bank Conference on Land and Poverty. The World Bank, Washington, DC, 24–27 March 2014

Nolte K, Chamberlain W, Giger M (2016) International land deals for agriculture fresh insights from the land matrix: Analytical Report II. Bern, Montpellier, Hamburg, Pretoria: Centre for Development and Environment, University of Bern; Centre de coopération internationale en recherche agronomique pour le développement; German Institute of Global and Area Studies; University of Pretoria; Bern Open Publishing

Onoja AO, Achike AI (2015) Large-scale land acquisitions by foreign investors in West Africa: learning points, Consilience. J Sustain Dev 14(2):173–188

Pearce F (2013) Splash and grab: the global scramble for water. New Scientist. 2906. http://www.newscientist.com/article/mg21729066.400-splash-and-grab-the-global-scramble-for-water.html#.Uq4uYqXGNjQ

Pignatelli V, Clamentel C (2006) I biocarburanti in Italia: ostacoli da superare e opportunità di sviluppo. ENEA, Unità Tecnico Scientifica Biotecnologie, Protezione della Salute e degli Ecosistemi Gruppo Sistemi Vegetali per Prodotti Industriali. http://www.qualenergia.it/UserFiles/Files/Rn_Bi_Ge_05_Panorama_Italia_ENEA_2006.pdf

Roberts RW (1992) Determinants of Corporate Social Responsibility Disclosure: an application of stakeholder theory. Acc Organ Soc 17(6):595–612

Scarlat N, Dallemand JF (2011) Recent developments of biofuels/bioenergy sustainability certification: a global overview. Energy Policy 39:1630–1646

Schoneveld GC (2014) The geographic and sectoral patterns of large-scale farmland investments in sub-Saharan Africa. Food Policy 48:34–50

Schut M, Soares NC, van de Ven G, Slingerland M (2014) Multi-actor governance of sustainable biofuels in developing countries: the case of Mozambique. Energy Policy 65:631–643

The Bioenergy Site News Desk (2009) Germany required to revise changes to biofuel law. February 2. Retrieved August 13, 2009, http://www.thebioenergysite.com/news/2961/germany-required-to-revise-changes-to-biofuellaw

Transparency International (2013) Corruption perceptions index 2013. http://www.transparency.org/cpi2013/results

Van Dam J, Junginger M, Faaij APC (2010) From the global efforts on certification of bioenergy towards an integrated approach based on sustainable land use planning. Renew Sust Energ Rev 14:2445–2472. https://doi.org/10.1016/j.rser.2010.07.010

Van Dam J, Junginger M, Faaij APC, Jürgens I, Best G, Fritsche U (2008) Overview of recent developments in sustainable biomass certification. Biomass Bioenergy 32:749–780

van Gelder JW and German L (2011) Biofuel finance: global trends in biofuel finance in forest-rich countries of Asia, Africa and Latin America and implications for governance. CIFOR infobriefs, 36

van Stappen F (2009) First implementations of sustainability schemes in European countries. Presentation held at the international workshop Sustainability Certification for Biofuels and Bioenergy in Brussels, January 29–30

Verhoeven H, Woertz E (2012) Mirage in the desert: the myth of Africa's land grab. CNN News, 09 July 2012

Weber M (2008) The business case for corporate social responsibility: a company-level measurement approach for CSR. Eur Manag J 26:247–261

White B, Borras SM, Hall R, Scoones I, Wolford W (2012) The new enclosures: critical perspectives on corporate land deals. J Peasant Stud 39:619–647

World Agronomy and VM Group (2011) Rush of investors forces spotlight on land tenure, World Agronomy, August

World Bank (2011) Rising global interest in farmland. Can it yield sustainable and equitable benefits? Washington DC. http://siteresources.worldbank.org/INTARD/Resources/ESW_Sept7_final_final.pdf

Zezza A (2013) Sustainability certification in the biofuel sector. Discussion Paper, Belfer Center for Science and International Affairs, Harvard University, Cambridge, Massachussets. http://belfercenter.ksg.harvard.edu/files/Zezza_Sustainability_Certification_Biofuel_DP_2013_web.pdf

Chapter 5
Determinants of Foreign Investment and International Aid for Meeting the Sustainable Development Goals in Africa: A Visual Cognitive Review of the Literature

Julia Lopes, Albert Novas Somanje, Esteban Velez, Rodolfo Dam Lam, and Osamu Saito

5.1 Introduction

Despite the significant progress made in the past decades, the attainment of equitable and sustainable development remains one of the most difficult global challenges, especially in developing countries. Poverty, for example, has been consistently decreasing since the early 1980s, when almost half of the global population lived below the international poverty line of USD 1.25 per day (Chen and Ravallion 2009; United Nations 2015). However, progress towards poverty alleviation has been variable across regions. While extreme poverty rates in Asia decreased rapidly (mainly in China), in Sub-Saharan Africa (SSA), the absolute number of people

J. Lopes (✉)
United Nations University, Tokyo, Japan

University of Brasília, Brasilia, Brazil

Brazilian Research Network on Global Climate Change (Rede CLIMA), Brasilia, Brazil

A. N. Somanje · E. Velez
United Nations University, Tokyo, Japan
e-mail: somanje@student.unu.edu

R. D. Lam
University of Tokyo, Tokyo, Japan

O. Saito
Institute for Global Environmental Studies (IGES), Hayama, Japan

Institute for Future Initiatives (IFI), The University of Tokyo, Tokyo, Japan
e-mail: o-saito@iges.or.jp

A. Gasparatos et al. (eds.), *Sustainability Challenges in Sub-Saharan Africa I*,
Science for Sustainable Societies, https://doi.org/10.1007/978-981-15-4458-3_5

living in poverty increased, largely due to population growth (Kates and Dasgupta 2007; Rivano 2014; Camfield et al. 2013) (Chap. 1 Vol. 1).

Most of the sustainability challenges in Africa require investments in (and financing of) key economic sectors (Adams 2009) (Chap. 1 Vol. 1). For example, external finance played a central role in promoting resilience to economic and financial shocks in the continent during the past decades (UNCTAD 2018). This includes finance obtained through foreign direct investment (FDI), aid in the form of official development assistance (ODA), short- and long-term loans, portfolio equity and remittances (UNCTAD 2018). FDI is currently the largest source of external financing in developing countries (39% of total financing), with ODA being the financing main mechanism for the least developed countries (LDCs), accounting for 36% of total financing compared to 21% from FDIs (UNCTAD 2018). Nonetheless, Kates and Dasgupta (2007) identify the failure of international institutions in providing sufficient investment and aid for overcoming multi-faceted sustainability challenges such as access to markets, education and lack of resources, among others (Chap. 1 Vol. 1).

New development agendas have sought to promote international collective action and coordinated efforts, with safeguarding or even increasing levels of FDI and ODA being a key priority (UN General Assembly 2015). Despite the geographically uneven progress, the Millennium Development Goals (MDGs) implementation process identified that international cooperation is an undeniably powerful mechanism for tackling sustainability challenges that are too great for any single nation to address unilaterally (United Nations 2015; Camfield et al. 2013). More specifically, coordinated action through FDIs played a key role towards the progress made in meeting some of the MDGs in Africa, and has been a key element in developing and implementing national development policies (Asongu and Nwachukwu 2017; Fowkes et al. 2016; Lim et al. 2016).

Following the MDGs, the sustainable development goals (SDGs) have culminated in a common framework to coordinate efforts for targets that have been commonly agreed both by developed and developing countries. With its 17 goals, 169 targets and 213 indicators, the SDGs pose an ambitious challenge, not the least due to their complex interlinkages (ICSU 2017; Zhou and Moinuddin 2017; IAEG-SDGI 2016) (see Chap. 1 Vol. 1). These interlinkages reflect the interconnectedness of the earth system and global political and socioeconomic processes, established in the Anthropocene (Steffen et al. 2004), as well as the multi-scalar nature of common pool resources (Duraiappah et al. 2014). There are six essential elements of SDGs for addressing sustainability challenges in Africa, namely (1) ensuring healthy lives, knowledge and inclusion of women and children; (2) ending poverty and fighting inequality; (3) growing strong, inclusive and transformative economies; (4) protecting ecosystems; (5) promoting safe and peaceful societies and strong institutions and (6) establishing global partnerships for sustainable development (United Nations 2014).

However, the rapid population growth, urbanization and other demographic and socioeconomic transitions continue to shape the dynamics of livelihoods in the region (UNDESA 2014) (Chap. 1 Vol. 1). Poorly planned solutions may result in higher transaction costs and inefficient investments (Adams 2009; Rasul and Sharma 2016; von Braun and Mirzabaev 2016) (see Chap. 1 Vol. 1; Chap. 4 Vol. 2). In particular, individual SDGs often address overlapping issues that are equally important for achieving more than one target. For example, the provision of improved sanitation and safe drinking water, which is a goal itself (SDG6), is equally important for ensuring good health and well-being (SDG3) and the sustainability of cities and local communities (SDG11) (Chap. 4 Vol. 2).

Nevertheless, it is not rare that such overlaps give rise to negative externalities and competing objectives. For example, the eradication of hunger (SDG2), is often approached through simplistic strategies related to enhancing food provision and production from improved agricultural systems (Chap. 3 Vol. 1; Chap. 1 and 2, Vol. 2), which often have direct trade-offs with SDGs 13 and 15 seeking to reduce ecosystem degradation and enhance biodiversity conservation (Chap. 10 Vol. 1; Chap. 5 Vol. 2).

Despite the ever-increasing number of publications addressing the SDGs and the UN Agenda in Africa (Aguayo-Tellez 2013; Asongu and Nwachukwu 2017; Fowkes et al. 2016; Lim et al. 2016), the aspects related to sustainable financing are still among the least explored (Asiedu 2002; Gui-Diby and Renard 2015; Cleeve et al. 2015). As already mentioned, the SDGs have essentially established an international commitment of "leaving no one behind" (UN General Assembly 2015). However, this all-inclusive and broad commitment, combined with the interconnectedness between SDGs, creates many different pathways to achieve effective international cooperation. For example, SDG17 (partnership for the goals) is entirely dedicated to promoting global partnership and cooperation (UNDP 2018; United Nations 2015). Furthermore, within each SDG, individual targets and indicators call for financial cooperation specifically aligned to the needs of each thematic issue (UN Social and Economic Council 2016).

The effective use of FDI and ODA can help solve many of the sustainability challenges faced by Africa (Adams 2009), including many of the issues discussed in these two edited volumes (Chap. 1 Vol. 1). However, strategically targeting key sustainable development priorities for funding, attracting sufficient amounts of FDI/ODA (Asiedu 2006; Dupasquier and Osakwe 2006), and effectively using these, presents a challenge in itself (Adams 2009). This is especially true in some of the least developed and resource-poor African countries that often lack appropriate institutions, capacity and/or infrastructure (Dupasquier and Osakwe 2006) (Chap. 1 Vol. 1).

The aim of this chapter is to systematize the recent literature related to FDI and ODA in African countries in the context of sustainable development. In particular, we analyse the academic literature and investment flows to unveil interactions, trends and the FDI and ODA contexts related to SDGs in Africa. By adopting an interlinkage approach, this chapter contributes to larger ongoing efforts to understand the connections between SDG targets and their potential to support Africa in achieving sustainable development (ICSU 2017; Zhou and Moinuddin 2017). Section 5.2 outlines the adopted methodology and Sect. 5.3 covers the main results and their implications. In particular, we (a) discuss how donors and academia have engaged with SDG priorities in Africa (Sect. 5.3.1); (b) identify the determinants of FDI and ODA in Africa (Sect. 5.3.2) and (c) identify pathways for transforming negative FDI and ODA determinants (Sect. 5.3.3). Section 5.4 highlights the policy implications and opportunities emerging from this analysis.

5.2 Methodology

5.2.1 ODA Flows and Linkages to SDGs in Africa

Section 5.3.1 focuses on how ODA and academic research have targeted and supported the attainment of the SDGs in Africa over the last decade. To achieve this, we assess which SDGs have been targeted by ODA funding (and how) between 2000 and 2013. We use information from the AidData[1] project dataset, which contains evidence about all past ODA as reported by OECD countries and international development agencies between 2000 and 2013 (Sethi et al. 2017). This database contains detailed information for over 435,000 development projects in Africa, and how the funding of each project was distributed between the 17 SDGs. Funding distribution is weighted relative to its contribution to each SDG, which is based on the amount of SDG targets that each ODA project addresses. Examples about the funding weight allocation method across SDG can be found in the AidData publications (Dilorenzo et al. 2017).

Subsequently, we conduct a systematic literature review to analyse how the academic literature has engaged with the SDGs during the past decade. This systematic review considered more than 30,000 documents identified through Elsevier Scopus (Boafo et al. 2018). We visualize SDG interlinkages by showing the number of ODA projects that have addressed more than one SDG within the same project, and how they connect between each SDG.

5.2.2 Factors Shaping FDI and ODA Landscapes in Africa

Sections 5.3.2 and 5.3.3 identify the most important factors and transformative pathways shaping the FDI and ODA landscapes in Africa. To achieve this, we conduct a literature review employing frameworks and ICT tools commonly used for the co-generation of interdisciplinary knowledge. In particular, we follow the three steps below:

- Identify and review relevant recent literature on FDI and ODA in Africa (Step 1).
- Generate conceptual maps (CM) to classify the main negative or positive determinants of ODA and FDI (Step 2).
- Use social network analysis (SNA) to interpret the relationships and patterns between the negative or positive determinants of ODA and FDI, and how frequently they are discussed in the reviewed studies (Step 3).

During Step 1, we identitfy literature relevant to the research question: "how does the academic literature discuss FDI and ODA in Africa?". We identify the relevant literature published between 2000 and 2018, through Elsevier Scopus and Google

[1]For more information, refer to: http://aiddata.org/financing-for-sustainable-development

Scholar using the following keywords: “investment in Africa”, “FDI”, “ODA”, “international cooperation”, “flow of aid” and “aid for Africa”. Date of publication and number of citations are used as initial filters. We use a second filter to select papers that address FDI or ODA at the regional level, thus excluding local and context-specific information. Finally, based on relevance to the topic and after reading the abstracts, we identify 35 papers for a more thorough analysis through concept mapping and social network analysis (see Steps 2–3).

During Step 2, we follow the literature-mapping process outlined by Hart (2018) that uses concept maps as visual representations of the relationship between concepts and processes (Sect. 5.3.2). Such concept maps can help transform declarative knowledge (i.e. what the topic is about) into procedural knowledge (i.e. understand the classification and relationships between the elements of the topic). By mapping the multiple ideas of different authors, it is possible to unveil common interactions, trends and landscapes, between them, thus enabling a better understanding of the relationships between individual studies (Hart 2018). The conceptual maps are discussed in more depth in Sect. 5.3.2. In summary, the coloured text represents the *concepts* and the *connectors* are linked to concepts with arrows. The relationship between concepts is determined by the direction of these arrows and is explained by the connectors between the arrows. Whenever a concept is identified as a positive determinant to attract FDI or ODA in the reviewed literature, it is highlighted in green. Conversely, concepts highlighted in red are identified in the literature as factors that drive away FDI and ODA. The conceptual map is generated using the free CMapTools software, developed by IHMC Public Cmaps.

During Step 3, we employ SNA to interpret the relationships between the factors outlined above (Sect. 5.3.3). In particular, SNA allows mapping how frequently these concepts relate to each other in the reviewed documents (i.e. how strongly authors agree on these relationships) through estimating the *degree centrality*. *Nodes* (i.e. the concepts identified in the literature) are represented by the labelled coloured circles, which are connected to each other through *ties* (i.e. directional links and relationships between concepts), which are represented by the lines and arrows (Sect. 5.3.3). The diameter of each node and the font size of its label represent the degree centrality of each concept (i.e. the number of connections). The social network analysis is developed using the open-source Gephi Graph Visualization and Manipulation software.

5.3 Results and Discussion

5.3.1 ODA Flows and Academic SDG Research Priorities in Africa

Between 2000 and 2013, a total of USD 465 billion in ODA was directed to Africa (USD 550 billion if debt-related assistance is considered). Approximately, 41% of

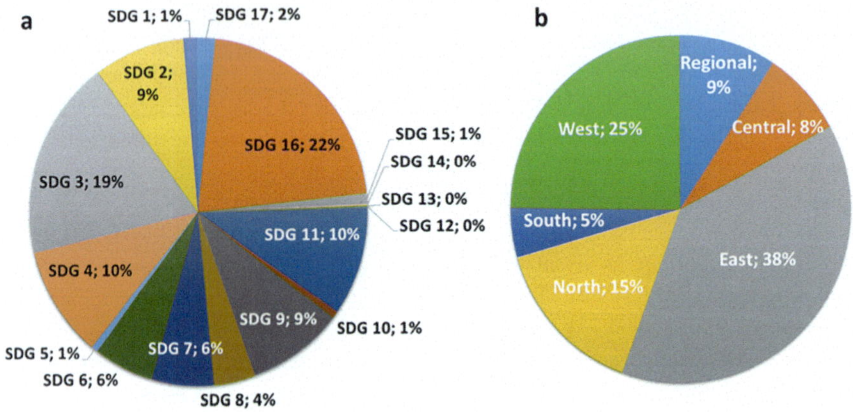

Fig. 5.1 Fraction of total ODA flows in Africa between 2000 and 2013 by SDG (**a**) and region (**b**). Note: Estimates exclude investments related to debt relief but include regional projects. Regional population distribution is: East Africa (33%); West Africa (30%); North Africa (19%); Southern Africa (5%); Central Africa (13%) (UNDESA 2017)

this ODA was concentrated in two goals, namely SDG 16 (peace, justice and strong institutions; 22% of the total funding), and SDG 3 (good health and well-being; 19% of the total funding), see Fig. 5.1a. However, despite some substantial successes, especially related to progress for SDG3 (good health and well-being), there is still more to be done. For example, even though some countries such as South Africa have invested heavily on controlling sexually transmitted diseases (STD) and HIV/AIDS (recording the largest decline in absolute number of new infections), they are still among the countries with the highest incidence of new infections globally (United Nations 2015) (Chap. 1 Vol. 1).

Overall, FDI and ODA flows seem to be equally distributed throughout Africa, with the regional funding allocation reflecting quite well the population distribution within the continent. However, West and Central Africa receive a slightly lower fraction of ODA compared to its population. Approximately 76% of the total FDI is directed to Sub-Saharan Africa, while 9% of the ODA supports African regional programs that span across multiple countries (Fig. 5.1b).

During the same period, the region experienced fundamental changes on how the external debt is handled (Sethi et al. 2017), which in turn precipitated higher amounts of allocated ODA. In particular, the sharp decline in debt-related ODA during the second half of the study period is related to debt relief (re)financing (Fig. 5.2). This decline reflects the decision of the Group of Eight (G8) major industrialized countries to forgo the remaining debt of heavily indebted poor countries (HIPC) through the Multilateral Debt Relief Initiative (MDRI) (Sethi et al. 2017). Overall, a total of USD 85 billion was allocated during this period for debt-related assistance with a steady increase in the first 7 years (2000–2006), peaking at 18 billion USD in investments (in 2006), most of which directed towards West African countries (Fig. 5.2). The second half of the period saw a sharp decline in debt-related ODA,

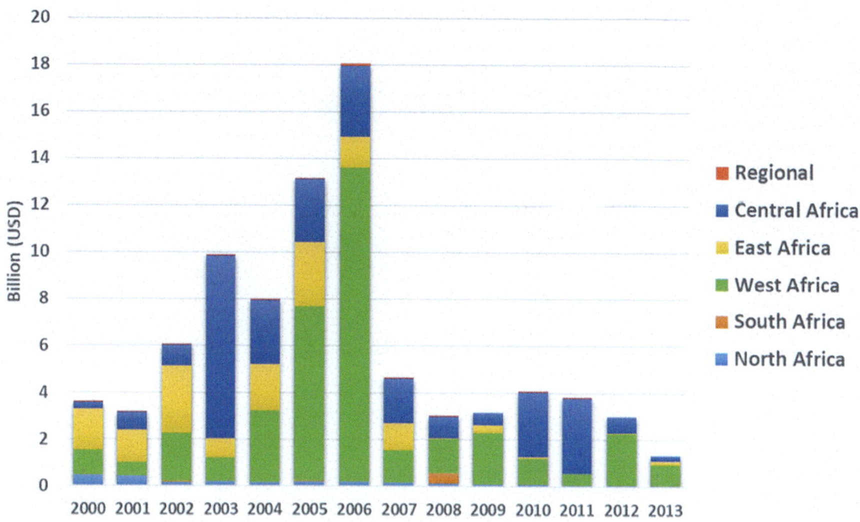

Fig. 5.2 Debt-related ODA between 2000 and 2013 by region

to less than USD 5 billion per year. Preliminary findings suggest that SSA countries free of external debt have experienced increases in total domestic revenue and public investments (Cassimon et al. 2015).

While the overall assistance directed to Africa is roughly distributed proportionally among the sub-regions (Fig. 5.1), certain donor countries tend to prefer specific regions to allocate ODA funds. This possibly reflects past colonial ties and present social and economic interests (Zanger 2000). For example, the top five donor countries[2] targeted mainly North and East Africa (Table 5.1). The United States and the United Kingdom focus their ODA primarily on East Africa (44% and 50%, respectively), while France focuses mainly on North Africa (43% of its total ODA). Germany and Japan, on the other hand, split their ODA between North and East Africa (52% and 70% of their total ODA respectively). Similarly, the top four international agencies in terms of ODA[3] focus predominately on East and West Africa (41% and 29%, respectively) (Table 5.1).

Figure 5.3a highlights the research priorities related to individual SDGs as captured by Boafo et al. (2018). There is extensive literature for SDG5 (Gender equality) (Chap. 1 Vol. 1; Chap. 4 Vol. 2), SDG13 (Climate action) (Chap. 1. Vol. 1; Chaps. 2 and 3 Vol. 2), and SDG1 (No poverty) (Chap. 1 Vol. 1). This is evident in Cluster 3 of Fig. 5.3a. However, the actual funding directed to those SDGs is relatively low (Fig. 5.3b). As a matter of fact, these research priorities reflect some

[2]The United States, France, Germany, Japan and the United Kingdom accounted for 36% of the total ODA flows in Africa between 2000 and 2013 (Table 5.1).

[3]The World Bank, European Communities, African Development Bank and Global Fund to Fight Aids, Tuberculosis and Malaria are responsible for 38% of the total ODA inflows to Africa (Table 5.1).

Table 5.1 Total ODA flows by the main donor countries and international agencies by region (in USD millions)

	North	South	West	East	Central	Regional	Total
Country total	***58,329***	***13,645***	***63,089***	***97,999***	***24,885***	***29,070***	***287,017***
European Communities (EC)	11,799	3582	11,621	15,022	6079	7529	55,633
France	15,451	1391	8716	5408	4201	479	35,646
Germany	6517	1960	4481	6457	2375	3100	24,890
Japan	6632	546	3095	6870	1132	1045	19,320
United kingdom	880	790	4534	9089	1050	1876	18,219
Global fund to fight Aids, tuberculosis and malaria (GFATM)	473	1354	4356	7493	1679	69	15,423
Canada	667	231	3603	4449	460	3299	12,708
Netherlands	679	624	2968	5319	186	2174	11,951
Norway	554	353	657	6248	459	1538	9810
Denmark	359	221	2459	5231	23	634	8927
Others	14,317	2592	16,598	26,413	7240	7328	74,489
International total	***79,673***	***28,803***	***108,299***	***180,981***	***35,499***	***46,447***	***479,737***
Joint United Nations programme on HIV/AIDS (UNAIDS)	–	58	78	130	46	64	399
African Capacity Building Foundation (ACBF)	0	8	47	62	43	65	225
Greece	87	9	16	29	18	58	218
African development Bank (AFDB)	20	13	26	38	17	14	129
Nordic development fund (NDF)	0	0	22	51	0	7	80
New Zealand	2	9	2	41	1	–	65
Czech Republic	2	0	1	12	2	0	18
World Trade Organization (WTO)	0	2	2	2	2	0	8
United Nations democracy fund (UNDEF)	0	0	2	2	1	2	8
Cyprus	1	–	2	0	–	–	6
Others	79,560	28,704	108,099	180,613	35,370	46,236	478,583

Note: The analysis is based on data from the AidData project (Sethi et al. 2017). Debt-relief-related investments are excluded

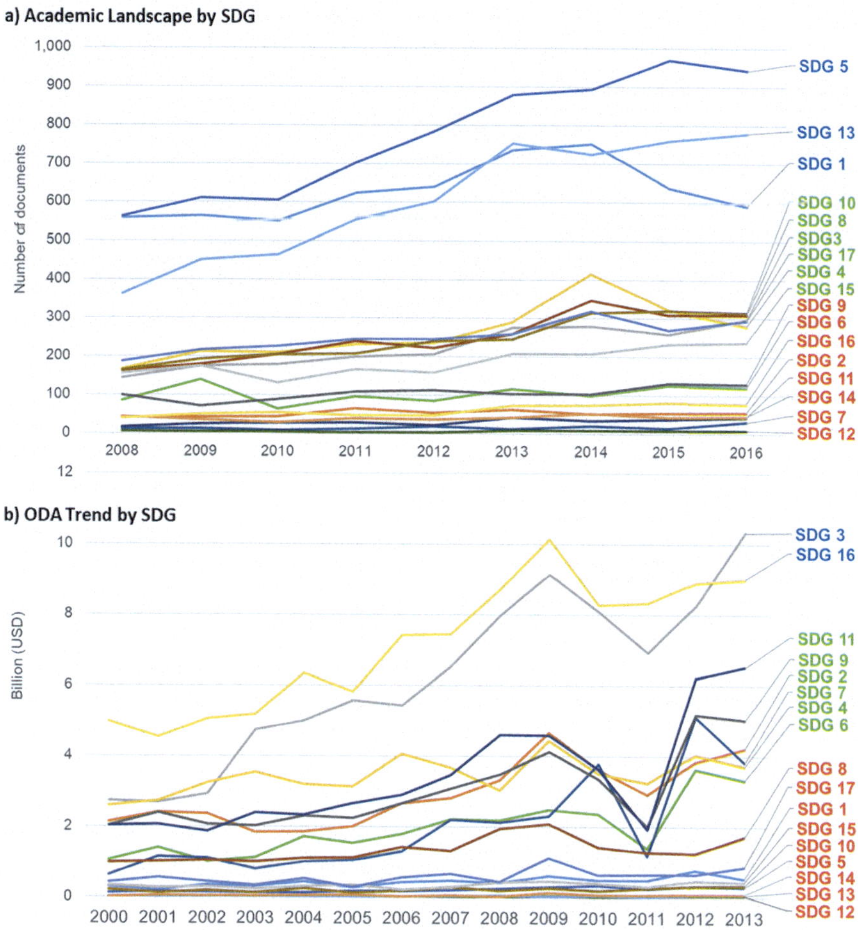

Fig. 5.3 Comparison of the focus of academic research and ODA funding by year; (**a**) academic publication trends in the last 10 years and (**b**) ODA funding between 2000 and 2013

of the SDGs for which the lowest amount of ODA has been allocated (see Cluster 1, Fig. 5.3b). Few studies have addressed those SDGs that have received the most funding, such as SDG 16 (peace, justice and strong institutions) (Fig. 5.1a), with little academic research focusing on this target (Fig. 5.3).

In this sense, there is a clear mismatch between academic research priorities (Fig. 5.3a[4]) and actual FDI and ODA flows in Africa (Fig. 5.3b). This mismatch represents a potential risk for the effective allocation of resources, as it can be argued that there is a lack of robust academic research to support the development of proper

[4]This is based on a systematic literature review of relevant publications over the past decade. This review included more than 30,000 documents identified through Elsevier Scopus (see Boafo et al. 2018).

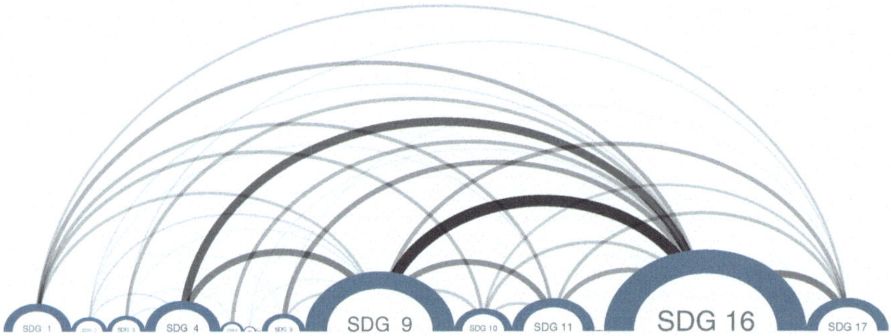

Fig. 5.4 SDG interlinkages within ODA projects in Africa between 2000 and 2013. Note: The size of each arc at the horizontal axis represents the number of projects that involve a component related to the SDG. The size of the links between arcs represents the number of projects that address those two SDGs within the same project. Debt-related ODA is included in the underlying calculations

mechanisms to enhance the effectiveness of FDI and ODA in meeting the SDGs (Asongu and Nwachukwu 2017; Fowkes et al. 2016).

As already mentioned in Sect. 5.1, there are extensive synergies and interlinkages between SDG targets (ICSU 2017; Zhou and Moinuddin 2017). Similar synergies and interlinkages can be found within ODA flows (Fig. 5.4). The results suggest that most ODA projects have some component related to SDG 16 and SDG 9 (industry, innovation and infrastructure). These multiple and strong connections to SDG 9 hint that it is assumed that national economic development can indeed contribute substantially to meeting the SDGs, while the importance of SDG 16 points to the underlying challenge of addressing weak institutions, conflicts and corruption (Mohamed et al. 2015; Asongu and Nwachukwu 2016) (Chap. 1 Vol. 1).

5.3.2 FDI and ODA Determinants in Africa

Our literature review suggests a general agreement that FDI and ODA can help solve many of the sustainability challenges in Africa, supplementing domestic savings to incentivize sustainable development and spur local value-added activities (Asiedu 2002; Gui-Diby and Renard 2015; Cleeve et al. 2015) (Chap. 1 Vol. 1). Nonetheless, both attracting FDI/ODA, and effectively using them to achieve sustainable development, resilience and the development of human capital, presents a challenge in itself, especially in African countries that are lacking appropriate institutions, capacity and/or infrastructure (Chap. 1 Vol. 1).

Figure 5.5 contains a conceptual map that visually represents the factors that influence the inflow and effective use of FDI and ODA in Africa. In particular, the linkages in Fig. 5.5 essentially delineate positive and negative cycles associated with the attraction of FDI and ODA. Green boxes indicate constructive aspects that

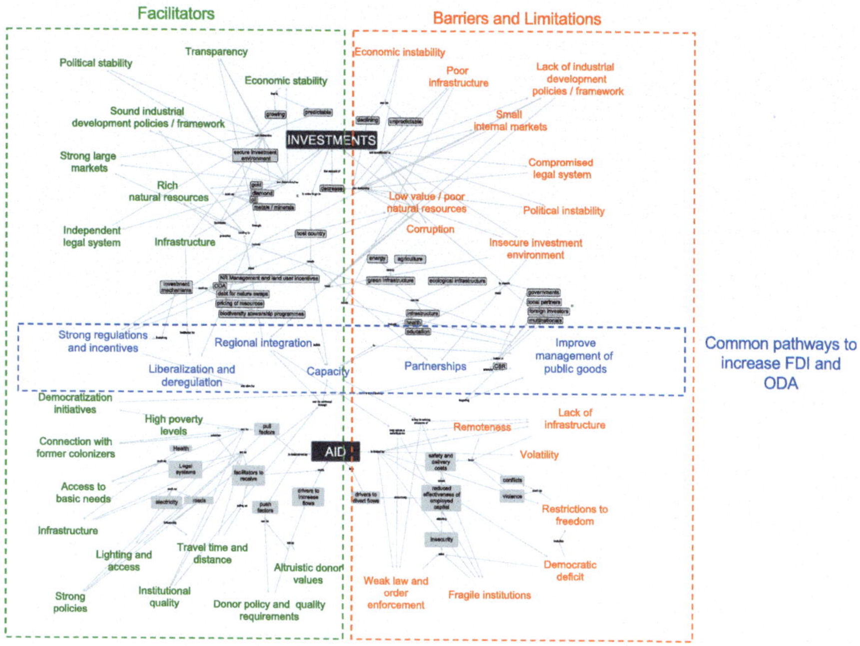

Fig. 5.5 Conceptual mapping of the literature on FDI and ODA in Africa

represent incentives for attracting, facilitating and fostering FDIs and ODA. Red boxes indicate negative aspects such as barriers (e.g. detrimental local characteristics to the investment environment) for attracting FDIs and ODA. We observe cyclic patterns that either determine these limitations to investment (i.e. negative spirals) or incentives to attract them (i.e. positive spirals). According to the reviewed literature, cycles in negative spirals can be reversed into positive spirals only if transformational interventions occur (represented below in blue) (Sects. 5.3.3 and 5.4). Figures 5.6 and 5.7 are essentially simplifications of the cycles identified in the conceptual map (Fig. 5.5), and are discussed in more detail subsequently.

When it comes to FDIs, the main factors that attract such investments include (a) stable economic, legal and democratic systems and (b) robust and open business environments with financial openness and good investment frameworks (Fig. 5.6). Such factors essentially shape safer and more conducive investment environments that, historically, have offered greater guarantees to achieve the expected returns on investment (Dupasquier and Osakwe 2006; Gui-Diby and Renard 2015). Other factors that seem to attract FDIs include natural resource availability (Asiedu 2006), monetary and market integration (Anyanwu and Yameogo 2015), as well as infrastructure that can facilitate further investments and logistics (Asiedu 2002) (Chap. 1 Vol. 1). However, these factors alone are not enough on their own to secure FDIs, as other characteristics are necessary to create enabling environments for attracting FDIs (Dupasquier and Osakwe 2006).

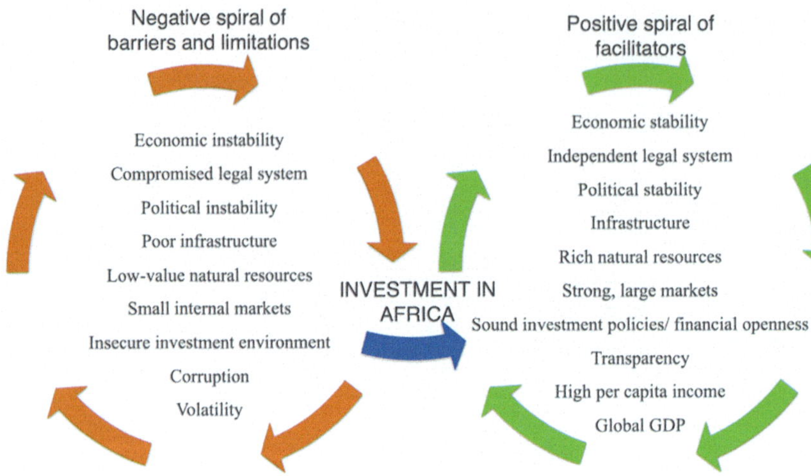

Fig. 5.6 Positive and negative spirals determining FDI flows in Africa

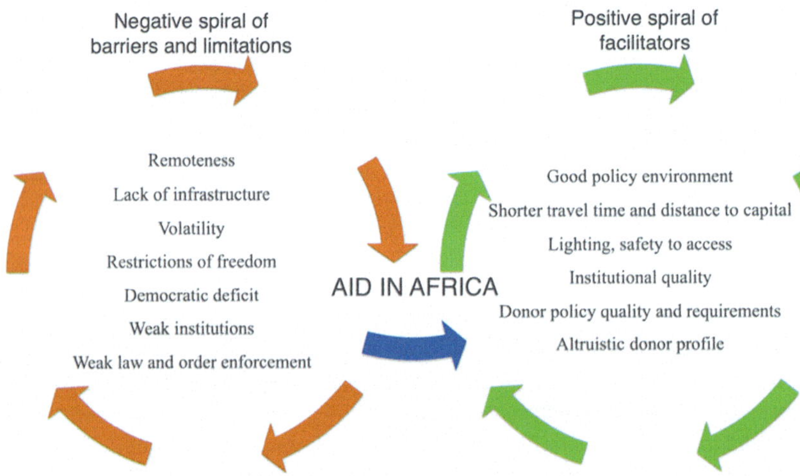

Fig. 5.7 Positive and negative spirals determining ODA flows in Africa

On the other hand, barriers to attract FDIs include corruption, lack of transparency and weak governance (Schoentgen and Gille 2017; Adams 2009) (Chap. 1 Vol. 1) (Fig. 5.6). Other barriers that are often mentioned but rarely addressed, include uncertainty due to *political instability* (e.g. due to war, military interventions, ethnic conflict); *macroeconomic instability* (e.g. due to currency crashes, inflation, budget deficits) and *lack of political transparency* (Asiedu 2006; Dupasquier and Osakwe 2006), as well as colonial legacies (Dunning 2004).

When it comes to ODA, countries with democratic regimes, high poverty and a colonial legacy have a higher chance of attracting ODA flows (Alesina and Weder

2002) (Fig. 5.7). Similar to FDI, countries that actively promote access to health and infrastructure (e.g. lighting, electricity, roads to access the targeted areas), and take steps towards developing functional legal systems, are more favourably considered when donors prioritize ODA-recipient countries. This is because such factors facilitate the effective implementation of ODA-funded projects. However, remote areas are often overlooked (even if they are highly vulnerable and in need of ODA) due to lack of infrastructure, secure and safe access and high delivery costs (Briggs 2018) (Chap. 1 Vol. 1).

Some of the factors discussed above can either act as push or pull factors to attract both FDI and ODA. Push factors drive capital flows from developed to developing countries due to external circumstances that are beyond the control of recipient countries, e.g. global economic growth, interest rates and investor risk aversion (Opperman and Adjasi 2017; Anyanwu and Yameogo 2015). On the contrary, pull factors reflect some of the characteristics of recipient countries, such as interest rates for investments (Opperman and Adjasi 2017; Anyanwu and Yameogo 2015). Donor intentions also positively or negatively influence ODA cycles (Fig. 5.7). At the same time, the strategic and political influences of donors and the economic deficits and political conditions in recipient countries, often determine the direction of ODA (Alesina and Dollar 2000). ODA outcomes thus often depend on the characteristics of the donors, their underlying motives/agendas and complementary policies (Minasyan et al. 2017).

Lastly, the volatility of FDIs and ODA plays a significant role on the direction of the cycles (Museru et al. 2014). The stability and predictability of ODA flows can promote higher internal confidence and trust in recipient countries, having a positive reinforcing effect on their ability to attract ODA (Dupasquier and Osakwe 2006). Conversely, volatile and variable ODA inflows can create insecurity and affect the internal circumstances in recipient countries, thus having a negative reinforcing effect on their ability to attract ODA (Minasyan et al. 2017). This can cause the diversion of financial resources directed towards research, science and technology, for example, to more urgent and basic needs, thus taking a toll on long-term economic development (Briggs 2018) (see Chap. 1 and 8 Vol. 1). Finally, large-scale and prolonged ODA initiatives can increase the dependency of recipient countries on foreign aid, which can lead to weak governance, institutions and state-building capacity (Bräutigam and Knack 2014).

5.3.3 Pathways for the Transformation of FDI and ODA Determinants

While the conceptual mapping outlined in Sect. 5.3.2 allows for the visualization of recurrent concepts and relationships within the literature, the SNA shows the *degree centrality* of these concepts (see below). Figure 5.8 outlines the frequency of interactions between concepts or how these concepts are discussed at the same time in the literature.

In Fig. 5.8, coloured circles represent *nodes* that are connected to each other through *ties* represented by lines. The green nodes represent those aspects that positively influence FDI and ODA flows, acting as drivers that increase such flows to recipient countries. Orange nodes represent aspects that drive away FDI and ODA flows. Transformative interventions (pathways) that either reduce the effect of barriers/limitations or maximize the effects of facilitators are identified in blue and are positioned in the inner circle. Factors that shape FDI and ODA flows (green and orange nodes) are positioned in the outer circle. Nodes with larger diameters/font sizes indicate that the particular concept is connected to a larger number of determinants or pathways. The thickness of the ties is proportional to the frequency with which each pair of concepts is addressed together in the literature. The arrows indicate the direction in which negative determinants are connected with positive determinants through pathways. Concepts without connections represent some factors that shape FDI and ODA flows and that are frequently identified in the literature, but they are not necessarily connected to pathways in the literature used in this SNA.

The current literature does not agree on a single set of determinants that incentivizes or hinders FDI and ODA flows in Africa. In fact, some authors suggest that what may work in one region or country may not work in another. Such an example

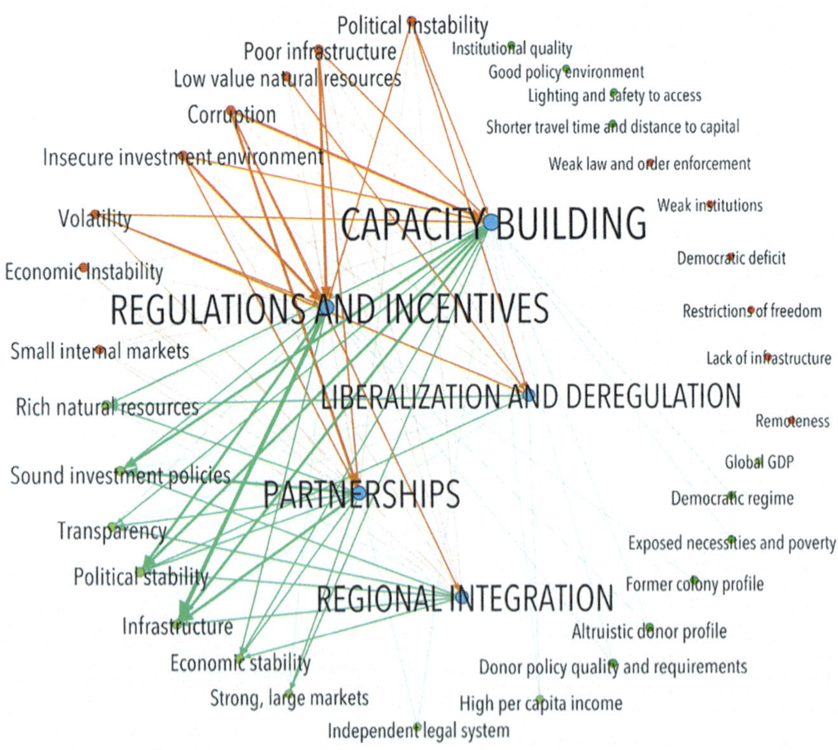

Fig. 5.8 Social network analysis of the FDI and ODA literature in Africa

is the infrastructure-related return on investment in Sub-Saharan Africa (SSA) and Northern Africa (Asiedu 2002). In such cases, some regional differences are captured as specific interventions may break negative cycles and provide significant shifts in the financial environment in some regions/countries.

It is worth mentioning that slightly different terms between different contexts are used to refer to these pathways. For the purpose of this analysis, we group and define them as a cluster of related actions with similar mechanisms. For example, components that relate to economies of scale are either clustered under regional integration or partnerships, according to their nature and scale. Both pathways can be associated with the development of large, strong and stable markets, and good access to natural resources and infrastructure (Chap. 1 Vol. 1), but are separated depending on the underlying mechanisms for implementation and stakeholder involvement (Dupasquier and Osakwe 2006; Opperman and Adjasi 2017). Below, we discuss the main nodes related to some pathways that can enhance FDI and ODA flows, namely partnerships, regional integration, capacity building, regulations/incentives and liberalization/deregulation.

Partnerships encompass the collaboration between stakeholders that have a stake, influence and interest in catalysing sustainable development at local, national and international levels (United Nations 2014). Yakovleva et al. (2017) highlight that the development of partnerships could support the reliable delivery of infrastructure, energy, water, health and education. Furthermore, the effective partnership with local stakeholders can also facilitate a better understanding of the context and local needs other areas targeted for FDIs and ODA. The thicker connections in Fig. 5.9 indicate that developing partnerships can be a good pathway to reduce investment uncertainties as partnerships can increase the credibility of transactions. Furthermore, partnerships can possibly enable more transparent decision-making processes, reduce corruption and catalyse the development of sound policies that favour investments in infrastructure. Public–private collaborations geared towards new technologies, human and financial resources are ideal for solving complex sustainability challenges, while bringing transparency and clear rules to financial interactions (Yakovleva et al. 2017; United Nations 2014) (Chap. 1 Vol. 1). Cooperation among African countries through regional bodies and partnerships is also vital for enhancing FDI and ODA flows (Dupasquier and Osakwe 2006; Gui-Diby and Renard 2015). To that end, South–South cooperation through partnerships for technology transfer, policy innovation and organizational change is an important response to failed efforts towards poverty eradication (Milhorance 2016).

Regional integration in the context of this chapter encompasses multi-national processes between countries that have common goals and needs, and share markets, resources and capacities to achieve sustainable development (Dupasquier and Osakwe 2006; Gui-Diby and Renard 2015; Yakovleva et al. 2017). Dupasquier and Osakwe (2006) stress that the relatively larger market sizes achieved through regional integration can attract and enhance FDI, as it reduces volatility and insecurity, while increasing political and economic stability (Fig. 5.10) (Chap. 2 Vol. 1). For example, the potential contribution of mining companies in achieving relevant SDGs in SSA increases with access to larger regional markets and the presence of

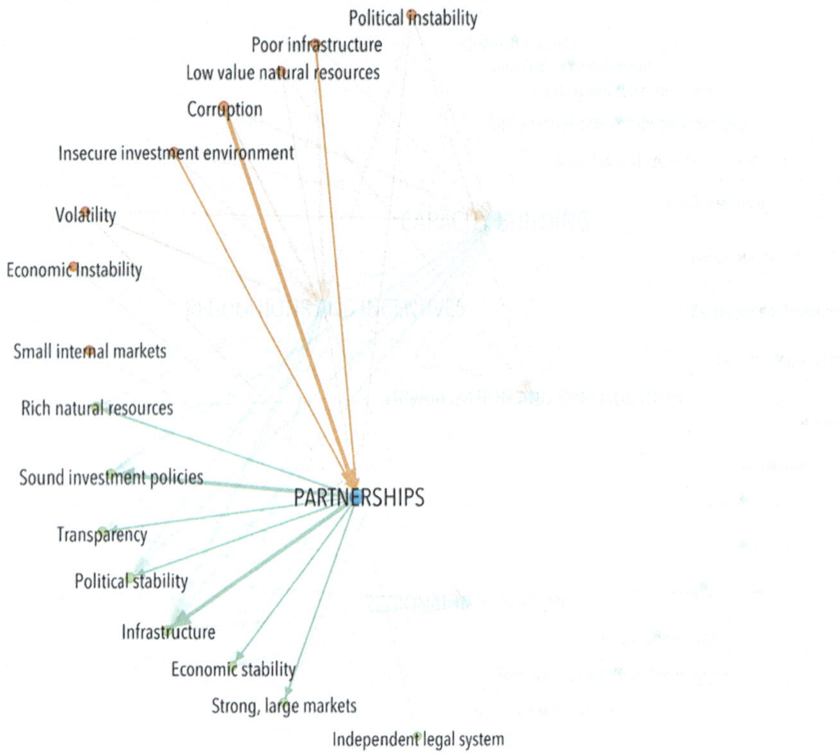

Fig. 5.9 Social network analysis for partnerships interventions

infrastructure jointly developed through tapping shared resources, thus having greater benefits for regional economies (Yakovleva et al. 2017) (see also Chap. 7 Vol. 2). On the other hand, due to the small size of most national economies in SSA, the industrialization of individual countries through FDIs was not very pronounced (Gui-Diby and Renard 2015) (Chap. 1 Vol. 1). In fact, a regional approach towards FDIs might have better industrialization outcomes (Gui-Diby and Renard 2015). Stemming from the above, due to the accelerated negotiations for international megatreaties in Africa, it is projected that FDIs will increase by 20% due to booming commodity prices and progress in operationalizing the African Continental Free Trade Area (AfCFTA) (UNCTAD 2018).

Capacity building has very varied definitions,[5] but there is a general agreement that capacity-building initiatives can have important ramifications for boosting FDI

[5]For the purpose of this chapter capacity is understood as sets of measures that institutions use to survive, grow and/or develop (Honadle 1981), or the ability of a system to use skills and resources to become more resilient, especially when competing for resources (UNISDR 2009; Pachauri et al. 2014).

Fig. 5.10 Social network analysis for regional integration interventions

and ODA flows in Africa. In particular, the literature views capacity building (mainly in terms of strengthening institutions and human capital) as a pathway to avoid/reduce corruption, thus having a positive effect on political stability, infrastructure development and the design and implementation of sound investment policies (Fig. 5.11). Capacity building thus entails the realization of actions that improve the ability of relevant stakeholders to make informed decisions, and identify policy/practice objectives and priorities using reliable/complete information and appropriate decision support tools (Honadle 1981). In the context of transformative interventions for improving the ability to attract FDI and ODA, capacity-building efforts should focus on improving management practices for (a) the delivery of public goods and services (e.g. infrastructure, health, education) (Cumming et al. 2017; Dupasquier and Osakwe 2006; Gui-Diby and Renard 2015; Yakovleva et al. 2017) and (b) resource mobilization (e.g. through tax system, development/implementation of robust policies) (Dupasquier and Osakwe 2006; Cumming et al. 2017; Dunning 2004).

Having in place strong domestic *regulatory frameworks and incentive* conveys an ability to acquire and use effectively FDIs and ODA, and thus increases the attractiveness of recipient countries. This is because such regulatory frameworks

Fig. 5.11 Social network analysis for capacity-building interventions

and incentives can send signals to funders about the transparency of transactions and indicate a safe investment environment. Figure 5.12 suggests, that even though these frameworks and incentives do not attract FDI and ODA in themselves (and are not the only motivations that influence donors decisions over resource allocation), they act to correct market failures, information asymmetries and structural deficiencies, promoting in the process guarantees to investors, local partners and the wider labour force (Barbour 2005; Dunning 2004). Donor quality and profile also influence the influx of ODA, which is subsequently affected by regulations that structure the means for such transactions. Scholars have suggested that mature and stable regulatory frameworks can (a) protect national and international investors, (b) encourage transparency against corruption, (c) promote enforcement mechanisms, (d) facilitate compliance through the elimination of bureaucracy and (e) allow for flexibility to address emerging issues (Schwerhoff and Sy 2017; Schoentgen and Gille 2017; UNCTAD 2018; Barbour 2005). Incentives may take several forms such as generic economic incentives (e.g. tax exemptions, cash grants) and sector-related mechanisms such as payment for ecosystem services (PES) schemes, debt-for-nature swaps, incentives to land users and stewardship programs (UNCTAD 2018; Barbour 2005; Cumming et al. 2017).

Fig. 5.12 Social network analysis for regulation and incentive-based interventions

Finally, measures related to *liberalization and deregulation* have been very popular globally in the recent past as a means of attracting FDI (Chap. 3 Vol. 1). For example, they have accounted for around 84% of all FDI-related measures and policies monitored in 65 countries (UNCTAD 2018). Africa is the second region that is most prone to investment liberalization, following Asia. Countries such as Morocco, Egypt, Tanzania, Zimbabwe, and Angola have adopted noteworthy liberalization measures that have diversified their respective economies from dependence on natural resources and commodities (UNCTAD 2018). Reducing bureaucracy and the role of the state in private transactions could facilitate the entry of new investors and donors (Cleeve et al. 2015). At the same time, dividing the separate stages of large infrastructure development projects among different investors may reduce uncertainties by sharing risks (Collier 2014; Yakovleva et al. 2017). This is especially relevant for facilitating transactions in countries with few natural resources and political instability (Fig. 5.13). However, international donors such as the World Bank are cautious towards financial openness, and specifically its risks for increasing market volatility, wage and gender inequality (Aguayo-Tellez 2013).

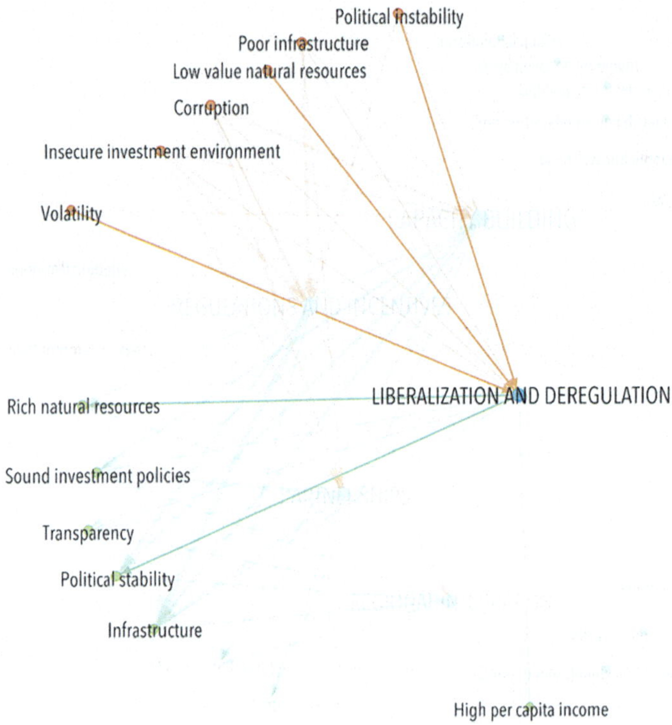

Fig. 5.13 Social network analysis for liberalization and deregulation interventions

5.4 Policy Implications and Recommendations

As discussed throughout this chapter, attracting and effectively utilizing funding in the form of FDIs and ODA are essential for achieving practically all SDGs in Africa (Sect. 5.1). Indeed the lack and ineffective use of funding has been identified as a major sustainability challenge in Africa (Chap. 1 Vol. 1). Section 5.3.3 outlined some of the main pathways for overcoming the barriers to attract and utilize effectively FDIs and ODA namely, (a) building capacity, (b) developing strong regulatory frameworks and incentives, (c) fostering multi-stakeholder partnerships, (d) promoting regional integration and (e) promoting deregulation and liberalization. Below, we discuss some of the main policy implications and opportunities to enhance these pathways and support FDI/ODA recipient countries to convey that they are safe and reliable funding destinations.

Policies seeking to build and improve capacity should go beyond a simple focus on financial capacity to encompass multiple other dimensions, including the development of capable institutions, infrastructure and human resources. First, capable institutions are critical for both developing and forming a conducive environment to attract and effectively use FDIs and ODA (Asiedu 2006). Policies should also facilitate the generation, sharing and management of trustworthy data and information that is usually required by investors and aid organizations to make well-informed funding decisions (United Nations Economic Commission for Africa 2010). This information should, to the extent possible, be generated and managed by independent institutions to avoid the corruption that is prevalent in many African countries, as a means of enhancing the transparency required by donors and investors (Chap. 1 Vol. 1). Second, as discussed accessible and reliable infrastructure (e.g. roads, ports, energy systems, ICT) can facilitate the implementation of projects financed through FDIs and ODA. Thus, the development of such critical infrastructure should become a priority policy domain in those countries seeking to absorb effectively FDIs and ODA (Chap. 1 Vol. 1). For example, Rwanda (a country with scarce natural resources) has invested heavily in ICT infrastructure to improve connectivity,[6] skills and institutional frameworks, partly through its Vision 2020 project to enhance investment flows. This has helped the country make impressive strides towards transitioning to a knowledge-based economy (Ntale et al. 2013). Third, policies should promote education-related actions for developing a qualified skill force to support value-addition in products/services, and more broadly to transition to knowledge-based economies (Asiedu 2002). The above would have ripple positive outcomes for multiple SDGs such as SDG3 (good health and well-being), SDG4 (quality education), SDG7 (affordable and clean energy) and SDG16 (peace, justice and strong institutions).

Although putting incentives may help attract to some degree FDIs and ODA, it is robust regulatory frameworks and policies that sit the core of a country's ability to receive and appropriately use such funds. Thus, any effort seeking to develop strong institutions and regulations to support FDIs and ODA should go beyond creating purely economic incentives. Instead, there should be a concerted effort to develop integrated policy frameworks to support good governance in other relevant policy domains such as the labour market, biodiversity conservation, landscape management, research and development, conflict resolution and infrastructure development, among others (Clark et al. 2018) (Chap. 1 Vol. 1). At the same time, the security of FDIs should be guaranteed through long-term economic policies that seek to ensure the stability of the local currencies (and the broader economic environment), and functional judicial systems that can enforce business contracts. Even though FDIs are critical for the viability of large investments in African countries that lack substantial domestic resources, there should be concomitant efforts to mobilize local financial resources for small and medium enterprises (SMEs), as well as support the financial inclusion of local communities. Institutional investments and local banks would be critical in making financial resources available to SMEs by granting them access to affordable credit (Chirambo 2016). The financial inclusion of local businesses (and the broader communities) may help boost economic growth

[6]Connectivity improves service delivery for businesses, and the broader agriculture and health sectors. For example, it facilitates diverse processes such as telemedicine, cashless mobile banking, money transfers and remittances, and financial inclusion (Allen 2017).

and employment generation, having thus knock-on effects for multiple other SDGs such as SDG8 (decent work and economic growth) and SDG9 (industry, innovation and infrastructure) (CGAP 2018) (Chap. 1 Vol. 1).

Fostering partnerships and international cooperation to secure FDIs and ODA can be an initial step towards more transformative change. In particular, successful sustainability interventions and investments in Africa would require both the thorough understanding of regional and local contexts (for which local stakeholders are indispensable) (Asiedu 2006) and the know-how, technical capacity and development experience of international actors. To the extent possible, appropriate policies should incentivize the development of foreign–local partnerships to raise and manage funding, as well as the adoption and transfer of technologies that are appropriate to the local contexts (Chirambo 2016). For example, cooperation plans such as the China–Africa Cooperation Beijing Action Plan (2019–2021) could provide financing for economic sectors related to energy and agriculture (FOCAC 2018). However, such partnerships should promote the financial inclusion of the local workforce and the training of a skilled workforce, in order to achieve broader societal benefits. Apart from being relevant to many of the SDGs mentioned above (and throughout this volume), fostering partnerships and international cooperation can directly contribute to the attainment of SDG17 (partnerships for the goals).

Stronger regional integration can boost the development of economies of scale, large/stable markets and diversified goods and services (Chaps. 1 and 2 Vol. 1). Regional integration[7] has been identified as one of the main factors for the observed increase in FDI and ODA flows in Africa, despite the loss of momentum in global mega-treaties and trade blocs (UNCTAD 2018). Conscious and coordinated efforts that harmonize policies across regions, simplify processes and optimize entry costs can in theory facilitate business growth and reduce barriers for attracting FDIs and private sector involvement across borders. At the national level, policy integration can reduce competition and the risk of conflict between countries, and at the same time increase their cooperation.

Liberalization and deregulation processes can provide signals for broader economic transformation within individual countries. Such processes can become an avenue for attract FDIs by speeding/simplifying relevant processes and reducing opportunity costs (United Nations Economic Commission for Africa 2010). Liberalization and deregulation processes, nonetheless, require at least some minimal governmental intervention through robust institutions that provide a conducive investment environment, and monitor and enforce regulations (see above). Thus, liberalization and deregulation policies should promote private investment and co-participation, while simplifying overly bureaucratic processes and leveraging

[7]Current regional integration processes in Africa include the Community for Sahel-Saharan States (CEN-SAD), the Common Markets for Eastern and Southern Africa (COMESA), the Eastern Africa Community (EAC), the Economic Community of Central African States (ECCAS), the Economic Community of West African States (ECOWAS), the Southern Africa Development Community (SADC), the Arab Maghreb Union (AMU) and the African Union (AU) (United Nations Economic Commission for Africa 2010).

governmental efforts in an efficient manner. Streamlining the registration of companies and facilitating the entry of foreign capital can be further assisted through the development of integrated institutions ("one-stop shops") and the ability to provide faster feedback on investment proposals.

However, we should note that despite the possible benefits of regional integration, liberalization and deregulation for attaining various SDGs and individual targets, they also pose many risks. This especially true in countries lacking a strong institutional capacity to undertake or engage effectively in such processes. In this sense, there is a need to have in place robust policy frameworks and capacity to navigate such processes (e.g. policies able to recognize market distortions timely and avoid any possible negative socioeconomic and environmental justice impacts). Labour rights, equality of opportunity and the recognition of environmental and social limits are only some of the aspects that should be considered/reflected in such policy frameworks in order to reduce to the extent possible any of the negative sustainability outcomes of broader regional integration, liberalization and deregulation processes, especially for vulnerable social groups.

5.5 Conclusions

By combining secondary data analysis and literature review, this chapter identified (a) the SDGs mainly targeted by ODA in Africa; (b) the research priorities related to the SDGs in Africa; (c) the factors that attract or drive away FDI and ODA in Africa. We systematized this information using conceptual mapping and social network analysis.

For (a) and (b), our results suggest that ODA flows seem to be equally distributed throughout the continent. Countries with functional democracies, high poverty and a colonial past have a higher chance in attracting ODA. However, there is a mismatch between academic research priorities related to the SDGs and actual SDGs targeted by ODA flows in Africa. Academic studies seem to focus more on SDGs1, 5 and 13, while ODA flows target interventions more related to SDGs 2, 11 and 16. This poses a potential risk for the effective allocation of ODA, as it implies a lack of robust academic research to inform those mechanisms seeking to enhance ODA effectiveness for meeting the SDGs in Africa.

For (c), we identified a series of push and pull factors that respectively drive FDI and ODA flows from developed to developing countries or are internal characteristics of developed countries that influence the attraction of FDI and ODA. Some of the factors having a positive influence include economic stability, robust legal systems, functional democratic systems, conducive and open business environments, functional infrastructure and market integration. The availability of natural resources and the size of the targeted markets are also advantages boosting the possibility of attracting FDIs. On the other hand, the quality and legitimacy of donor intentions also positively or negatively influence ODA acquisition. For example, donors tend to prioritize ODA to countries that promote access to basic needs such as health, legal systems and infrastructure.

Finally, we identified five pathways that can help transform some of the negative factors into positives that can enhance the ability to secure (and effectively use) FDIs and ODA, namely capacity-building liberalization and deregulation, incentives and regulations, partnerships and regional integration. Policies should aim to strengthen these areas in order to enhance the influx and effective use of FDIs and ODA in African countries.

Acknowledgements This chapter was developed as part of the Development Strategy for Urban Sustainability in Africa on the Basis of SDGs Interlinkage Analysis (USiA) project led by research teams from the United Nations University Institute for the Advanced Study of Sustainability (UNU-IAS), the Integrated Research System for Sustainability Science (IR3S) at the University of Tokyo, the Institute for Global Environmental Strategies (IGES), the University of Ghana, the University of Malawi and the University of Cape Town. The project was supported and funded by the Japan International Cooperation Agency (JICA).

References

Adams S (2009) Foreign direct investment, domestic investment, and economic growth in sub-Saharan Africa. J Policy Model 31:939–949

Aguayo-Tellez E (2013) *The impact of trade liberalization policies and FDI on gender inequalities: a literature review*. World Bank

Alesina A, Dollar D (2000) Who gives foreign aid to whom and why? Working paper 6612. J Econ Growth 5(1):33–63

Alesina A, Weder B (2002) Do corrupt governments receive less foreign aid? Am Econ Rev 92 (4):1126–1137

Allen JP (2017) Technology and inequality: concentrated wealth in a digital world. Palgrave Macmillan, San Francisco

Anyanwu JC, Yameogo ND (2015) What drives foreign direct investments into west africa? An empirical investigation. Afr Dev Rev 27(3):199–215

Asiedu E (2002) On the determinants of foreign direct investment to developing countries: is Africa different? World Dev 30(1):107–119

Asiedu E (2006) Foreign direct investment in Africa: the role of natural resources, market size, government policy, institutions and political instability. World Economy 29:63–77

Asongu SA, Nwachukwu JC (2016) Foreign aid and governance in Africa. Int Rev Appl Econ 30 (1):69–88. https://doi.org/10.1080/02692171.2015.1074164

Asongu SA, Nwachukwu JC (2017) Foreign aid and inclusive development: updated evidence from Africa, 2005-2012∗. Soc Sci Q 98(1):282–298. https://doi.org/10.1111/ssqu.12275

Barbour P (2005) An assessment of South Africa's investment incentive regime with a focus on the manufacturing sector. Economics and Statistics Analysis Unit—Overseas Development Institute, London

Boafo YA, Lam RD, Juju DB, Saito O et al (2018) The role of sustainability science in addressing Africa's sustainable development challenges: a bibliometric survey. In: Constructing sustainability science in developing countries. UNESCO, Paris

Bräutigam DA, Knack S (2014) Foreign aid, institutions, and governance in sub-Saharan Africa. Econ Dev Cult Chang 52(2):255–285

Briggs RC (2018) Poor targeting: a gridded spatial analysis of the degree to which aid reaches the poor in Africa. World Dev 103:133–148

Camfield L, Crabtree A, Roelen K (2013) Editorial: poverty, vulnerability and resilience in a Post-2015 world. Soc Indic Res 113(2):599–608

Cassimon D, Van Campenhout B, Ferry M, Raffinot M (2015) Africa: out of debt, into fiscal space? Dynamic fiscal impact of the debt relief initiatives on African heavily indebted poor countries (HIPCs). Int Econ 144:29–52. https://doi.org/10.1016/j.inteco.2015.04.007

CGAP (2018) Donors and investors: funding instruments. Retrieved 01 10, 2018, from Consultative Group to Assist the Poor – CGAP: Focus Areas: http://www.cgap.org/topics/funding-instruments

Chen S, Ravallion M (2009) The developing world is poorer than we thought, but no less successful in the fight against poverty. The World Bank Development Research Group, Washington

Chirambo D (2016) Moving past the rhetoric: policy considerations that can make Sino-African relations to improve Africa's climate change resilience and the attainment of the sustainable development goals. Adv Clim Chang Res 7(4):253–263

Clark R, Reed J, Sunderland T (2018) Bridging funding gaps for climate and sustainable development: pitfalls, progress and potential of private finance. Land Use Policy 71:335–346

Cleeve EA, Debrah Y, Yiheyis Z (2015) Human capital and FDI inflow: an assessment of the African case. World Dev 74:1–14

Collier P (2014) Attracting international private finance for African infrastructure. J Afr Trade 1:37–44

Cumming TL, Shackleton RT, Förster J, Dini J, Khan A, Gumula M, Kubiszewski I (2017) Achieving the national development agenda and the sustainable development goals (SDGs) through investment in ecological infrastructure: a case study of South Africa. Ecosyst Serv 27:253–260

Dilorenzo M, Ghose S, Turner J (2017) Estimating baseline aid to the sustainable development goals. AidData, Williamsburg, VA

Dunning T (2004) Conditioning the effects of aid: cold war politics, donor credibility, and democracy in Africa. Int Organ 58:409–423

Dupasquier C, Osakwe PN (2006) Foreign direct investment in Africa: performance, challenges, and responsibilities. J Asian Econ 17(2):241–260

Duraiappah AK, Asah ST, Brondizio ES, Kosoy N, O'Farrell PJ, Prieur-Richard A-H et al (2014) Managing the mismatches to provide ecosystem services for human Well-being: a conceptual framework for understanding the new commons. Curr Opin Environ Sustainability 7:94–100

FOCAC (2018) Beijing declaration, action plan adopted at FOCAC summit. Retrieved 09 05, 2018, from Forum on China-Africa Cooperation: https://focacsummit.mfa.gov.cn/eng

Fowkes FJ, Draper BL, Hellard M, Stoové M (2016) Achieving development goals for HIV, tuberculosis and malaria in sub-Saharan Africa through integrated antenatal care: barriers and challenges. BMC Med 14(1):202

Gui-Diby SL, Renard M-F (2015) Foreign direct investment inflows and the industrialization of African countries. World Dev 74:43–57

Hart C (2018) Doing a literature review: releasing the social science research imagination. Sage Publications, London

Honadle BW (1981) A capacity-building framework: a search for concept and purpose. Public Adm Rev 41:575–580

IAEG-SDGI (2016) *Report of the Inter-Agency and Expert Group on Sustainable Development Goal Indicators.* United Nations, Economic and Social Council. New york: United Nations Secretariat

ICSU (2017) A guide to SDG interactions: from science to implementation. International Council for Science (ICSU), Paris

Kates RW, & Dasgupta P (2007, 10 27) African poverty: a grand challenge for sustainability science. PNAS, 104 16747-16750

Lim SS, Allen K, Bhutta ZA, Dandona L, Forouzanfar MH, Fullman N et al (2016) Measuring the health-related sustainable development goals in 188 countries: a baseline analysis from the global burden of disease study 2015. Lancet 388(10053):1813–1850

Milhorance C (2016) Le rôle du Sud dans la fabrique du développement : l'internationalisation des instruments des politiques publiques brésiliennes pour le secteur rural. Paris: Thèse de doctorat. Université Paris Saclay, Paris

Minasyan A, Nunnenkamp P, Richert K (2017) Does aid effectiveness depend on the quality of donors? World Dev 100:16–30

Mohamed MR, Kaliappan SR, Ismail NW, Azman-Saini W (2015) Effect of foreign aid on corruption: evidence from sub-Saharan African countries. Int J Soc Econ 42(1):47–63. https://doi.org/10.1108/IJSE-04-2013-0089

Museru M, Toerien F, Gossel S (2014) The impact of aid and public investment volatility on economic growth in sub-saharan Africa. World Dev 57(1966):138–147

Ntale A, Yamanaka A, Nkurikiyimfura D (2013) The metamorphosis to a knowledge-based society: Rwanda. In: Bilbao-Osorio B, Dutta S, Lanvin B (eds) The global information technology report 2013: growth and jobs in a hyperconnected world. World Economic Forum and INSEAD, pp 119–126

Opperman P, Adjasi CK (2017) The determinants of private capital flow volatility in sub-Saharan African countries. Res Int Bus Financ 42:312–320

Pachauri RK, Allen MR, Barros VR, Broome J, Cramer W, Christ R, . . . van Ypserle JP (2014) *Climate Change 2014: Synthesis Report. Contribution of Working Groups I, II and III to the Fifth Assessment Report of the Intergovernmental Panel on Climate Change [Core Writing Team R.K. Pachauri and L. A. Meyer (eds)].* IPCC, Geneva, Switzerland

Rasul G, Sharma B (2016) The nexus approach to water–energy–food security: an option for adaptation to climate change. Clim Pol 16(6):682–702

Rivano NS (2014) African power in the 21st century and beyond. Panoeconomicus 5(Special Issue):585–596

Schoentgen A, Gille L (2017) Valuation of telecom investments in sub-Saharan Africa. Telecommun Policy J 41:537–554

Schwerhoff G, Sy M (2017) Financing renewable energy in Africa—key challenge of the sustainable development goals. Renew Sust Energ Rev 75:393–401

Sethi T, Custer S, Turner J, Dilorenzo M, Latourell R (2017) Realizing agenda 2030: will donor dollars and country priorities align with global goals? AidData at the College of William & Mary, Williamsburg, VA

Steffen W et al (2004) Global change and the earth system: a planet under pressure—executive summary. IGBP Secretariat, Stockholm

UN General Assembly (2015) Transforming our world: the 2030 agenda for sustainable development. United Nations, New York

UN Social and Economic Council (2016, 03 08) *United Nations Statistical Comission—47th Session (2016)—Chairs of past sessions*. Retrieved 05 24, 2016, from United Nations Statistics Division: http://unstats.un.org/unsd/statcom/47th-session/

UNCTAD (2018) World investment report 2018: investment and new industrial policies. United Nations Conference on Trade and Development, New York and Geneva

UNDESA (2014) World urbanization prospects. Undesa. https://doi.org/10.4054/DemRes.2005.12.9

UNDESA (2017) *Population and vital statistics report.* United Nations Department of Economic and Social Affairs, UNDESA. United Nations, New York

UNDP (2018). *Sustainable development goals.* Retrieved 04 24, 2018, http://www.undp.org/content/undp/en/home/sustainable-development-goals.html

UNISDR (2009) 2009 UNISDR terminology on disaster risk reduction.. United Nations International Strategy for Disaster Reduction. UNISDR, Geneva

United Nations (2014) The road to dignity by 2030: ending poverty, transforming all lives and protecting the planet. Synthesis report of the Secretary-General on the post-2015 Agenda. United Nations, New York

United Nations (2015) The millenium development goals report. United Nations Secretariat, Department of Economic and Social Affairs. United Nations, New York

United Nations Economic Commission for Africa (2010) *Assessing Regional Integration in Africa IV Enhancing Intra-African Trade*. https://doi.org/e-ISBN-13:978-92-1-055278-3
von Braun J, Mirzabaev A (2016) Nexus scientific research: theory and approach serving sustainable development. In: Dodds F, Bartram J (eds) The water, food, energy and climate nexus: challenge and an agenda for action. Routledge, London, New York, pp 58–71
Yakovleva N, Kotilainen J, Toivakka M (2017) Reflections on the opportunities for mining companies to contribute to the United Nations sustainable development goals in sub-saharan Africa. Extr Ind Soc 4(3):426–433
Zanger SC (2000) Good governance and European aid: the impact of political conditionality. Eur Union Polit 1(3):293–317. https://doi.org/10.1177/1465116500001003002
Zhou X, Moinuddin M (2017) Sustainable development goals interlinkages and network analysis: a practical tool for SDG integration and policy coherence. Instiutute for Global Environmental Strategies (IGES), Kanagawa

Part II
Western Africa

Chapter 6
Perceived Community Resilience to Floods and Droughts Induced by Climate Change in Semi-arid Ghana

Yaw Agyeman Boafo, Osamu Saito, Godfred Seidu Jasaw, Gerald A. B. Yiran, Rodolfo Dam Lam, Geetha Mohan, and Gordana Kranjac-Berisavljevic

6.1 Introduction

The large-scale scientific assessments of climate change point out that many communities and households around the world are already facing severe negative impacts from extreme weather events, sea level rise and abnormal temperature and

Y. A. Boafo (✉)
Centre for Climate Change and Sustainability Studies, University of Ghana, Accra, Ghana
e-mail: yboafo@ug.edu.gh

O. Saito
Institute for Global Environmental Studies (IGES), Hayama, Japan

Institute for Future Initiatives (IFI), The University of Tokyo, Tokyo, Japan
e-mail: o-saito@iges.or.jp

G. S. Jasaw · G. Kranjac-Berisavljevic
University for Development Studies, Wa and Tamale Campuses, Tamale, Ghana
e-mail: gjasaw@uds.edu.gh

G. A. B. Yiran
Department of Geography and Resource Development, University of Ghana, Accra, Ghana
e-mail: gyiran@ug.edu.gh

R. D. Lam
Graduate School for Frontier Studies, The University of Tokyo, Tokyo, Japan

G. Mohan
United Nations University Institute for the Advanced Study of Sustainability (UNU-IAS), Tokyo, Japan

Institute for Future Initiatives, The University of Tokyo, Tokyo, Japan

A. Gasparatos et al. (eds.), *Sustainability Challenges in Sub-Saharan Africa I*,
Science for Sustainable Societies, https://doi.org/10.1007/978-981-15-4458-3_6

precipitation patterns (IPCC 2014) (Chaps. 2 and 9, Vol. 2). There is consensus within the scientific community that climate change exacerbates the risk, frequency, intensity, spatial extent, duration and timing of extreme weather and climate-related hazards (IPCC 2014). This is consistent across most regions and biomes around the world, with a rather strong consensus between scientists.

Sub-Saharan Africa (SSA) is expected to face some of the most severe impacts of climate and environmental change due to a combination of environmental and socio-economic factors (Serdeczny et al. 2016) (see Chap. Vol. 1). Many of the countries at a greater risk are concentrated in SSA, where the capacity to prevent and cope with the adverse impacts of climate change is poor (Chen et al. 2015). Since 2014, for instance, annual and unprecedented droughts have been threatening biodiversity, ecosystem services and the quality of life of rural and urban households across southern and eastern SSA (IPBES 2018). Remarkably, the vulnerability of communities and households to disasters related to climate and ecosystem change is highly disproportional across SSA sub-regions, as well as within individual countries and localities (Narain et al. 2011; World Bank 2012; IPCC 2014). Furthermore, due to the generally weak adaptive capacity across much of SSA, entire livelihood systems, biodiversity and ecosystem services are threatened (Niang et al. 2014) (see also Chap. 10 Vol. 1; Chaps. 2 and 3 Vol. 2). Similarly, poorer rural communities experience the highest negative effects from such events, as they depend directly on agriculture and ecosystems for their livelihoods (IRIN 2017) (Chap. 1 Vol. 1; Chaps. 2 and 3 Vol. 2).

The effects of climate change are possibly more pronounced in the semi-arid areas of SSA that span an estimated 5073 km^2 (UNDP/UNSO 1997). In these areas, the livelihood resilience of most people is continuously weakened due to floods and droughts, which are contributing to water stress and hindering food production exacerbating food insecurity (see also Chaps. 2 and 3 Vol. 2).[1] Several scenarios and modelling studies show that tropical savannas and grasslands will be significantly threatened by climate change in the next decades (Higgins and Scheiter 2012; Lehmann et al. 2014).

Many studies have been focusing on the ability of vulnerable communities and households to adapt to and/or recover from the effects of potentially harmful natural and human-induced disasters in a manner that protects livelihoods, accelerates/sustains recovery and supports economic and social development (Manyena 2006; Gaillard 2007; Turnbull et al. 2013; Zhang and Shay 2019). Despite an emerging consensus on the fundamental tenets of the resilience concept and the proliferation of relevant studies (especially in the last two decades), key definitions, interpretations and measurement approaches are still contested (Sect. 6.2.1). Current definitions suggest that resilience is a rather complex subject, with its different theoretical and

[1]Rockström et al. (2009) point out that water scarcity and rainfall variability (coupled with human pressure on water resources for food production) in the savanna semi-arid regions could lead to floods, droughts and dry spells, and why there is a need for the adoption of effective water strategies to increase the resilience of local social-ecological systems.

empirical aspects captured within the different studies (Tobin 1999; Adger 2000; Pelling 2003; Buckle 2006; Janssen et al. 2006) (Chap. 7 Vol. 2). The prevalence and use of diverse indicators and parameters to assess the resilience of vulnerable groups have contributed significantly to conceptual and practical developments within the disaster risk management discourse.

Reducing the vulnerability (and enhancing the resilience) of local communities to climatic hazards and disasters is a major sustainability challenge in SSA, and especially in its semi-arid regions (AGRA 2017). This is further complicated by a series of factors including the prevailing deficit in resources, knowledge and infrastructure in these regions (Chap. 1 Vol. 1). A key prerequisite of resilience building at the local scale is the wide understanding and appreciation among stakeholders about the resilience needs and coping/adaptive conditions of communities vulnerable to climatic hazards (NASAC 2015). Community resilience assessments can help identify gaps, raise awareness and propose potential interventions and strategies to enhance resilience. Such studies can build on the verifiable, fact-based and consensual data on the causes, risks and potential negative impacts of climate and ecosystem changes in SSA on current and future generations.

In the context of the above, this study aims at assessing community and stakeholder perceptions of resilience to droughts and floods in selected locations of semi-arid Ghana. This reflects the proliferation of strategies at the national, regional and local levels in semi-arid Ghana over the last five decades that seek to alleviate the effects of climate change (and associated socio-economic impacts) and build resilient communities (Songsore 2011). Even though multiple stakeholders (including central government, international development agencies, non-governmental organizations) have attempted to provide solutions for enhancing the resilience of households and communities through poverty reduction, sustainable infrastructure development and alternative livelihood strategies, they appear to have yielded limited results so far (Pasteur 2011).

In line with existing studies evaluating community resilience to climatic hazards and disasters, our study acknowledges the role of diverse elements and indicators in mediating households' and individuals' sensitivity and exposure to floods and droughts (Sect. 6.2.1). This can be a first step towards the identification, development and operationalization of reactionary and anticipatory actions for current and future events, respectively. A central element of this study is eliciting the perceptions of stakeholders/households/communities (Sect. 6.2.1), as such perceptions can affect behavioural response in relation to strategies and interventions for managing hazards and disasters linked to environmental change (Whitmarsh 2008). The study highlights the practicality and importance of an inclusive and participatory approach that goes against purely top-down approach to resilience assessment in disaster risk management efforts (Argon 2014).

Section 6.2 outlines the methodology by delving into the research approach, study sites, field data collection and data analysis approaches. Section 6.3 presents the community resilience perceptions by community, locality and age group. Section 6.4 examines the implications of the study findings. It focuses especially on how to facilitate the uptake of the proposed methodology, as well as the policy relevance for achieving sustainable development.

6.2 Methodology

6.2.1 Research Approach

The main starting points of our conceptualization of resilience are the works of Cutter et al. (2008) and UNISDR (2012). Cutter et al. (2008: 599) define disaster resilience as, "the ability of a social system to respond and recover from disasters and includes those inherent conditions that allow the system to absorb impacts and cope with an event, as well as post-event adaptive processes that facilitate the ability of the social system to reorganize, change, and learn in response to a threat". The United Nations International Strategy for Disaster Reduction (UNISDR) defines resilience as "the ability of a system, community or society exposed to hazards to resist, absorb, accommodate to and recover from the effects of a hazard in a timely and efficient manner, including through the preservation and restoration of its essential basic structures and functions" (UNISDR 2012: 3).

According to Folke et al. (2003), in order to build communities that are resilient to disasters within their respective socio-ecological systems, there is a need for: (a) social learning developed through a memory of past events and the abandonment of the notion of stability; (b) different forms of social and ecological community diversity that can enable coping with shocks and stresses; (c) different types of knowledge that can stimulate learning and innovation and (d) opportunities that can enable self-organization and the formation of cross-scale linkages.

The concept of resilience has four inter-related dimensions, namely technical, organizational, social and economic (Bruneau et al. 2003). Simpson (2006) thoroughly examined hazards, community assets, social capital, infrastructure quality, planning, social services and population demographics as key elements and indicators for resilience. Further advances to community resilience indicators include ecological, social, economic, institutional, infrastructure and community capacity dimensions (Cutter et al. 2008). Earlier studies led to the development of the Social Vulnerability Index (Cutter et al. 2008) and identified economic development, social capital and community competence as the key elements in measuring community resilience (Norris et al. 2008). Collective action, community resources and strategic action have also been emphasized as key dimensions in the assessments of community resilience (Magis 2010). Similarly, Burton (2015) used six indicators (i.e. economic resilience, social resilience, infrastructure resilience, institutional resilience, environmental resilience, community resilience) as proxies to assess the resilience of groups vulnerable to natural disasters.

The above suggest that most of the indicators and parameters commonly used for the assessment of resilience fit seamlessly into top-down approaches, rather than bottom-up approaches. Top-down approaches to resilience assessment tend to rely on secondary data such as government census reports and administrative data in designing proxies and indicators of resilience (Cutter et al. 2008; UNU 2014). However, the limited degree of community involvement in top-down resilience assessments means that they are usually undertaken at international or national level. Even though many factors (e.g. scale, costs, end use of the output), influence the adoption of top-down approaches it can be argued that they are insufficient to offer fine-grained information as there is little (or no) space for integrating the lived experiences, viewpoints and local knowledge systems and practices of diverse stakeholders.

We therefore argue in this study that attention needs to shift more to bottom-up assessment approaches. Our argument is further strengthened by the fact that the occurrence of climatic disasters (and resilience/adaptive capacity to them) varies across space and time (Yiran and Stringer 2016). The proxies commonly used in top-down assessment frameworks often fail to adequately involve, capture and integrate diverse interest groups and stakeholders' perspectives in the assessment (Burton 2015). Thus, some interventions seeking to reduce the vulnerability and/or increase the resilience to natural disasters might fail because they often do not reflect local contexts and needs.[2]

Conversely, participatory resilience assessments in disaster-prone settings of SSA can provide a useful knowledge base for policy- and decision-makers at the local, regional and national level. Hence, this study focuses on the development of a community-based resilience assessment approach through the input of diverse and relevant community groups and stakeholders in flood and drought-prone semi-arid regions of Northern Ghana. The overall research approach is outlined below and draws from Antwi et al. (2014), which was conducted in the same research setting under the auspices of the CECAR-Africa project.

Step 1 entails a comprehensive review of the scientific and grey literature on the concept of disaster resilience, focusing on human–environment interactions. In particular, the review aims at identifying the broader concepts, dimensions, elements and variables related to community resilience to floods and droughts. Based on this review, we identify three key dimensions of community resilience, namely ecological, engineering and socio-economic resilience. Ecological resilience reflects those elements and actions within local communities that enhance tolerance to droughts and floods, as well as the ability of socio-ecological system to return to a desirable state after such events (Gunderson and Holling 2002; Walker et al. 2004; Norris et al. 2008). Engineering resilience reflects the attributes of local communities that enable the entire socio-ecological system to adapt to and absorb the shocks and stresses of perennial floods and droughts (e.g. anticipation, consistency, efficiency,

[2]Such an example comes from the Afar landscape in north-eastern Ethiopia, where households and communities are both highly exposed and sensitive to perennial droughts and warming, which negatively affect water resources, livestock production, and agricultural productivity (Magnan et al. 2016). The promotion of regional adaptation policies such as irrigation agriculture and non-pastoral livelihoods systems, generated conflicts and widened social disparities due to competition over land and other natural resources (Eriksen and Marin 2015).

predictability, preparedness) (Holling 1973; Ludwig et al. 1997). Socio-economic resilience focuses on the capability of social and economic elements to maintain the social–ecological system after disturbances from flood and drought (e.g. household livelihoods, assets, social capital, social services). Socio-economic resilience stresses the availability, access, exposure, diversity and functionality of the key resources and services needed to meet basic needs (Adger 2000; Langridge et al. 2006; Perrings 2006).

Step 2 involves the identification of appropriate parameters and elements to serve as proxies for measuring ecological, engineering and socio-economic resilience. This required a second round of literature review aiming to identify the necessary elements for measuring resilience within the context of the three dimensions outlined above. We relied on both peer-reviewed and grey literature to explore issues of vulnerability and resilience in disaster-prone settings. We also sourced secondary data from relevant government agencies, public organizations, NGOs and international development agencies related to droughts and floods in our study areas (Sect. 6.2.2).

We initially identify approximately 73 elements of resilience across the ecological, engineering and socioeconomic dimensions outlined above. We then select the major elements based on the (a) extent of application in similar socio-ecological systems, regions and contexts and (b) scalability in terms of the ability to apply the different elements within the study communities. This resulted in 52 elements, which are grouped into parameters that are relevant to each dimension (Table 1). We should note that even though this process for identifying and selecting resilience elements has been detailed and thorough, the current list is by no means exhaustive, nor transferable directly to other contexts (e.g. socio-ecological, disaster, geographical).

Step 3 tests the draft community[3] resilience matrix developed through the extensive review process. We undertook a preliminary field survey to the study communities (Sects. 6.2.2–6.2.3) to identify the possible applicability of the matrix locally. Based on the collected information, we refine the community resilience matrix, which involved the addition, omission or revision of the 52 elements of community resilience (see above).

Step 4 validates the selected elements through community and stakeholder meetings (Sect. 6.2.3). These meetings aimed at (a) identifying the concerns of local stakeholders and community members related to droughts and floods in the study area and (b) incorporating scalable elements of resilience, which fitted into the local socio-ecological, economic and traditional context. We synthesize the input from all study communities to compile the final community resilience assessment matrix. Overall, this phase was extremely important in the context of this study, as it

[3]The term 'community' in this study refers to the people that share a common physical and/or geographical space. This includes access to (and utilization of) common ecosystem goods and services, whilst interacting regularly and being connected through common socio-cultural beliefs and norms.

Table 1 Resilience dimensions and elements identified during Stage 2

Ecological resilience	
1	Protected and conservation areas
2	Understanding of functioning of local environment
3	Biodiversity richness
4	Quality of vegetation cover and soil
5	Level of dependence on ecosystem services
6	Secured land tenure system
7	Available and enough rangelands
8	Availability of productive arable lands
9	High landscape elevation minimizes flood impacts
10	Drought-tolerant crops cultivation
11	Flood-resistant crop cultivation
12	Functioning irrigation system
13	Access to crop storage system in community
14	Seascape (water bodies) diversity
15	Land area with wetland decline
16	Land area with primary forests
17	Percentage of degraded forest
18	Land area not erodible
19	Existence of natural windbreaks
Engineering resilience	
20	Access to community-managed resources
21	Knowledge and understanding of flood impacts
22	Community recovery potential after flood and drought
23	Suitable area for flood evacuation
24	Good and accessible roads
25	Flood-resistant household unit
26	Functioning and available health centres
27	Electricity accessibility
28	Portable water for domestic use
29	Platform for effective dissemination of DRR knowledge information
30	Effective communal risk and cost-sharing system
31	Reliable early warning and forecast system for drought
32	Reliable early warning and forecast system for flood
33	Trained local population to manage immediate disaster effects
34	Local DRR is integrated into livelihood systems
35	Knowledge and understanding of climate change impact
36	Knowledge and understanding of climate change impact
37	Established local disaster management network
38	Local compliance with disaster management strategies
39	Knowledge and understanding of drought impacts
40	Diversity of local food systems
41	Value addition to local produce

(continued)

Table 1 (continued)

42	Community-managed landscape
43	Use of local ecosystem for ecotourism
44	Age groups awareness of resilient local knowledge
45	Mechanisms to transfer knowledge
Socio-economic resilience	
46	Security in land tenure
47	Collective memories of past disaster
49	Diversified livelihood strategies
50	Access to money transfers and remittances from cities
51	Access to savings and credit schemes
52	Equity in wealth distribution
53	Livestock income
54	Small-scale businesses (milling, provision stores)
55	Livestock and poultry holdings
56	Secured food supply during disaster
57	Available and access to local markets and trade links
58	Active labour force in community
59	Secured water for domestic use
60	Access to electricity
61	Access to common-pool resources
62	Active social support and network system
63	Intra-community DRR information cooperation
64	Inter-community DRR information cooperation
65	Coordination and collaboration with external agencies
66	Household savings per capita
67	Share of population in formal employment
68	Ability to obtain credit
69	Change in dietary pattern
70	Unemployment rate
71	Existence of self-help groups
72	Women voices heard in decision-making
73	Youth involvement in decision-making

enabled the research team to explain the concept of resilience, receive feedback and generate community interest for the rest of the study.[4]

[4]The aforementioned process was a central element of the CECAR-Africa project, which aimed at developing and implementing a community-based resilience enhancement framework in rural communities in Ghana that are vulnerable to floods and droughts (Sect. 6.2.2). This framework drew on the participation of relevant stakeholders at the local, district, regional, national and global levels, and is referred to as the 'Ghana Model' (Saito et al. 2018b).

6.2.2 *Study Sites*

Semi-arid Ghana formally comprises of three political and administrative regions, namely Northern, Upper East and Upper West (Saito et al. 2018a). The three regions comprise approximately 41% (97,702 km^2) of Ghana's landmass (Fig. 1). The Northern and Upper West regions are located in the Guinea Savanna agro-ecological zone (Peprah 2015) that is characterized by semi-arid climatic conditions and savanna grassland vegetation. Northern Ghana is noted for its high sensitivity and susceptibility to climate and ecosystem change, making it the most ecologically vulnerable region in Ghana (Yiran et al. 2017). Some of the well-documented indications of climate change include rainfall decline and the prevalence of extreme high temperatures over the past five decades (Stanturf et al. 2011). Anthropogenic activity and natural disasters such as wildfires, storms, floods, droughts, pests and diseases increase the vulnerability of households and local communities in the region (Armah et al. 2010).

In recent years, the legal and illegal harvesting and exporting of rosewood trees (Ohene 2019), and the production of charcoal have contributed significantly to the decline of biodiversity and ecosystem resilience across the semi-arid regions of

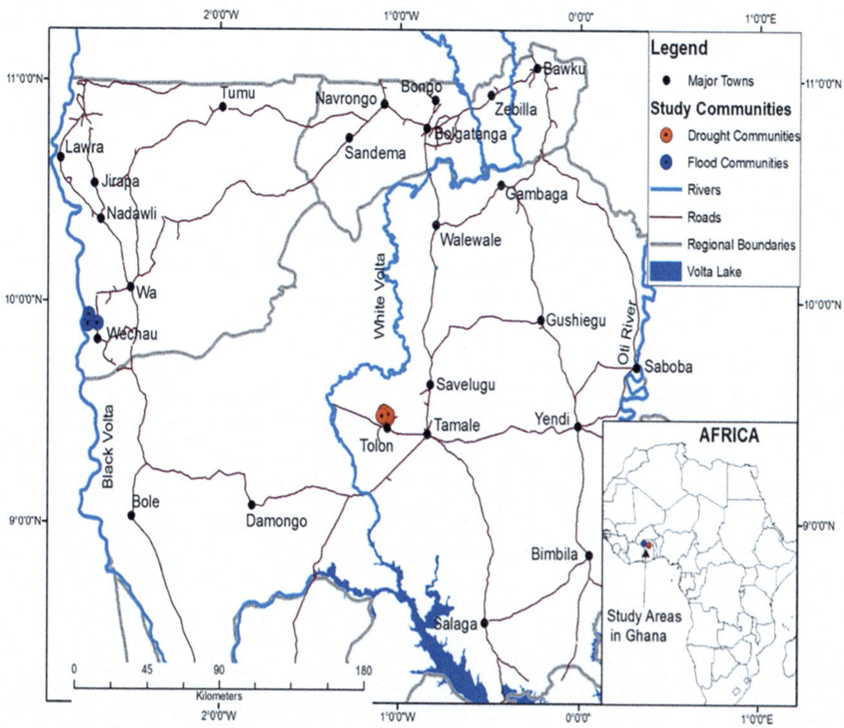

Fig. 1 Location of study communities

Ghana (Jasaw et al. 2017) (see Chap. 2 Vol. 1 for similar situation in Uganda). Semi-arid Ghana is also noted for being socio-economically isolated, characterized by poor social support systems and physical infrastructure, and the highest incidence of poverty in the country (World Bank 2011; Songsore 2011; Ghana Statistical Service (GSS) 2014). The seasonal and permanent youth out-migration to cities in the southern parts of Ghana is commonly used as a coping and adaptation strategy (Schraven 2010).

Based on past studies, reports of disaster management organizations and preliminary field surveys under the CECAR-Africa project, we selected Tolon and Wa West Districts as the specific study sites (see Fig. 1). These districts have high incidence of droughts and floods, which have substantial impacts on social, economic, cultural and environmental systems (Musah and Oloruntoba 2013; Owusu et al. 2016). Although both study locations are drought and flood prone, droughts are more severe than floods in Tolon, while floods are more severe than droughts in Wa West (Antwi et al. 2014). Thus, in this chapter, we consider the Tolon area as "drought-prone", and the Wa West area as "flood prone".

Subsequently, we purposely selected ten rural communities due to their proneness to droughts and floods. Six drought-prone communities were selected from Tolon district while four flood-prone communities were selected from Wa West district (Table 1). The number of communities were more in Tolon due to the fact that drought is more pronounces and frequent. The population of the studied communities ranges from 3751 people (Yoggu) to 195 people (Cheshagu), with also some variation in some other key demographic characteristics (Table 2).

An overwhelming proportion of the population in both districts live in rural settings, with over 90% of the workforce being involved in smallholder rain-fed farming and animal husbandry (Ghana Statistical Service 2012). Staple crops and vegetables produced include maize, sorghum, millet, yam, groundnut, tomatoes and cowpea, and are grown by almost every household and local community in the study districts. Food insecurity, linked to droughts and floods, is an annual challenge for communities in both districts, as well as many other local communities across semi-arid Ghana (Armah et al. 2011).

Torrential rainfall is the major cause of flooding in both study districts, which further causes significant damage to infrastructure, building, farmlands and other natural land uses (Acheampong et al. 2014; Armah et al. 2010). Annual flooding in Tolon can be characterized as small-scale and short-lived (i.e. within 1 hour), as no major rivers flow through the communities. On the other hand, the studied communities in Wa West district are located in close proximity to the Black Volta River, and thus some of these communities have experienced some of the most severe and catastrophic floods in recent memory in semi-arid Northern Ghana.

Droughts often precede the periodic floods in the studied sites, and have many negative impacts on food crop yields, water availability and livelihood security (EPA 2012) (see also Chap. 2 Vol. 2). The reliance of local communities on rain-fed agriculture for subsistence and income generation means that the irregular rainfall patterns, coupled with sustained droughts, can result in poor food crop yields and food insecurity. Linked to droughts is the issue of regulated and unregulated

Table 2 Demographic characteristics of study communities

Study community	Geographic coordinates	Population (people)	Population (households)	Illiteracy (%)	Population over 60 years old (%)	Female-headed households (%)
Drought-prone communities						
Yoggu	9°28′N, 1°04′W	3751	216	92.1	44.5	12.6
Cheshagu	9°04′N, 1°06′W	195	26	80.8	33.7	8.5
Kpalgun	9°28′N, 1°05′W	1573	111	81.1	48.2	12.6
Daboshe	9°05′N, 1°04′W	430	32	96.9	52.3	11.6
Zagua	9°05′N, 1°08′W	561	44	84.1	39.5	6.8
Fihini	9°04′N, 1°07′W	655	38	97.8	56.2	5.1
Flood-prone communities						
Bankpama	9°54′N, 2°34′W	799	79	89.5	35.7	9.3
Baleufili	9°53′N, 2°41′W	802	91	85.3	40.1	11.3
Chietanga	9°92′N, 2°69′W	438	46	96.5	55.3	8.2
Zowayili	9°90′N, 2°70′W	231	28	98.3	59.1	6.5

Note: These demographic characteristics were derived through other surveys conducted during the CECAR–Africa project. We use this as background information for the purpose of this chapter, but the interested reader is diverted to other publications for further results and the methodological protocols (e.g. Antwi et al. 2014; Boafo et al. 2014) bushfires, which are commonly used by locals to access vital ecosystem services such as bushmeat, fodder, firewood and charcoal (see also Chap. 10 Vol. 1). This contributes to the overexploitation and further decline of some of the already scarce plant and animal species (Boafo et al. 2014). It is worth mentioning that when such events occur, the external assistance from national and international disaster management agencies is the primary adaptive strategy (NADMO 2011).

6.2.3 Data Collection

For Step 3 (see Sect. 6.2.1), we performed preliminary fieldwork during the rainy period when floods are expected (8–25 August 2013), and during the long dry season (12–27 April 2014). These preliminary fieldworks involved direct observation, transect walks, farmland visits and informal interviews with opinion leaders and local households to understand the main characteristics of the communities, the hazards and disasters they experience and how they cope with them. Participants included local community members, agriculture extension officers and disaster management agencies from the Tolon and Wa West District Assemblies.

For Step 4 (see Sect. 6.2.1), we conducted a total of 30 focus group discussions (FGDs) across the ten study communities between July–August 2015 and April–May 2016. Specifically, in each community, we conducted three FGDs, stratified into elderly men (50 years old and above), elderly women (50 years old and above) and young adults (18–49 years old) (Table 3). These age groupings were identified and applied purposely for this study due to their distinctive characteristics and inherent differences, as they relate to the recognition, perception, experience and knowledge of the local social, economic and environmental conditions. For example, studies have suggested the proclivity of the young adults to migrate from rural semi-arid Ghana to urban centres in Southern Ghana in search of better livelihoods

Table 3 Participants of focus group discussions across communities and stakeholder groups

	Elderly men	Elderly women	Young adults
Drought-prone			
Yoggu	12	7	8
Cheshagu	10	8	9
Kpalgun	9	9	11
Daboshe	12	10	7
Zagua	10	5	10
Fihini	8	10	8
Flood-prone			
Bankpama	9	8	7
Baleufili	12	9	8
Chietanga	9	10	9
Zowayeli	7	8	6

(Awumbila et al. 2015; Songsore 2009). Similarly, women in the study area tend to have limited control and ownership of productive resources such as land and capital (Rademacher-Schulz and Mahama 2012). Women also participate less in decision-making, although they contribute significantly to labour and housekeeping. In fact, in the study area, men have the major say in household decision-making and practically control most productive resources and assets (Zakaria et al. 2015; Zakaria 2017). Participation in the FGDs varied from 6 to 12 people (Table 3). Even though we sought to obtain a representative sample of each target group in each community, the actual participation was ultimately influenced by the availability and willingness of invitees to participate.

During the FGDs, we aimed to elicit perceptions about the performance of each age group across the different resilience elements. To elicit a qualitative estimate, we used a Likert-type scale where 1 = very low performance, 2 = low performance, 3 = moderate performance, 4 = high performance and 5 = very high performance. We clarified to all participants that their answer should represent their group's performance in that specific resilience element, with lower scores denoting poor performance in a specific element, whilst higher scores denoting a better performance. Research assistants, who were native to the study areas and proficient in the local dialects, translated the resilience elements to the local dialects. Prior to each FGD, the scoring rules and approach were explained to all participants using five rounded, equally sized stones.

The FGDs generated dialogue among participants and sought to capture consensus opinions. Thus, each of the value elicited during Step 4 characterizes the entire group, and not the individual respondent. When the group reached an agreement for a given resilience element, the moderator moved on to the next element. When participants failed in reaching a consensus immediately, the moderator brought up a shared experience of an impact associated with a flood or drought event, as a means of resolving the lack of consensus over the score. For most resilience elements, participants were asked to provide tangible and practically relevant evidence to support their proposed performance scores.

6.2.4 Data Analysis

The FGD results are aggregated at three levels: (a) social group (i.e. elderly men, elderly women, young adults), (b) location (i.e. disaster proneness, community) and (c) resilience category (i.e. dimension, sub-dimension, element). We calculate the mean resilience scores across communities and interest groups and use the outcomes from each FGD to estimate the perceived resilience score.

We obtain the scores for each sub-dimension through averaging the scores for each resilience element. Finally, the resilience scores for each resilience dimension (i.e. ecological, engineering, socio-economic) are calculated as the average of each sub-dimension. The average of averages is used to obtain the resilience score for each resilience dimension and sub-dimension to provide the same weight for each parameter and sub-dimension when aggregated to the higher grouping level (Sect. 6.3.1).

We also calculate the scores for each age group as the averages by each resilience element (Sect. 6.3.2). Subsequently, the average of each element is used to obtain the resilience score for each sub-dimension. Finally, each sub-dimension is averaged to obtain the resilience dimension score and an overall resilience score that averages all three dimensions for each community. Once again, the average of averages is used to provide the same weight for each of the estimated resilience level.

Lastly, we assess interrater agreement (IRA) for each resilience sub-dimension score to identify the level of consensus for each group. The IRA explores how each rater (i.e. FGD participants in this case) assign similar values for each item (Gisev et al. 2013). Therefore the IRA is a valuable analytical tool that allows for the quantification of levels of consensus and dissent with regards to a particular resilience element (Burke and Dunlap 2002; O'Neill 2017). To estimate the consensus and dissent for each resilience element, we use the critical values[5] for the 5-point scale (Smith-Crowe et al. 2013). The IRA is limited to the resilience element since higher levels are calculated and not directly derived from the FGD participants.

6.3 Results

6.3.1 Perceived Community Resilience by Study Community

Table 4 contains the overall mean resilience scores across each of the ten communities. Generally, our findings suggest that the perceived community resilience to floods and droughts is "very low" to "low" across all study communities. The mean resilience scores in drought-prone communities (1.99) are slightly higher compared to flood-prone communities (1.96). When stratified by community, Yoggu (drought-prone) and Baluefili (flood-prone) have the highest overall resilience scores, with both communities recording a mean value of 2.37. On the otherhand, Daboshe (drought-prone) and Zowayili (flood-prone) have the lowest overall resilience scores with mean values of 1.64 and 1.69, respectively. However, there is no considerable variation in the resilience scores across the ten communities. In terms of resilience dimensions, the perception of ecological resilience is higher (drought-prone = 2.10; flood-prone = 2.07) compared to socio-economic resilience (drought-prone = 2.07; flood-prone = 1.950) and engineering resilience (drought-prone = 1.83; flood-prone = 1.86) (Table 4). These results strongly point to the fact that respondents rate their community resilience to floods and droughts as rather weak. One reason

[5]Critical values are assigned depending on the distribution skewness, with slight skew at 0.69, moderate skew at 0.49, and heavy skew at 0.42. Level of skewness are measured with absolute values ranging between 0–0.5, as slight skew; 0.5–1 as moderate skew; and > 1 as heavy skew (Bulmer 1979).

Table 4 Perceived community resilience by location and study community

	Mean community resilience score			
	Overall	Ecological	Engineering	Socio-economic
Drought-prone	*1.99*	*2.10*	*1.83*	*2.05*
Yoggu	2.37	2.43	2.39	2.30
Cheshagu	1.87	1.66	1.76	2.18
Kpalgun	1.94	2.14	1.62	2.05
Daboshe	1.64	1.79	1.43	1.70
Zagua	2.16	2.51	1.91	2.06
Fihini	1.99	2.10	1.89	1.98
Flood-prone	*1.96*	*2.07*	*1.86*	*1.95*
Bankpama	2.05	2.22	1.98	1.95
Baleufili	2.37	2.40	2.26	2.46
Chietanga	1.73	1.90	1.57	1.72
Zowayili	1.69	1.75	1.63	1.68

accounting for this is inability to recall since the impacts of the events are not recorded.

FGD participants in drought-prone communities such as Yoggu, Fihini and Daboshe constantly drew our attention to how scarce and unproductive arable land has resulted in the adoption of intensive and uninterrupted farming practices. Most of the discussants revealed that, maize is extensively cultivated largely due to the availability of drought-resistant varieties, lower labour requirement and quick maturation compared to other staple crops such sorghum and yam. Although there is no substantial spatial variation in ecological resilience levels across locations, some communities such as Chietanga and Zowayili (flood-prone) and Daboshe and Fihini (drought-prone) appear to have the lowest overall ecological resilience score (Table 5). For example, an elderly woman in Daboshe mentioned that:

> Every year, our land becomes infertile and our vegetables do not survive. All this is happening because the weather is becoming hotter and hotter and the rains are not coming as we expect. The small dam is dry all year round so we cannot even plant anything in the dry season. There is nothing we can do about the ongoing situation so please help us.

6.3.2 *Perceived Community Resilience by Dimension, Sub-dimension and Element*

Table 5 contains the mean resilience scores by community type for the resilience dimensions and sub-dimensions. The outcomes reveal a consensus among communities on their low resilience to floods and droughts. The ecological dimension has the highest mean score (2.1) among the three resilience dimensions in both the drought- and flood-prone areas. However, the average score still ranges between "very low" and "low", with the mean scores for the three sub-dimensions

Table 5 Perceived community resilience by resilience dimension, sub-dimension and element

	Drought-prone communities	Flood-prone communities
Ecological	**2.1**	**2.1**
Environmental management	***2.2***	***2.0***
Protected and conservation areas	2.3	2.3[a]
Environment functions effectively	2.2[a]	1.9[a]
Species diversity richness	1.9[a]	2.1
Quality soil and vegetation	2.1[a]	1.8[a]
Ecosystem services access	2.3[a]	1.8[a]
Land use and cover	***2.1***	***2.2***
Land tenure	2.3[a]	2.3
Healthy rangelands	2.1[a]	2.4[a]
Cultivable lands	2.3[a]	2.2
Landscape elevation	1.9	2.0
Agriculture systems	***2.0***	***2.0***
Drought-tolerant crop	1.7	1.7[a]
Flood-resistant crop	1.8	2.3[a]
Irrigation system	2.0[a]	2.3[a]
Crop storage system	2.3	2.0[a]
Dry season farming	2.3[a]	2.1[a]
Fodder and forage availability	2.2	2.1[a]
Food supply	1.7	1.8[a]
Engineering	**1.8**	**1.9**
DRR knowledge and preparedness	***1.9***	***1.8***
Climate change impacts	*2.3*[a]	*2.0*[a]
Local disaster management	*1.8*[a]	*1.7*[a]
Local DRR compliance	*1.9*	*2.1*
Drought impacts	*1.6*	*1.8*[a]
Flood impacts	*1.8*[a]	*1.9*[a]
Recovery potential	*1.7*	*1.6*
Physical infrastructure	***1.8***	***1.8***
Flood evacuation	*1.9*[a]	*1.8*[a]
Road access	*1.9*	*1.8*[a]
Flood resist house	*1.7*	*1.8*[a]
Health centres	*1.6*[a]	*1.8*[a]
Electricity access	*1.6*	*1.9*[a]
Potable water	*2.0*[a]	*2.1*[a]
Planning systems	***1.9***	***1.9***
DRR dissemination	*1.9*[a]	*1.9*[a]
Risk and cost sharing	*1.6*	*1.9*[a]
Early warning drought	*1.9*[a]	*2.2*[a]
Early warning flood	*1.8*[a]	*1.7*[a]
Trained local population	*2.0*	*1.7*[a]
DRR in livelihoods	*2.0*[a]	*2.0*

(continued)

Table 5 (continued)

Socio-economic	**2.1**	**1.9**
Social capital and risk management	***2.2***	***1.9***
Livelihood diversification	2.1[a]	1.8[a]
Social support network	2.1[a]	1.7[a]
Intra-community DRR	2.2[a]	1.8[a]
Inter-community DRR	2.2[a]	1.6[a]
Effective leadership	2.2[a]	2.1[a]
External agency	2.3[a]	2.0[a]
Disaster memory	2.6	1.8[a]
Active labour force	2.3[a]	2.1[a]
Financial instrument	***1.9***	***2.0***
Money transfers	1.7[a]	2.1[a]
Savings and credit	1.7	2.0[a]
Livestock income	1.9[a]	2.0[a]
Ecosystem income	2.3	1.6
Petty commerce	2.0	2.1[a]
Other incomes	1.8[a]	2.2[a]
Livelihood asset base	***2.0***	***2.0***
Livestock holdings	2.0[a]	1.7
Secured food	2.0[a]	2.0[a]
Market and trade	2.2	2.7
Secured water	1.8[a]	1.8[a]

Note: [a]Denotes consensus between communities

(i.e. environmental management, land use and cover and agriculture productivity) being still "very low". This suggests that most of these resilience elements are of major concern, including protected/conservation areas, understanding of local ecosystem functions, access to critical ecosystem services, rangeland availability, arable land availability, secure food supply and nutrition and cultivation of flood- and drought-tolerant crops.

With regards to the perceived engineering resilience, the overall mean score of 1.8 and 1.9 for drought- and flood-prone locations can be considered as quite low. The mean scores for the three sub-dimensions of engineering resilience (i.e. DRR knowledge and preparedness, physical infrastructure and planning systems parameters) are very low, with a widespread consensus among households (Table 5). This low score was due to the perceived weaknesses of the listed engineering resilience elements assessed, such as (a) the lack of critical infrastructure such as flood evacuation zones, well-maintained and accessible roads, and functioning health centres and (b) the lack of reliable early warning systems and trained locals to assist in disaster management. As a result, the study communities need a very long time to recover following droughts and floods, and in some cases, they experience new disasters while being in the process of recovery.[6]

The overall mean scores of the socio-economic resilience dimension are not markedly different to the score of ecological and engineering resilience (Table 4). FGD participants scored sub-dimensions such as social capital and risk elements (i.e. livelihood diversification, social support system, disaster memory) slightly higher than financial instrument (i.e. money transfers and remittances, savings and credits, livestock income, petty commerce) and the livelihood asset base (i.e. livestock holding, secured food during disaster, secured water for domestic use) (Table 5).

The weak socioeconomic resilience is reflected in the lack of alternative income sources and low asset ownership, which both suggest the low capacity to meet household needs and eventually the inability to enhance community resilience to disasters. Related to the low livelihood asset base is the fact that households in the study areas prefer to keep livestock, such as cattle as a means of savings, a safety net for contingencies, a self-reproducing asset and a source of income.[7]

6.3.3 Perceived Community Resilience by Age Group

Figures 2–4 show the mean perceived community resilience scores and consensus ratings among the three age groups (i.e. elderly men, elderly women and young adults). To highlight more themes and compare consensus among different groups on resilience elements, we analyse below scores by resilience dimension and sub-dimension through the summation of scores for various resilience elements. Figures 2–4 help to identify and classify the elements of resilience with mean scores greater than 2 (i.e. scores closer to moderate), whilst also exploring for possible commonalities among groups.

Figure 2 suggests that all three age groups perceive community resilience to droughts and floods as "very low" and "low". Elderly men (drought = 2.3); (flood = 2.4) and elderly women (drought = 2.1; flood = 2.0) recorded slightly higher mean resilience scores compared to young adults (drought 1.8; flood 1.8). The further analysis of the three sub-dimensions of ecological resilience (i.e. environmental management, land use and cover, agricultural productivity) reveals similar patterns and trends for the different age groups (i.e. elderly men report higher mean scores, followed by elderly women, and young adults). These

[6]It is well known that the synergistic effects of different disasters can have even higher adverse outcomes (Yiran and Stringer 2016), and reducing further the resilience of local communities in semi-arid Ghana.

[7]Even though not specifically mentioned, cattle can also support farm intensification, as they are a major source of nutrients for agriculture in the form of manure and crop residues from feed (Mortimore and Adams 2001).

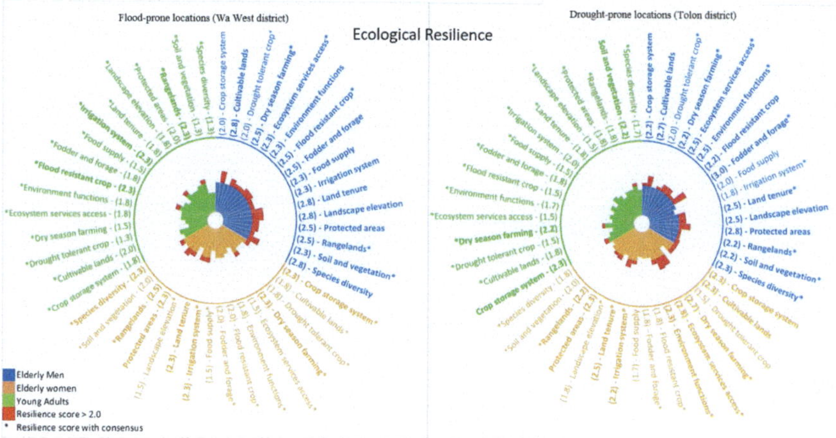

Fig. 2 Ecological resilience scores and consensus by age group

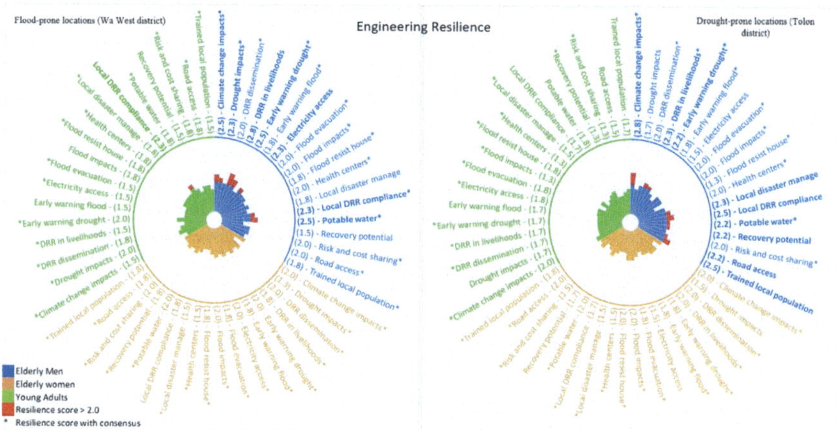

Fig. 3 Engineering resilience scores and consensus by age group

results indicate a consensus among all social groups about the low community resilience against droughts and floods (Fig. 2).

The overall mean engineering resilience scores is lower compared to ecological resilience for all groups. Engineering resilience scores ranged between 1.6 and 2.1, with the aggregated mean scores in drought-prone locations being 1.8, and in flood-prone communities being 1.9. The differences in mean resilience scores are minor between age groups, although elderly men scores are higher compared to elderly women, and young adults in both drought- and flood-prone communities. Much like

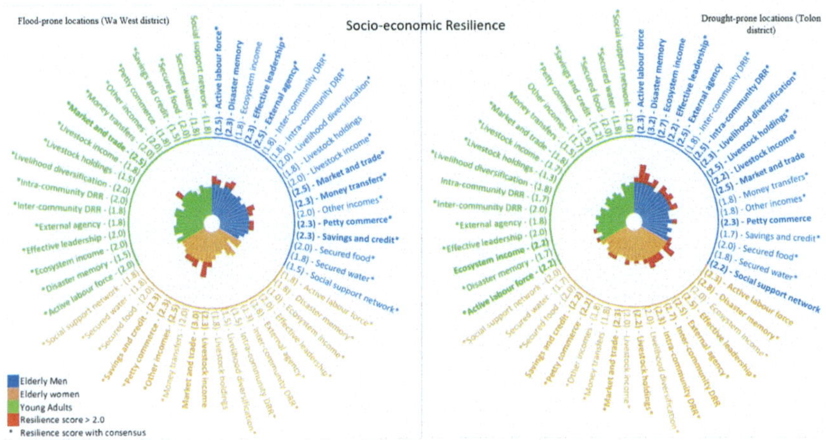

Fig. 4 Socio-economic resilience scores and consensus by age group

ecological resilience, elderly men in flood- and drought-prone communities consistently report higher scores for all the three sub-dimensions of engineering resilience, including DRR knowledge and preparedness, physical infrastructure and planning systems. Elderly women report moderately higher scores compared to young adults (Fig. 3). One of the resilience elements mentioned by FGD participants includes the decline in use of local farming technologies. As a FGD participant from Kpalgun mentioned:

> I have been employing for many years the *zai* method, which involves digging pits in the soil to store water and compost. Even though, this local soil improvement technology has served many people in this community well since our forefathers, its use is in decline partly because of the prolonged drought.

Community perceptions follow similar patterns for socio-economic resilience as for ecological and engineering resilience. All age groups within drought-prone (2.1) and flood-prone (1.9) communities perceive their socio-economic resilience to be rather low (Fig. 4). Within the Tolon district, elderly men (2.2) and elderly women (2.2) both perceive a "low" to "moderate" socio-economic resilience compared to young adults that perceive a "low" (1.8) socioeconomic resilience. In flood-prone communities (Wa West district), elderly men report a higher score (2.1), with elderly women and young adults reporting "low" socioeconomic resilience (1.9). Finally, there appears to be a minor difference between elderly men and elderly women (but not younger adults) for the three sub-dimensions of socioeconomic resilience (i.e. social capital and risk management, financial instrument and livelihood asset base).

6.4 Discussion

6.4.1 Synthesis of Findings

The results of this study reveal that local communities in semi-arid Ghana have low resilience to floods and droughts, and thus a limited ability to deal adequately with their impacts (Table 4, Figs. 2–4). This low perceived community resilience is largely consistent between groups and localities and further suggests the high vulnerability of local communities in semi-arid Ghana to the interplay of multiple natural and human shocks and stresses (Songsore 2011). Furthermore, they corroborate other studies in semi-arid Ghana (Antwi et al. 2014; Nyantakyi-Frimpong and Bezner-Kerr 2015), providing further evidence about context-specific vulnerabilities to climate hazards and the underlying biophysical conditions and socio-economic disparities (Acheampong et al. 2014; Antwi-Agyei et al. 2012).

The, generally low community resilience, combines with the low adaptive capacity and high exposure and sensitivity to seasonal flooding to explain the very negative effects that such hazards have for local livelihoods (Musah and Oloruntoba 2013; Musah and Akai 2014; Owusu et al. 2016; Armah et al. 2011; Wossen and Berger 2015). For example, a large fraction of the more than 325,000 people negatively affected by the devastating 2007 floods in semi-arid Ghana were located in the study areas (NADMO 2011). For similar reasons, the persistent droughts across the study sites have been identified as a major reason behind the prevailing food insecurity, malnutrition and labour outmigration (Jarawura 2013; Lolig et al. 2014; Kusakari et al. 2014) (see Chap. 2 Vol. 2 for similar case in Uganda). All these findings reinforce partly the widely accepted proposition that the semi-arid areas of Ghana lag behind the rest of the country despite their potential due to such economic, social and ecological vulnerabilities (Oteng-ababio et al. 2017).

However, despite the general consensus in resilience, there are some interesting variations between age groups for some resilience dimensions, sub-dimensions and elements (Figs. 2–4). Overall, elderly respondents (both men and women) tend to report relatively higher scores for more resilience elements compared to young adults (Figs. 2–4), possibly due to the fact that many young adults may fail to appreciate and value the ecological knowledge, practices and norms that could help them cope with floods and droughts. This might reflect their comparatively limited engagement with (and experience of) the local agro-ecosystem due to modernization, and especially the influence of western education and religion (Boafo et al. 2016). A second, and closely related, reason may be the widespread outmigration of young adults from the semi-arid regions of Ghana to the larger cities of southern Ghana such as Kumasi and Accra (Songsore 2011; Awumbila et al. 2014).

However, the low perceived community resilience among young adults could be one of the factors influencing their outmigration. This further suggests that in order to improve socioeconomic resilience in the rural communities of semi-arid Ghana, young adults must be provided with tangible opportunities for decent, stable and viable occupations. These livelihood options should go beyond the prevailing smallholder-based farming systems.

6.4.2 Possible Practical Applications

Considering the similar findings across all groups and resilience dimensions/elements (Sect. 6.4.1), there is a dire need to increase the adaptive capacity of the vulnerable local communities. We argue that both the data generated from this study and the underlying research approach can be used to inform the development of interventions for increasing community resilience to floods and droughts.

First, our observations during fieldwork and the study findings suggest that local communities have indeed a rather good understanding of their (lack of) resilience for specific sub-dimensions and elements. Such insights can be used to identify priority thematic areas for interventions. At the same time, this makes the case that collective and participatory processes can inform the development of local adaptation options for floods and droughts. Second, the knowledge about the disparities among groups and localities can provide finer-grain information to develop more tailored interventions for particularly vulnerable regions and groups.

Our overall methodology was transdisciplinary, participatory and context-specific, relying heavily on baseline information from vulnerable socio-ecological systems. This bottom-up approach, when employed properly, allows for incorporating diverse community viewpoints and lived experiences in the resilience assessment process. Apart from providing useful information for the development of appropriate interventions, such information can also assist in during the decision-making. By involving communities in this manner, it gives offers them greater power and a voice in decisions that truly affect them. This bodes well with recommendations that interventions seeking to increase the resilience of local communities to climatic hazards should allow for the participation of vulnerable groups during the design, implementation and evaluation of such interventions (UNFCCC 2018; Norris et al. 2008). One added benefit of this methodology is that it is rapid, requires little expertise and can be implemented with few resources. In this sense, it can act as a quick pre-planning tool to identify broad community patterns and priority intervention areas. Furthermore, it is rather flexible and can offer complementary information to seasonal prediction models and other more technical tools aiming to build resilience to climatic hazards.

However, despite their strengths, our results and underlying methodology have limitations. First, there seems to be a trade-off between speed/cost-effectiveness and accuracy. This trade-off is quite common for analytical techniques that depend on self-reported and qualitative data like the one reported in this study. Apart from the general methodological issues, there should be some caution when generalizing the actual results of this study. We used non-probabilistic sampling, which possibly resulted in the unequal representation of some respondents in the FGDs. Second, in our attempt to reduce the number of FGDs per site (and thus the needed time/resources), we did not include a distinct FGD group consisting of respondents above 70 years old, a group that is considered as extremely vulnerable to climatic hazards in the study area (Zimmerman et al. 2007; Haq et al. 2008). Hence, there is large variation in respondent ages within the elderly groups, which has possibly caused some of the large perception variation in these groups and the generally higher resilience scores.

6.4.3 Policy Implications and Recommendations

Overall, our study reflects the current discourse pointing to the need for identifying and implementing context-specific interventions when seeking to enhance local resilience to climatic hazards. Such interventions, if designed and implemented properly, can have positive ripple outcomes for national development priorities and the progress to meet the sustainable development goals (SDGs). For example, Ghana's medium-term development agenda (Ghana Shared Growth and Development Agenda II) has explicitly identified climate variability and climatic hazards such as floods and droughts to be a "major threat to national development" (NDPC 2014:82). Similarly, enhancing resilience and building adaptive capacity to climatic hazards has been encapsulated in SDG13 (Take urgent action to combat climate change and its impacts), and especially Target 13.1 on "urgent action to combat climate change and its impacts including strengthening resilience and adaptive capacity to climate-related hazards and disasters". To all intents and purposes, building resilience to climate change (and its impacts) is a cross-cutting theme across multiple SDGs (Yonehara et al. 2017), and especially SDG1 (no poverty), SDG6 (clean water and sanitation) and SDG11 (sustainable cities and communities), among others (UN 2015).

Integrative, collaborative and participatory approaches to resilience planning could thus have a positive effect in the ongoing efforts to meet such policy objectives. Below we discuss four possible recommendations for enhancing the resilience of local communities to climatic hazards. Even though these recommendations are mostly relevant to the specific context of this study (i.e. droughts and floods in semi-arid Ghana), they can have some applicability to other SSA contexts where climate change is a major driver of environmental and socioeconomic change (e.g. Chaps. 2 and 3 Vol. 2).

First, there is a need to introduce and improve the predictive power of early warning systems. The outputs of such systems should be combined with bottom-up information about the adaptation needs/priorities of local communities and the factors affecting the resilience to climatic hazards at the district, community and household level. This information should be combined for the development, monitoring and evaluation of intervention strategies, while combining the viewpoints of different stakeholders (Parnewll 2011; Yonehara et al. 2017; Saito et al. 2018a, b). To forge a broader understanding of such techniques and information it would be important to build capacity among stakeholders and researchers. A key step would be to incorporate resilience concepts into educational policies and curricula to equip multiple stakeholders with the practical knowledge and necessary skillsets to conduct and appreciate the outputs of resilience assessments (Apronti et al. 2015).

Second, it should be ensured that national fiscal, planning and implementation mechanisms towards enhancing resilience and reducing disaster risk target vulnerable communities and groups. Relevant government departments and agencies such as the Ministry for Finance, the Environmental Protection Agency and the Ministry for Food and Agriculture that have clear mandates and defined roles within the national development agenda should be responsive and proactive in this regard.

Third, there should be efforts to promote alternative livelihood options in order to reduce the negative effects of climate variability on vulnerable groups and particularly women, elderly and the youth. A key element of such efforts should be to diversify from primary economic sectors such as agriculture, livestock and forestry, as many studies have shown that the ongoing climate variability and change reduce the capacity of semi-arid ecosystems to supply such ecosystem services that are essentially the basis of local livelihoods (Boafo et al. 2014; Armah et al. 2011). This would require coordinated policy actions between the national and local government, as well as with broader stakeholder groups from the private sector, civil society and international organizations and donors.

Fourth, a practical approach towards disaster prevention and reduction could be to integrate ecosystem-based adaptation (EbA) approaches into community resilience adaptation and disaster risk-reduction strategies. EbA approaches, with their strong focus on landscapes and ecosystem services, could provide multiple benefits to enhance the rather low socioeconomic, ecological and engineering resilience of the vulnerable local communities, while directly contributing to several SDGs including SDG1 (no poverty), SDG2 (zero hunger) and SDG15 (life on land). Due to their multi-functionality EbA approaches could also reduce the investments needed for disaster response, thus offering cost-effective solutions to enhance resilience to climatic hazards.

6.5 Conclusions

This chapter presented the results of a community resilience assessment to floods and droughts using a bottom-up participatory approach that elicited the perceptions of different groups in semi-arid Ghana. The results suggest that community resilience is consistently low between the studied groups and localities, as well the individual dimensions and elements of resilience. This overwhelming agreement clearly shows that local communities and households in the studied areas (and semi-arid Ghana more generally) are far from capable to withstand the negative outcomes of climatic hazards, now and in the future. Considering the ongoing climate change and variability, it becomes ever so urgent to improve their resilience across all studied dimensions (i.e. ecological, engineering, socioeconomic), and underlying elements. Such a holistic understanding and bottom-up evaluation of community resilience are imperative to assist the design and implementation of effective intervention strategies. This is not the least because bottom-up approaches can capture better the experiences and needs of local communities that have intimate experience with such climatic hazards. However, we should also point that our results are also based on self-reported community perceptions and are thus highly subjective and prone to different biases. For a more in-depth and comprehensive investigation of the status of community resilience to floods and droughts, the factors affecting it and the possible interventions, these results should be triangulated with other available sources of information.

Acknowledgements The research within this chapter was conducted within the framework of the "*Enhancing Resilience to Climate and Ecosystem Changes in Semi-Arid Africa: An Integrated Approach*" project (CECAR-Africa). We acknowledge financial support from the Japan International Cooperation Agency (JICA) and the Japan Science and Technology Agency (JST).

References

Acheampong E, Ozor N, Owusu E (2014) Vulnerability assessment of Northern Ghana to climate variability. Climate Change 126(12):31–44

Adger W (2000) Social and ecological resilience: are they related? Prog Hum Geogr 24(3):347–364

AGRA (2017) Africa agriculture status report: the business of smallholder agriculture in Sub-Saharan Africa. Nairobi, Kenya: Alliance for a Green Revolution in Africa (AGRA), (Issue 5). https://doi.org/http://hdl.handle.net/10568/42343

Antwi E, Otsuki K, Saito O et al (2014) Developing a community-based resilience assessment model with reference to Northern Ghana. J Integr Disaster Risk Manage 4. 10.5595

Antwi-Agyei P, Fraser EDG, Dougill AJ, Stringer LC, Simelton E (2012) Mapping the vulnerability of crop production to drought in Ghana using rainfall, yield and socioeconomic data. Appl Geogr 32:324–334

Apronti PT, Saito O, Otsuki K, Kranjac-Berisavljevic G (2015) Education for disaster risk reduction (drr): linking theory with practice in Ghana's Basic Schools. Sustainability 7(7):9160–9186. https://doi.org/10.3390/su70x000x

Argon P (2014) Developing a model and tool to measure community disaster resilience. Aust J Emerg Manage 29:12–16

Armah FA, Odoi JO, Yengoh GT, Obiri S, Yawson DO, Afrifa EKA (2011) Food security and climate change in drought-sensitive Savannah Zones of Ghana. Mitig Adapt Strateg Glob Chang 16(3):291–306

Armah FA, Yawson DO, Genesis TY, Odoi JO, Ernest K, Afrifa A (2010) Impact of floods on livelihoods and vulnerability of natural resource dependent communities in Northern Ghana. Water 2(2):120–139

Awumbila M, Darkwah A, and Teye JK (2015) Migration, intra-household dynamics and youth aspirations. Unpublished Paper. Migrating out of Poverty, University of Sussex

Awumbila M, Owusu G and Teye JK (2014). Can rural-urban migration into slums reduce poverty? Evidence from Ghana. Migrating out of Poverty Working Paper 13. University of Sussex

Boafo YA, Osamu S, Takeuchi K (2014) Provisioning ecosystem services in rural savannah landscapes of Northern Ghana: an assessment of supply, utilization and drivers of change. J Disaster Res 9(4):501–515

Boafo YA, Saito O, Jasaw GS, Otsuki K, Takeuchi K (2016) Provisioning ecosystem services-sharing as a coping and adaptation strategy among rural communities in Ghana's semi-arid ecosystem. Ecosyst Serv 19:92–102

Bruneau M, Chang S, Eguchi R, Lee G, O'Rourke T, Reinhorn A, Shinozuka M, Tierney K, Wallace W, von Winterfelt D (2003) A framework to quantitatively assess and enhance the seismic resilience of communities. EERI Spectra J 19(4):733–752

Buckle P (2006) Assessing social resilience. In: Paton D, Johnston D (eds) Disaster resilience: An integrated approach. Charles C. Thomas, Springfield, IL, pp 88–103
Bulmer MG (1979) Principles of statistics. Dover, New York
Burke MJ, Dunlap WP (2002) Estimating Interrater Agreement with the Average Deviation Index: A User's Guide. Organ Res Methods 5:159–172
Burton CG (2015) A validation of metrics for community resilience to natural hazards and disasters using the recovery from Hurricane Katrina as a case study. Ann Am Assoc Geogr 105 (2015):67–86
Chen C, Noble I, Hellmann J, Coffee J, Murillo M, Chawla N (2015) University of notre dame global adaptation index: country index. University of Notre Dame Global Adaptation Index Country: Country index technical report, 46
Cutter SL, Barnes L, Berry M, Burton CG, Evans E, Tate EC, Webb J (2008) A place-based model for understanding community resilience to natural disasters. Glob Environ Chang 18:598–606
EPA (Environmental Protection Agency) (2012). Flood and drought risk mapping in Ghana; EPA: Accra, Ghana
Eriksen S, Marin A (2015) Sustainable adaptation under adverse development? Lessons from Ethiopia. In: Inderberg TH, Eriksen SH, O'Brien K, Sygna L (eds) Climate change adaptation and development: changing paradigms and practices. Routledge, Oxford, pp 178–199
Folke C, Colding J, Berkes F (2003) Synthesis: building resilience and adaptive capacity in social–ecological systems. In: Berkes F, Colding J, Folke C (eds) Navigating social–ecological systems: building resilience for complexity and change. Cambridge University Press, Cambridge, UK, pp 352–387
Gaillard J (2007) Resilience of traditional societies in facing natural hazards. Disaster Prev Manag 16(4):522–544
Ghana Statistical Service (2012) 2010 Population and housing census: summary report of final results. Ghana Statistical Service, Accra
Ghana Statistical Service (GSS) (2014) Ghana living standards survey round 6. GSS, Accra, Ghana
Gisev N, Bell JS, Chen TF (2013) Interrater agreement and interrater reliability: key concepts, approaches, and applications. Res Soc Adm Pharm 9(3):330–338
Gunderson LH, Holling CS (eds) (2002) Panarchy: understanding transformations in human and natural systems. Island Press, Washington, DC
Haq G, Whitelegg J, Kohler M (2008) Growing old in a changing climate: meeting the challenges of an ageing population and climate change. Stockholm Environment Institute, Stockholm
Higgins S, Scheiter S (2012) Atmospheric CO_2 forces abrupt vegetation shifts locally, but not globally. Nature 488(7410):209–212
Holling CS (1973) Resilience and stability of ecological systems. Annu Rev Ecol Syst 4:1–23
IPBES (2018) Summary for policymakers of the assessment report on land degradation and restoration of the Intergovernmental science-policy platform on biodiversity and ecosystem services. Intergovernmental Science-Policy Platform on Biodiversity and Ecosystem Services, Bonn, Germany
IPCC (2014) Contribution of Working Groups I, II and III to the Fifth Assessment Report of the Intergovernmental Panel on Climate Change (2014). Climate Change 2014: Synthesis Report. IPCC, Geneva, Switzerland: Cambridge University Press
IRIN (2017) Recovery pledges for hurricane-ravaged Caribbean are a drop in the ocean. https://www.irinnews.org/analysis/2017/11/24/recovery-pledgeshurricane-ravaged-caribbean-are-drop-ocean
Janssen MA, Schoon ML, Ke W, Börner K (2006) Scholarly networks on resilience, vulnerability and adaptation within the human dimensions of global environmental change. Glob Environ Chang 16(3):240–252
Jarawura FX (2013) Drought and Migration in Northern Ghana. Doctor of Philosophy Thesis Presented to The University of Ghana Centre for Migration Studies, Accra

Jasaw G, Saito O, Gasparatos A, Shoyama K, Boafo Y, Takeuchi K (2017) Ecosystem services trade-offs from high fuelwood use for traditional shea butter processing in semi-arid Ghana. Ecosyst Serv 27:127–138

Kusakari Y, Asubonteng K, Jasaw G, Dayour F, Dzivenu T, Lolig V, Donkoh S, Obeng KF, Gandaa B, Kranjac-Berisavljevic G (2014) Farmer-perceived effects of climate change on livelihoods in Wa West District, Upper West Region of Ghana. J Disaster Res 9:516–528

Langridge R, Christian-Smith J, Lohse KA (2006) Access and resilience: analyzing the construction of social resilience to the threat of water scarcity. Ecol Soc 11(2):18

Lehmann C, Anderson T, Sankran M, Higgins S, Archibald S, Hoffmann W et al (2014) Savanna vegetation-fire-climate relationships differ among continents. Science 343(6170):548–552

Lolig V et al (2014) Households' coping strategies in drought- and flood-prone communities in Northern Ghana. J Disaster Res 9(4):542–553

Ludwig D, Walker B, Holling CS (1997) Sustainability, stability, and resilience. Conserv Ecol. [online] 1(1):7

Magis K (2010) Community resilience: an indicator of social sustainability. Soc Natur Resour 23 (5):401–416

Magnan AK, Skipper ELF, Burkett M, Bharwani S, Burton I, Eriksen S, Ziervogel G (2016) Addressing the risk of maladaptation to climate change. Wiley Interdiscip Rev Clim Chang 7 (5):646–665. https://doi.org/10.1002/wcc.409

Manyena S (2006) The concept of resilience revisited. Disasters 30(4):434–450

Mortimore MJ, and Adams WM. (2001). Farmer adaptation, change and 'crisis' in the Sahel. Global Environ Change—Human and Policy Dimensions 11: 49–57

Musah BAN, Akai CY (2014) Effects of flood disasters on livelihood coping mechanism in Tolon/ Kumbumgu district of northern region of Ghana. Int J Agric Policy Res 2(1):033–040

Musah BAN, Oloruntoba A (2013) Effects of seasonal floods on households' livelihoods and food security in Tolon/Kumbungu of the northern region, Ghana. Am J Res Commun 1(8):160–171

NADMO (National Disaster Management Organisation). Annual Report 2011; National Disaster Management Organisation: Accra, Ghana

Narain U, Margulis S, Essam T (2011) Estimating costs of adaptation to climate change. Clim Pol 11(3):1001–1019. https://doi.org/10.1080/14693062.2011.582387

NASAC (2015) Climate change adaptation and resilience in Africa. Recommendations to policymakers, Network of African Science Academies

NDPC (National Development Planning Commission) (2014) Medium-term national development policy framework: Ghana shared growth and development agenda II 2014–2017: vol II: policy framework. www.ndpc.gov.gh/downloads

Niang I, Ruppel OC, Abdrabo MA, Essel A Lennard C, Padgham J and Urquhart P (2014) Africa. In: Climate Change 2014: Impacts, Adaptation, and Vulnerability. Part B: Regional Aspects. Contribution of Working Group II to the Fifth Assessment Report of the Intergovernmental Panel on Climate Change. Cambridge University Press, Cambridge, UK

Norris FH, Stevens SP, Pfefferbaum B, Wyche KF, Pfefferbaum RL (2008) Community resilience as a metaphor, theory, set of capabilities, and strategy for disaster readiness. Am J Community Psychol 41:127–150

Nyantakyi-Frimpong H, Bezner-Kerr R (2015) The relative importance of climate change in the context of multiple stressors in semi-arid Ghana. Glob Environ Chang 32:40–56

O'Neill TA (2017) An overview of interrater agreement on likert scales for researchers and practitioners. Front Psychol 8:777

Ohene E (2019) Rosewood, the timber to kill for. Retrieved from https://www.graphic.com.gh/features/features/ghana-news-rosewood-the-timber-to-kill-for.html. June 27, 2019

Oteng-ababio M, Mariwah S, Kusi L (2017) Is the underdevelopment of northern Ghana a case of environmental determinism or governance crisis? Ghana J Geogr 9(Special Issue):5–39

Owusu AB, Jakpa JT, Awere KG (2016) Smallholder farmers' vulnerability to floods in the Tolon District, Ghana. Interdiscip Environ Rev 17(3/4):286–311

Parnewll M (2011) Agropolitan and bottom-up development. In: Desai V, Potter RB (eds) The companion to development studies, 2nd edn. Chennai Micro Print, Chennai, India

Pasteur K (2011) From Vulnerability to Resilience: A framework for analysis and action to build community resilience. Vasa 128. http://practicalaction.org/media/preview/9654/lng:en

Pelling M (2003) The vulnerability of cities: natural disasters and social resilience. Earthscan, London

Peprah K (2015) Land degradation is indicative: proxies of forest land degradation in Ghana. J Degrad Min Lands Manage 3(1):477–489. (2015): 477

Perrings C (2006) Resilience and sustainable development. Environ Dev Econ 11(4):417–427

Rademacher-Schulz C and Mahama ES (2012). "Where the rain falls", Project case study: Ghana. Results from Nadowli District, Upper West region, Ghana. (Report No. 3). United Nations University Institute for Environment and Human Security (UNU-EHS), Bonn

Rockström J, Falkenmark M, Karlberg L et al (2009) Future water availability for global food production: The potential of green water for increasing resilience to global change. Water Resour Res 45:1–16

Saito O, Boafo YA, Jasaw GS et al (2018a) The Ghana Model for resilience enhancement in Semiarid Ghana: conceptualization and social implementation. In: Saito O, Kranjac-Berisavljevic G, Takeuchi KA, Gyasi E (eds) Strategies for building resilience against climate and ecosystem changes in Sub-Saharan Africa. Science for sustainable societies. Springer, Tokyo

Saito O, Kranjac-Berisavljevic G, Takeuchi K, A. Gyasi E. (eds) (2018b) Strategies for building resilience against climate and ecosystem changes in Sub-Saharan Africa. Science for Sustainable Societies. Springer, Tokyo

Schraven B (2010) Irrigate or migrate? Local livelihood adaptation in Northern Ghana in response to ecological changes and economic challenges (PhD dissertation, University of Bonn, Germany)

Serdeczny O, Adams S, Baarsch F, Coumou D, Robinson A, Hare W et al (2016) Climate change impacts in Sub-Saharan Africa: from physical changes to their social repercussions. Reg Environ 17(6):1585–1600. https://doi.org/10.1007/s10113-015-0910-2

Simpson DM (2006) Indicator issues and proposed framework for a disaster preparedness index (Dpi), Unpublished Draft report to the Fritz Institute Disaster Preparedness Assessment Project

Smith-Crowe K, Burke MJ, Kouchaki M, Signal SM (2013) Assessing interrater agreement via the average deviation index given a variety of theoretical and methodological problems. Organ Res Methods 16:127–151

Songsore J (2009) Regional Development in Ghana: The theory and the reality. Woeli Publishing Services, Accra, Ghana

Songsore J (2011) Regional development in Ghana: the theory and reality. Woeli Publishing Services, Accra

Stanturf JA, Warren Jr., ML, Charnley S, Polasky SC, Goodrick SL, Armah F and Nyako YA (2011) Ghana: climate change vulnerability and adaptation assessment. United States Agency for International Development (USAID)

Tobin G (1999) Sustainability and community resilience: the holy grail of hazards planning? Global Environ Change B Environ Hazard 1(1):13–25

Turnbull M et al (2013) Toward resilience: a guide to disaster risk reduction and climate change adaptation. Practical Action Publishing Ltd.. http://reliefweb.int/sites/reliefweb.int/files/resources/ECB-toward-resilience-Disaster-risk-reduction-Climate-Change-Adaptation-guide-english.pdf

UN (2015) Transforming our world: the 2030 agenda for sustainable development. https://Sustainabledevelopment.Un.Org/Content/Documents/7891Transforming%20Our%20World.pdf. doi:https://doi.org/10.1007/s13398-014-0173-7.2

UNDP/UNSO (1997) Aridity zones and dryland populations: an assessment of population levels in the world's drylands with particular reference to Africa. UNDP Office to Combat Desertification and Drought (UNSO), New York

UNFCCC (2018) Progress, experience, best practices, lessons learned, gaps, needs and support provided and received in the process to formulate and implement national adaptation plans. https://unfccc.int/sites/default/files/resource/inf01.pdf

UNISDR (2012) How to make cities more resilient: a handbook for local government leaders. Geneva: Switzerland

United Nations University (UNU), World Risk Report 2014, United Nations University Institute for Environment and Human Security and Alliance Development Works, Berlin, Germany, 2014

Walker B, Holling CS, Carpenter SR, Kinzig A (2004) Resilience, adaptability and transformability in social–ecological systems. Ecol Soc 9(2):5

Whitmarsh L (2008) Are flood victims more concerned about climate change than other people? The role of direct experience in risk perception and behavioural response. J Risk Res 11 (3):351–374

World Bank (2011) Global facility for disaster reduction and recovery (GFDRR) Country Report Ghana, http://gfdrr.org/ctrydrmnotes/Ghana.pdf [12 December 2013]

World Bank (2012) World development report 2012: gender equality and development. World Bank. © World Bank. https://openknowledge.worldbank.org/handle/10986/4391 License: CC BY 3.0 IGO

Wossen T, Berger T (2015) Climate variability, food security and poverty: Agent-based assessment of policy options for farm households in Northern Ghana. Environ Sci Policy 47:95–107

Yiran GAB, Stringer LC (2016) Spatio-temporal analyses of impacts of multiple climatic hazards in a savannah ecosystem of Ghana. Clim Risk Manag 14:11–26

Yiran GAB, Stringer LC, Attua EM, Evans AJ, Challinor AJ, Gyasi EA (2017) Spatial Mapping of Vulnerability of Multi-sector to Multi-hazards in the Savanna Ecosystem in Ghana. Reg Environ Chang 17(3):665–676

Yonehara A, Saito O, Hayashi K, Nagao M, Yanagisawa R, Matsuyama K (2017) The role of evaluation in achieving the SDGs. Sustain Sci 12:969–973

Zakaria H, Abujaja AM, Adam H, Yakub WS (2015) Does gender makes any difference in livelihoods diversification? Evidence from Northern Ghana. Int J Agric Extension Rural Dev Stud 1(1):36–51

Zakaria H (2017) The drivers of women farmers' participation in cash crop production: the case of women smallholder farmers in Northern Ghana. J Agric Educ Ext 23(2):141–158

Zhang XA, Shay R (2019) An examination of antecedents to perceived community resilience in disaster postcrisis communication. J Mass Commun Q 96(1):264–287. https://doi.org/10.1177/1077699018793612

Zimmerman, R., Restrepo C.E., Nagorsky B and Culpen M.A. (2007). Vulnerability of the elderly during natural hazard events. In *Proceedings of the Hazards and Disasters Researchers Meeting, pp.* 38–40. Boulder, Colo., July 11–12

Chapter 7
Linking Rural Livelihoods and Fuelwood Demand from Mangroves and Upland Forests in the Coastal Region of Guinea

Boubacar Siddighi Balde, Alice Karanja, Hajime Kobayashi, and Alexandros Gasparatos

7.1 Introduction

Access to reliable, affordable and modern energy is a critical precondition for enhancing human well-being and achieving the sustainable development goals (SDGs). For example, access to sustainable energy has been enshrined in a dedicated SDG that aims at "*ensuring universal access to affordable, reliable, sustainable and modern energy for all by the year 2030*" (SDG7) (UN 2015). However, apart from being a key element of SDG7, access to sustainable energy is also central to achieving other SDGs pertaining to poverty alleviation (SDG1), good health (SDG3), gender equality (SDG5), ecosystem conservation (SDG15) and climate action (SDG13), among others (IEA et al. 2019; Rosenthal et al. 2018).

However, most of the rural population in sub-Saharan Africa (SSA) relies extensively on traditional energy sources such as fuelwood, both for household use and small-scale livelihood activities (World Bank 2017) (see also Chap. 2 Vol. 1; Chap. 5 Vol. 2). About 85% of households across SSA rely on traditional biomass such as fuelwood, charcoal and animal dung in inefficient traditional cookstoves for their daily cooking and heating (IEA et al. 2019). Generally, wood-based fuel such as charcoal and fuelwood are the main energy sources, accounting for 92% of total wood consumption in SSA (World Bank 2011; IRENA 2018).

Wood-based fuel consumption is expected to further increase in SSA in the coming decades alongside with the growing urban demand for charcoal (IEA 2017) (Chap. 2 Vol. 1; Chap. 5 Vol. 2). However, it is not clear whether this

B. S. Balde (✉) · A. Karanja · A. Gasparatos
Institute for Future Initiatives (IFI), The University of Tokyo, Tokyo, Japan
e-mail: karanja.alice@ifi.u-tokyo.ac.jp; gasparatos@ifi.u-tokyo.ac.jp

H. Kobayashi
Tottori Study Center, Open University of Japan, Tottori, Japan
e-mail: khb10625@nifty.com

A. Gasparatos et al. (eds.), *Sustainability Challenges in Sub-Saharan Africa I*,
Science for Sustainable Societies, https://doi.org/10.1007/978-981-15-4458-3_7

continuous wood-based fuel demand can be met in a sustainable manner in the continent. First of all, there are substantial knowledge gaps, as most of the current and future fuelwood consumption estimates are typically based on fuel preferences and per capita consumption averages (Drigo and Nzabanita 2011; IEA 2014). Yet, there are significant discrepancies between such estimates, which are often drawn from different sources such as household surveys and simulation modelling based on secondary data (IRENA 2018). In any case, in 2011, SSA had the world's highest per capita fuelwood consumption among global regions, standing on average at 0.69 m^3/year, compared to the global average of 0.27 m^3/year (Iiyama et al. 2014). At the same time, there is a lack of reliable and updated data about the distribution of fuelwood demand among the main economic sectors in SSA. A sectoral analysis of fuelwood consumption identified that households accounted for more than 86% (537 Mm^3) of the total consumption in SSA in 1994 (FAO n.d.). The industrial sector (consisting mainly of traditional industries such as furniture production and housing construction) accounted for about 10% of the total consumption (59.4 Mm^3) and other sectors just 4% (25.9 Mm^3) (FAO n.d.).

Various studies have documented the many different negative and positive environmental and socioeconomic outcomes of fuelwood harvesting and use in SSA (Arnold et al. 2006; Bailis et al. 2015; Cerutti et al. 2015; Karanja and Gasparatos 2019). For instance, fuelwood and charcoal value chains generate employment and income to producers and traders in many SSA countries (Iiyama et al. 2014; Smith et al. 2017; Woollen et al. 2016). Approximately 200–350 jobs/TJ are generated across the charcoal value chain (FAO 2017b), while commercial biomass energy value chains employ about 13 million people across the continent (Openshaw 2010).

However, the unsustainable harvesting of trees for fuelwood and charcoal increasingly contributes to deforestation, ecosystem degradation, biodiversity loss and greenhouse gas (GHG) emissions (FAO 2007; World Bank 2017; IPBES 2018) (see Chap. 2 Vol. 1; Chap. 5 Vol. 2). Unsustainable fuelwood harvesting has been linked to deforestation and the loss of ecosystem services both at a regional level (IPBES 2018) and at many different specific local contexts (Drigo and Nzabanita 2011; Drigo 2005; Bolognesi et al. 2015; Sedano et al. 2016). For instance, Kenya loses 10.3 Mm^3 of wood from its forests every year due to overexploitation and unsustainable charcoal and fuelwood use (GoK 2013). Furthermore, unsustainable fuelwood harvesting and use accounts for a major part of GHG emissions in SSA, estimated at 1.0–1.2 Gt CO_2e/year (1.9–2.3% of global emissions) (Bailis et al. 2015).

Studies have also linked high fuelwood reliance to many negative socioeconomic outcomes related to poverty, energy insecurity, gender inequality, and unhealthy living conditions (Karanja and Gasparatos 2019). Usually women and girls undertake fuelwood procurement and collection at the household level (and often for other livelihood activities), especially in rural areas (Shankar et al. 2015; Jasaw et al. 2017). Several studies have also pointed to the health implications of biomass fuel use through traditional and inefficient cookstoves due to the high emissions of indoor air pollutants such as fine particulate matter (e.g. $PM_{2.5}$), carbon monoxide

(CO) and nitrogen oxides (NO_x) (WHO 2014; Ezzati and Kammen 2002; Bailis et al. 2007). Women, girls and young children are particularly exposed to such pollutants, as they spend substantial amounts of time cooking in poorly ventilated kitchens (Shankar et al. 2015). In fact, the World Health Organization (WHO) reported an estimated 600,000 annual premature deaths in SSA due to health complications from traditional cooking practices (WHO 2014).

Many organizations have raised alarming calls about the need to reverse the traditional reliance on fuelwood, especially for household energy use (WHO 2016; ESMAP 2015; GACC 2014; SE4All 2012). However, although biomass use is not, in itself, a cause for concern (Chap. 2 Vol. 1), it is the unsustainable current harvesting practices and energy conversion technologies (for both household use and livelihood activities) that raise these major concerns about the consequences for health, economic development and the environment. At the same time, imbalances on the demand, supply and use of biomass have been observed in many SSA countries, raising further concerns about the viability of the current biomass energy system in the long run (Chap. 2 Vol. 1; Chap. 5 Vol. 2). Often, policy responses to the prevailing unsustainable fuelwood procurement and use practices have aimed at banning fuelwood harvesting altogether in some contexts (FAO 2001). However, the evidence suggests that such restrictions are often ineffective in enhancing the sustainability of the energy system, particularly if national governments have low enforcement capacity and fail to take measures to promote and facilitate access to alternative household fuels (FAO 2017a, b).

The above suggest the strong linkage between energy use, traditional livelihoods, and environmental conservation in SSA. This interface presents a major sustainability challenge for many countries in the continent. Balancing such diverse objectives is indeed challenging but can catalyse progress for many different Sustainable Development Goals (SDGs) such as SDG1 (No poverty), SDG2 (Zero hunger), SDG 7 (Affordable and clean energy), SDG 9 (Industry, innovation and infrastructure) and SDG15 (Life on land). However, the dynamics between these different policy objectives are not well understood in many countries across the region, considering the large variation in ecosystems, local livelihoods and energy demand and supply practices.

Guinea is a country in western Africa that exemplifies this interface of high fuelwood reliance for domestic use and livelihood activities, with environmental degradation. With a population of 12.7 million people in 2013 (World Bank 2017), only less than 2% of the population has access to non-biomass cooking fuels (IEA et al. 2019). Indeed, more than 74% of households use fuelwood for cooking and about 24% charcoal (Republic of Guinea 2012). Only 34% have access to electricity (IEA et al. 2019), and although the fraction of households with access increased slightly in rural areas between 2007 (1.4%) and 2012 (2.6%), it declined markedly in urban areas (65.7% and 55.5% respectively) (AFDB 2010; IMF 2013a, b).

At the same time, fuelwood consumption grew from about 8.7 Mm^3 in 1980 to about 12.2 Mm^3 in 2000, largely due to population growth (Broadhead et al. 2001). However, there is a dearth of accurate estimates on the actual biomass potential, with estimates ranging between 8.5 and 14 Mm^3 in terms of accessible biomass volume

(REEEP 2012). A significant fraction of fuelwood in Guinea comes from mangrove forests located in the coastal region (Sect. 7.2.1). Mangrove forests are the main sources of fuelwood and charcoal for household cooking and small-scale agricultural processing (e.g. palm oil, fish smoking, salt making) (ESMAP 1994; Ewel et al. 1998; Balmford et al. 2002) (Sect. 7.2.1). Furthermore, mangrove areas host many other traditional livelihood activities that also end up degrading the forest. For example, rice farming in cleared mangroves accounts for 16% of the total rice crop area and 18% of the total rice production in the country (RoG 2009). Considering its extent and high-yield variability (1.5–3.5 t/ha), mangrove rice production is considered particularly important for national and local food security. However, these coastal mangroves provide important ecosystem services that are gradually degraded such as erosion protection for more than 300 km of the Guinean coast and habitat for biodiversity (USAID 2012).

The above imply the important trade-offs associated with energy use, traditional livelihoods and ecological conservation in the coastal areas of the country. These trade-offs are regulated by a constellation of government agencies, which further complicates their effective management. For example, the Ministry of Energy and Hydraulics (MEH) is responsible for regulations, incentives and legislative frameworks in the energy sector. The Ministry of Environment is in charge of forestry issues, thus playing an important role in (a) regulating the biomass sector and (b) ensuring the conservation of threatened ecosystems. Regarding the former, there are plans to protect the national biomass resources while at the same time reducing deforestation and the national contribution to global climate change. Towards this end, the Intended Nationally Determined Contributions (INDCs) aim at reducing fuelwood and charcoal demand by 50% per capita by 2030 through the introduction of improved cookstoves and charcoal production kilns (RoG 2015). Regarding the latter, the escalating conservation concerns over mangrove loss led to the revision of the Forestry Action Plan of 1992 that includes strategies for management and development of mangroves in coastal Guinea (Samoura et al. 2007). At the same time, the coastal region is important for national development (RoG 2004). In the past decades, a series of initiatives promoted by different government agencies have sought to achieve a pragmatic coastal development through multiple policies, strategies and investment programmes.

The aim of this chapter is to explore the interface between energy use, rural livelihoods and mangrove degradation in the coastal region of Guinea. In particular, we identify common fuelwood access regimes in four study districts for households that interact directly with mangrove forests, namely, loggers, rice farmers and salt makers. Section 7.2 outlines the study sites and the data collection and analysis methods. Section 7.3 assesses household energy consumption patterns, expenditure and cooking device utilization and fuelwood access regimes. Section 7.4 synthesizes the main findings at the interface of rural livelihoods and mangrove forest exploitation and identifies possible future trajectories and policy implications for the region.

7.2 Methodology

7.2.1 Study Site

The coastal region of Guinea spans an area of approximately 1 Mha, which stretches 300 km between Guinea-Bissau and Sierra Leone (Rue and Fontana 1995). Many rivers originating from the Fouta Djallon highland traverse this area from east to west. These rivers create an estuary covered by an extensive mangrove swamp forest, which covers a large portion of the coastal region. These forests are dominated by Rhizophoracea and Avicenniaceae plant species (RoG 2009) and play a crucial role for the livelihoods of local communities (Ellison 1997; Naylor and Drew 1998; Dahdouh-Guebas et al. 2000) (Sect. 7.1). The region consists of three broad zones with very different ecological features: (a) a coastal plateau; (b) a tidal marsh and (c) a coastal-marine area.

First, the coastal plateau is about 20 km wide and lies in the eastern parts of the coastal region, adjacent to the foothills of the Fouta Djallon highland. This area contains most of the road network and settlements and is covered by highly degraded and infertile sandy soils. The area contains areas of upland forest, which are increasingly degraded, particularly through timber extraction and agriculture. Slash-and-burn agriculture is practiced extensively in the region for many crops, such as upland rice, peanuts, fonio and oil palm.

Second, the tidal marsh area extends over 400,000 ha in an alluvial estuary that contains a highly biodiverse mangrove forest. The mangrove forest occupies a total area of approximately 270,000 ha and is roughly subdivided into three landscapes types: (a) the upper estuary, (b) the middle estuary, and (c) the waterfront (c). The mangrove upper estuary is located along the riverbanks and extends in the lowland coastal plateau for over 120,000 ha. The waterfront extends parallel to the shore for over 38,000 ha and is crossed by barrier beaches. The middle estuary is located between the upper estuary and the waterfront. It covers roughly 180,000 ha and is characterized by many channels and complex hydrological circulation patterns. Many human activities affect the tidal marsh areas such as timber extraction, agriculture, salt production, livestock farming, oyster gathering, fishing and fish smoking (see below). Mangrove forests are cleared extensively throughout the tidal marsh area in order to establish rice fields on its fertile soils, with coastal mangrove rice fields spanning an estimated 80,000 ha.

Third, the coastal-marine area extends over the large estuary formed by the mangrove head. It is characterized by the seaward extension of coastal mudflats that have an average width of 20 km. At its limits, the water depth is around 15–20 m. The area hosts intense fishing activities, both by artisanal fishers operating from approximately 100 docks and by large-scale industrial operators.

Mangrove forests in the region have experienced significant degradation in the past decades. The main pressures come from the increasing population and "informal" industrialization (e.g. logging, fish smoking, salt production from small-scale enterprises), agriculture and mining (Valiela et al. 2001; Wilkie and Fortuna 2003;

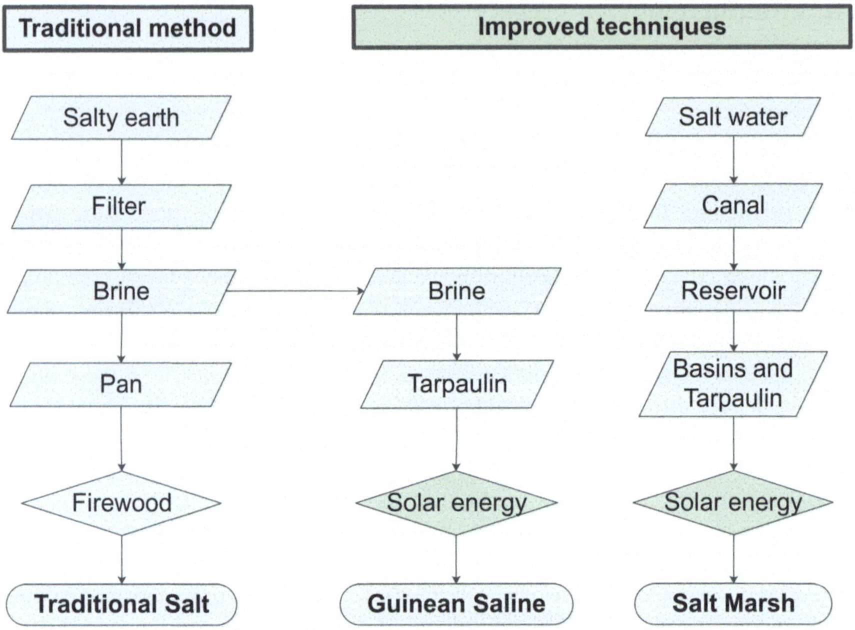

Fig. 7.1 Schematic representation of salt production practices in coastal Guinea. (Source: Adapted from (Balde et al. 2013b))

Duke et al. 2007; Feka and Manzano 2008). For example, Balde et al. (2014a) found extensive land use change due to agricultural activities at a site located in the Dubreka prefecture. In particular, they found that 41.7% of the 5099 ha have underwent substantial changes during between 1990 and 2010, with the upland forest area affected substantially due to local population growth and farmer migration (Balde et al. 2014a).

Extensive areas of mangrove forest and abandoned shrimp farms in the Makinsi district have been converted to paddy fields, with two mangrove rice production techniques prevalent in the region (Balde et al. 2014b). Traditional mangrove rice (TMR) farming is undertaken in cleared mangrove forestland, usually through slash-and-burn. Small embankments enclose these farms, with rice planted on ridges as a means of controlling the inflow of seawater and protecting the crop from crabs (Balde et al. 2014b). The Guinean government has introduced some improved mangrove rice (IMR) farming techniques performed in irrigated perimeters surrounded by large embankments separated by dikes and floodgates (Balde et al. 2014b). The gates are opened during the dry season to allow seawater intrusion for weed control and soil acidification prevention, while they remain closed during the cropping season to prevent seawater intrusion into the rice fields (Balde et al. 2014b).

Some of the other common livelihood activities also affect mangrove forest through fuelwood extraction. For example, fish drying and smoking consume an estimated 58,000 tons of fuelwood per year,[1] while the salt extraction consumes 90,000 tons (Diallo 1993; Republic of Guinea 2002; Samoura and Diallo 2003; Feka and Ajonina 2011). Salt extraction is a particularly important economic activity for many villages along the coast, providing subsistence to more than 8000 households. Salt is produced though different techniques (Box 7.1, Fig. 7.1), with some of the traditional salt-making techniques consuming a considerable amount of wood for preparing (cooking) the brine. On the contrary, salt-making techniques based on sun-drying crystallization do not consume fuelwood, are highly productive, but are perceived as less efficient by local communities.

Box 7.1: Main Salt Production Practices in Coastal Guinea. Source (Balde et al. 2013b)

Traditional salt production (TSP) is seasonal and practiced for around 4 months (mid-February to mid-May) in seasonal campsites. Earth with high salt content is scraped and collected in coastal areas or drained swamps. It is then transferred into a series of filters, which are filled up with saline water (leaching stage). The saline water dissolves the salt crystallized within the earth, with the resulting brine leachate collected into containers placed below the filters (single or multiple filters). The periodic addition of saline water can maintain a continuous leaching process, until the concentration of the leachate becomes not much different than that of the saline water. At that point, the salty earth is replaced with fresh one, to continue the leaching process. The brine obtained through this filtering process is poured into pans and boiled to evaporate the liquid and extract the salt. This latter process requires significant amounts of fuelwood collected from mangrove and/or upland forests.

The Guinean saline (GS) process is similar to the TSP process up to the brine evaporation stage. The main difference is that evaporation is not achieved through cooking but on a plastic canvas (tarpaulin). By practically removing the cooking stage his technique reqires no fuelwood and has significantly lower financial (for pans, fuelwood) and time costs leading to improved productivity. However, this technique requires building capacity through training and providing tarpaulins to producers.

The salt marsh (SM) is an improved technique introduced in 1999 by the French NGO, Charente Maritime. Seawater is collected in reservoirs and then filtered through a series of salt basins where, via evaporation and crystallization, salt crystals are deposited. These basins are always coated with tarpaulins allowing for the storage of a significant volume of seawater. Similar to the GS process, evaporation does not require fuelwood and is associated with

(continued)

[1]Mangrove wood is preferred for fish smoking due to its high calorific value, antimicrobial properties and the fact that it provides a golden-brown color to the smoked fish that enhances marketability (Feka et al., 2008).

Box 7.1 (continued)
significant productivity gains. However, similar to the GS process, it requires capacity building and the provision of necessary materials.

In order to explore the interface between livelihood activities, energy use and fuelwood procurement in coastal Guinea, we focus on the Koba sub-prefecture of the Boffa prefecture (Fig. 7.2). Koba covers an area of 1026 km^2 and has a total population of 52,720 inhabitants. It borders to the north Tanene sub-prefecture (Dubreka prefecture) (Balde et al. 2013a), the Mara Island in the west and the Atlantic Ocean to the southeast. It is one the major areas in Guinea in terms of mangrove rice and salt production and hosts the Koba agronomic research center (CRAK), which specializes in the experimentation of mangrove and freshwater rice varieties.

Based on information obtained from key informants and rapid exploratory studies, we targeted four districts. These districts were purposively selected to capture the main livelihood activities outlined above and Table 7.1, namely: (a) salt production (TSP, GS, SM); (b) logging (WE) and (c) mangrove rice production (TMR, IMR). The selected districts include Balessourou (TSP, GS, SM), Taboria (WE), Makinsi (TMR) and Bentya (IMR).

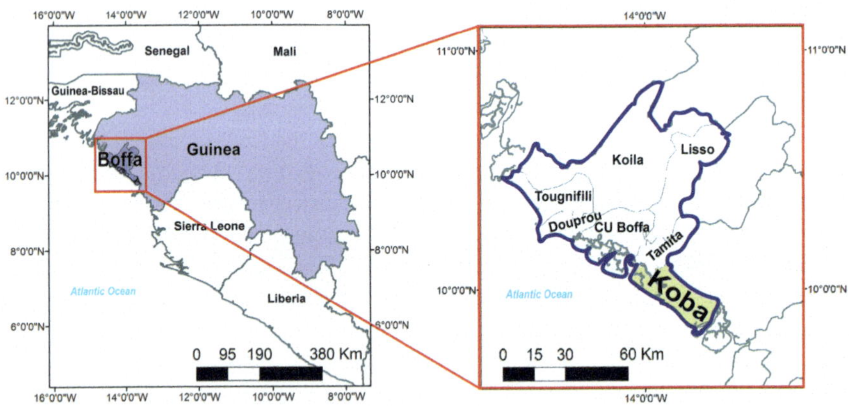

Fig. 7.2 Study locations in Koba region, Boffa prefecture

Table 7.1 Study groups by livelihood activity

Study groups	Code	Sample size	Study district
Traditional/semi-traditional salt producers	TSP/GS	60	Balessourou
Traditional mangrove rice producers	TMR	50	Makinsi
Irrigated mangrove rice producers	IMR	50	Bentya
Improved salt producers	SM	40	Balessourou
Wood loggers	WE	20	Taboria

7.2.2 Data Collection and Analysis

Primary qualitative and quantitative data were collected in March and April 2013 in different districts of the Koba region that contain mangrove social–ecological systems (Sect. 7.2.1). Quantitative and qualitative information from household surveys is used to elicit the actual fuelwood procurement practices and energy use patterns (Sects. 7.3.2–7.3.4). In particular, we conducted 220 surveys, with randomly selected households within five study groups that represent the main local livelihood options intersecting with mangrove forests (Sect. 7.2.1). In particular, the survey captured information regarding the fuelwood procurement and energy use patterns (Sects. 7.3.2–7.3.4). The survey data are analysed through crosstab statistics and one-way ANOVA tests. The Crosstabs procedure employs two-way and multi-way tables and provides various tests and measures of association for two-way tables.

Qualitative information was collected through expert interviews and field observations. This information is used to both complement/triangulate the household survey and understand the access mechanisms and possible effects of fuelwood procurement and consumption from mangrove and upland forests (Sect. 7.3.1). Expert interviews were conducted in the local language with forest-user groups and other local stakeholders. The interviews were first translated into French and then subsequently transcribed into English. The English transcripts were coded for key themes and sub-themes to establish the fuelwood procurement patterns and access mechanisms in the coastal mangroves and upland forests.

7.3 Results

7.3.1 Fuelwood Access Regimes

Generally speaking, in the study area, fuelwood is sourced from both mangroves and upland forests. It is harvested either by household members or by professional/commercial loggers. Access to (and interaction with) these forests differs according to the type of livelihood activity. Loggers extract timber products such as poles and chopped wood from the mangrove forest for sale to construction operators, fish processors and fuelwood traders (Sect. 7.3.2). Salt producers using traditional techniques (TSP) harvest directly wood from the upland forest to prepare and condition the brine before extracting the salt (Sect. 7.2.1). Traditional (TMR) and improved (IMR) mangrove rice producers also contribute to mangrove forest logging, either through direct demand for fuelwood or land clearing (i.e. mangrove forest conversion to rice paddies) (Sect. 7.2.1). Overall, we identify two different fuelwood access regimes that have been implemented historically and are currently operational in the study area.

The first regime entails zero restriction or ban on accessing the mangrove forest for wood. This regime is still operational in some areas, while it has been partly discontinued in others (see below). For example, in parts of Taboria and Makinsi districts, there is no mangrove logging ban according to some respondents. The surveyed loggers (WE) operate in remote sites that are accessible only by canoes from the main inland Koba and sell mangrove wood in Taboria that is the principal port of Koba. The remoteness and difficult access to logging sites, combined with the quick regeneration of the logged mangroves could explain why the local government has not imposed any restriction on mangrove logging. For example, according to an interviewed logger "*there are no worries over replenishment because there is a quick regeneration of almost 98% of logged mangrove areas*".

The second regime entails the partial restriction of mangrove logging in some of the surveyed areas that affect fuelwood access and use for some groups such as traditional salt producers (TSP). For example, in Balessourou district, traditional salt producers indicate that the local government of the Koba sub-prefecture imposed in 2004 the ban of wood extraction from the mangrove forest. According to the salt producers, this restriction was imposed following severe flooding in the area, which apart from affecting the broader local communities also damaged the salt production campsites and shortened the salt production season (Sect. 7.2.1). Interviews with traditional and improved salt producers confirmed that the ban of mangrove logging significantly reduced flooding due to the regeneration of the mangrove forest. However, due to this ban, traditional salt producers depend much more on fuelwood from upland forests and have to invest more to transport it to the ever-increasing distances of the salt production campsites (Sect. 7.2.1). Reportedly, before the ban, traditional salt producers in Balessourou district collected more than 95% of the fuelwood needed for salt production from the mangrove forest (see below).

In Bentya district, the improved mangrove rice farmers (IMR) also reported a ban on mangrove clearing for rice production following the irrigation development undertaken in 1999 (Sect. 7.2.1). Most of these rice farmers belong to the Darabo union, whose main objectives includes the protection of mangrove forest areas between the irrigated farm perimeters and the inlets of seawater (Sect. 7.2.1). Farmers believe that by protecting these mangrove forest areas, they may prevent floods and damages to their rice fields, ensuring thus their food security. Furthermore, they also believe that mangrove forest cutting could reduce rainfall and cause droughts, which could affect rice yields.

7.3.2 Timber Collection by Loggers and Salt Producers

First, we assess the average amount of wood harvested by traditional salt producers (TSP) and loggers (WE). Salt producers collect on average of 40.41 tons per year, mainly to be used for boiling the brine water to extract the salt (Sect. 7.2.1). This wood is usually collected in the form of bunches either from deadwood (salt producers, TSP) or cutting live trees (loggers, WE).

Loggers harvest wood and timber mainly in two forms: "Type A" and "Type B". Type A timber is processed from logs that are about 6–7 m long, and is sold to clients involved in construction in urban areas. On average loggers harvest 7.88 tons/month of "Type A" timber. Type B timber comes in the form of chopped logs that are 1.1–1.2 m long, and is sold to bakers, fish smokers and barwomen. On average, the interviewed loggers collected 367.5 tons/month of chopped-off Type B wood.

7.3.3 *Household Energy Consumption*

Household energy consumption was assessed through binary questions that captured the type and source of biomass fuel. This included (a) fuelwood from mangrove forest, fuelwood from upland forest and charcoal and (b) modern fuels including kerosene, batteries and electricity.

Results for the pooled sample indicate that 91%, 44% and 40% of the respondents rely on fuelwood extracted from the upland forest, charcoal and fuelwood collected from the mangrove forest, respectively (Table 7.2). However, there are some interesting variations among the individual study groups. For example, all loggers use fuelwood extracted from mangroves for cooking. This is perhaps due to (a) close proximity to the mangrove forest; (b) mangrove wood extraction being their main livelihood activity and (c) lack of restriction on the mangrove wood extraction in their area (Sect. 7.3.1).

On the other hand, despite the close proximity of salt production sites to the mangrove forest, all of the interviewed salt producers (TSP, GS and SM) reportedly consume fuelwood sourced from the upland forest. This is perhaps attributable to (a) the long distance of their homesteads to mangrove forests and (b) the ban on mangrove wood logging in the Balessourou district where all salt producers operate.

Table 7.2 Source of household energy among study groups

Districts		Taboria	Balesourou		Makinsi	Bentya		
Fuel types		WE	TSP/ GS	SM	TMR	IMR	Overall	*P* value
Fuelwood from mangrove forest (%)	No	0	25.9	18.2	4.1	11.8	60.0	0.000***
	Yes	9.1	1.4	0	18.6	10.9	40.0	
Fuelwood from upland forest (%)	No	0	0	0	8.6	0.9	9.5	0.000***
	Yes	9.1	27.3	18.2	14.1	21.8	90.5	
Charcoal (%)	No	7.7	5.9	10.9	9.1	22.7	56.4	0.000***
	Yes	1.4	21.4	7.3	13.6	0	43.6	
Electricity (%)	No	9.1	27.3	18.2	22.7	22.7	100	NA
	Yes	0	0	0	0	0	0	
Kerosene (%)	No	9.1	27.3	18.2	21.8	10.0	86.4	0.000***
	Yes	0	0	0	0.9	12.7	13.6	
Batteries (%)	No	5.5	27.3	12.3	20.5	5.0	70.5	0.000***
	Yes	3.6	0	5.9	2.3	17.7	29.5	

Note: No statistical analysis was performed for electricity as it is a constant variable

Table 7.3 Adoption of biomass stoves among study groups

Stove types		WE (%)	TSP/ GS (%)	SM (%)	TMR (%)	IMR (%)	Overall (%)	p-value
Charcoal stove	No	9.1	12.3	11.4	11.8	22.7	67.3	0.000***
	Yes	0	15.0	6.8	10.9	0	32.7	
Rocket stove (improved fuelwood stove)	No	3.6	23.6	4.5	14.1	6.8	52.7	0.000***
	Yes	5.5	3.6	13.6	8.6	15.9	47.3	
Tripod (traditional, open three-stone fire)	No	2.3	4.5	13.2	11.4	15.9	47.3	0.000***
	Yes	6.8	22.7	5.0	11.4	6.8	52.7	

Note: $***p < 0.01$

It is interesting to note that only one of the TSP respondents reported to use fuelwood from mangrove forests. However, this fuelwood is mainly purchased from commercial fuelwood markets procured from other areas outside Balessourou district.

Traditional mangrove rice producers (TMR) and improved mangrove rice producers (IMR), reportedly consume fuelwood from mangrove forest (18.6% and 10.9% of the total surveyed households, respectively) (Table 7.2). Furthermore 14.1% and 21.8% of the total surveyed households collect fuelwood from the upland forests, respectively.

When it comes to other cooking fuels, charcoal is used by 21.4%, 13.6%, 7.3% and 1.3% of the TSP/GS, SM, WE and TMR, respectively (Table 7.2). Approximately 14% of households use kerosene for cooking, while none of the households reported the use of electricity. For cooking devices, most households use inefficient traditional three-stone stoves (52.7%). On average, the adoption and use of improved biomass cookstoves such as rocket-firewood stove and charcoal stove was reported by 47.3% and 32.7% of the surveyed households, respectively (Table 7.3).

7.3.4 Monthly Household Expenditures

Table 7.4 contains the monthly expenditures for household energy consumption. Overall, the mean expenditure is statistically significant ($p < 005$) indicating a difference between households. For the pooled sample, the average monthly expenditure on upland forest fuelwood is greater than any other energy source, representing 122,077 Guinean Franc (GNF). This is followed by expenditures for fuelwood from mangrove forest (56,945 GNF) and charcoal (22,886 GNF). The average monthly expenditure for modern household energy (kerosene and batteries) is lower than biomass fuels at GNF 1614 and 1900 respectively. This low consumption of modern fuel can be attributed to fuel stacking with fuelwood (see Tables 7.2 and 7.3).

Table 7.5 and Fig. 7.3 present the household monthly expenditures on other household budget expenses such as food, ceremonies, transportation, education and health care. The average monthly expenditure (479,091 GNF) on food is the highest,

Table 7.4 Monthly energy expenditures by energy source and respondent group

		Taboria	Balesourou		Makinsi	Bentya		
		WE	TSP/GS	SM	TMR	IMR	Overall	*P* value
Fuelwood from mangrove forest (GNF)	Mean	250,000	22,667	0.00	67,760	55,600	56,945	0.000***
	Std. error	13,650	14,710	0.00	10,063	12,307	7069	
	Minimum	160,000	0.00	0.00	0.00	0.00	0.00	
	Maximum	370,000	800,000	0.00	270,000	300,000	800,000	
Fuelwood from upland forest (GNF)	Mean	272,600	123,667	235,500	41,360	49,940	122,077	0.000***
	Std. error	10,020	13,272	9352	8794	4899	7346	
	Minimum	189,000	60,000	130,000	0.00	0.00	0.00	
	Maximum	360,000	800,000	360,000	230,000	150,000	800,000	
Charcoal (GNF)	Mean	4000	48,000	19,200	26,140	0.00	22,886	0.000***
	Std. error	2224.27	3434.62	5150.21	4599.29	0.00	2083.17	
	Minimum	0.00	0.00	0.00	0.00	0.00	0.00	
	Maximum	30,000	75,000	150,000	120,000	0.00	150,000	
Kerosene (GNF)	Mean	0.00	0.00	0.00	620	6480	1614	0.000***
	Std. error	0.00	0.00	0.00	445	1191	338	
	Minimum	0.00	0.00	0.00	0.00	0.00	0.00	
	Maximum	0.00	0.00	0.00	19,000	30,000	30,000	
Batteries (GNF)	Mean	13,900	2500	15,575	13,620	5400	9100	0.057*
	Std. error	4091	1477	5410	5992	606	1794	
	Minimum	0.00	0.00	0.00	0.00	0.00	0.00	
	Maximum	45,000	60,000	200,000	180,000	15,000	200,000	

Note: $***p < 0.01$, $*p < 0.1$

Table 7.5 Other household expenditures by respondent group

		WE	TSP/GS	SM	TMR	IMR	Overall	*P* value
Food (GNF)	Mean	380,000	599,833	373,625	540,500	39,6800	479,091	0.000***
	Std. error	26,388	31,080	32,367	68,220	24,111	20,470	
	Minimum	200,000	360,000	190,000	100,000	30,000	30,000	
	Maximum	600,000	18,000,00	15,750,00	25,000,00	630,000	250,0000	
Children education (GNF)	Mean	91,750	70,583	73,200	125,760	83,700	88,505	0.014**
	Std. error	12,994	5130	2932	23,067	9100	6073	
	Minimum	30,000	0.00	30,000	0.00	0.00	0.00	
	Maximum	240,000	200,000	120,000	800,000	360,000	800,000	
Contribution to ceremonies (GNF)	Mean	65,400	71,817	104,375	60,660	90,600	78,886	0.001***
	Std. error	3585	2366	6114	13,406	6403	3757	
	Minimum	30,000	40,000	40,000	5000	30,000	5000	
	Maximum	86,000	120,000	300,000	450,000	200,000	450,000	
Health care (GNF)	Mean	125,000	73,000	153,875	206,620	102,020	129,395	0.000***
	Std. error	12,107	8040	14,617	34,139	5988	9238	
	Minimum	30,000	30,000	50,000	25,000	50,000	25,000	
	Maximum	200,000	400,000	600,000	950,000	210,000	950,000	
Transportation (GNF)	Mean	85,650	51,333	99,000	128,200	92,400	89,923	0.005***
	Std. error	10,850	1817	15,973	23,724	14,223	7197	
	Minimum	30,000	30,000	40,000	12,000	15,000	12,000	
	Maximum	210,000	80,000	660,000	960,000	400,000	960,000	
Others (GNF)	Mean	0.00	167	0.00	14,960	0.00	3445	0.000***
	Std. error	0.00	167	0.00	4200	0.00	1038	
	Minimum	0.00	0.00	0.00	0.00	0.00	0.00	
	Maximum	0.00	10,000	0.00	150,000	0.00	150,000	

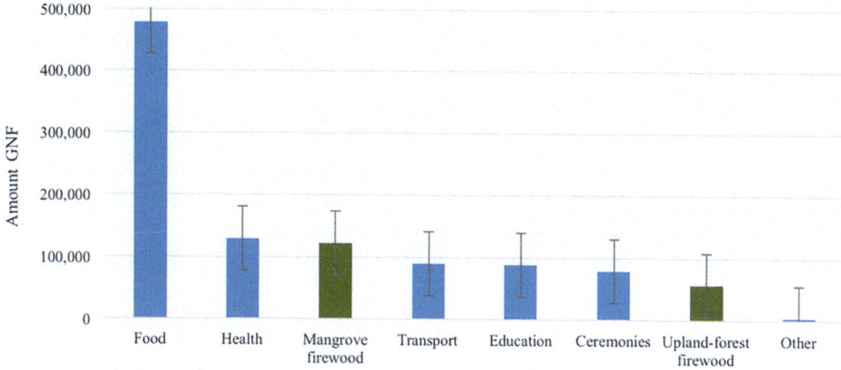

Fig. 7.3 Average monthly household expenditures for all respondent groups

followed by health care (129,395 GNF). Interestingly, as most respondents rely on traditional biomass fuels for cooking, this could cause health problems to household members involved in cooking activities, possibly increasing health-care costs (Table 7.5).

7.4 Discussion

7.4.1 Synthesis of Findings and Future Trajectories

Household energy use patterns indicate that the surveyed households rely mainly on fuelwood derived from both mangrove forests (40% of respondents) and upland forests (91% of respondents) (Sect. 7.3.3). However, the utilization of mangrove fuelwood is limited in some areas due to their large distance from mangrove areas and the partial restriction/ban, especially for traditional salt production (Sect. 7.3.1). As a result, many of the surveyed traditional salt producers (TSP) switched to fuelwood procured from upland forest (Sect. 7.3.1).

Furthermore, more than half of the surveyed households use inefficient traditional stoves that require large fuelwood quantities (Sect. 7.3.3), which can have detrimental impact on mangrove and upland forests (Sect. 7.3.1). Furthermore, due to the partial ban on mangrove forest logging (Sect. 7.3.1), most of the charcoal consumed in the study area is derived from fuelwood extracted in upland forest, particularly in the Balessourou and Bentya districts (Sects. 7.3.3–7.3.4). Such energy procurement and use practices may exacerbate deforestation and land degradation in the study areas, with all their subsequent environmental impacts (Sect. 7.1). Many studies across SSA have identified the high reliance on traditional fuels and stoves not only for household use (Drigo and Nzabanita 2011; Bailis et al. 2004; IRENA 2018; IEA et al. 2019) but also agricultural processing technologies and livelihood activities such as for example shea processing in Ghana (Jasaw et al. 2017).

Due to this need for large fuelwood quantities, households relying on commercial fuelwood markets tend to spend substantial amounts of money to purchase fuelwood, constituting fuelwood as the third largest household expense category among all study groups following food and health care (Sect. 7.3.4). Many studies have reported the significant economic burden of high fuelwood reliance on household budgets in many rural parts of SSA (Sola et al. 2016; ESMAP 2015). However, it is interesting to point that the overall monthly expenditure on fuelwood from upland forests is two times higher compared fuelwood sourced from mangrove forests (Sect. 7.3.4). A possible explanation behind this difference might be the restriction on mangrove wood extraction in some locations and the remoteness of mangrove forests from many villages (Sect. 7.3.1). Both these factors might end up increasing final costs as found in other parts of SSA (Bolognesi et al. 2015; Maurice et al. 2017).

Another major finding is that households engaging in prevalent livelihood options in the study area also tend to interact in different ways with the mangrove forest. Although in most areas mangrove fuelwood extraction and land conversion used to be practically unregulated, the situation has recently changed partly due to stricter regulation and efforts aiming at enhancing disaster prevention and food security (Sect. 7.3.1). In fact, many studies from different parts of SSA have identified how stricter regulation and other socioeconomic imperatives have catalysed changes in prevailing fuelwood extraction practices (Arnold et al. 2006; Lukumbuzya and Sianga 2017; Maurice et al. 2017). However, the ban on mangrove fuelwood is precarious and possibly depends on many economic and cultural factors as discussed below.

For example, irrigated rice fields in Balessourou and Bentya are not always maintained properly, as farmers lack the means to undertake such tasks that require usually large and recurrent investments. Failure to maintain the irrigated perimeters of improved rice farms (Sect. 7.2.1) may have negative impacts on the surrounding mangrove forest when such perimeters become unproductive (e.g. due to salinity, acidification). For example, possible crop yield decreases due to the underperformance of these irrigation systems may lead to agricultural expansion and the conversion of new mangrove areas to rice fields as has been identified in different parts of SSA (FAO 2009; UNEP-WCMC 2007; Juo et al. 2003).

With respect to salt production, only traditional salt producers (TSP) rely on fuelwood for brine preparation and conditioning before salt extraction (Sect. 7.2.1). The interviewed salt producers reported that fuelwood costs for salt production under traditional methods have increased substantially due to the remoteness of the upland forest from the campsites used for salt extraction (Sect. 7.3.1). However, two reasons may lead to the possible shift of fuelwood extraction back to the mangrove forest. The first reason is that fuelwood overexploitation in the upland areas might cause fuelwood scarcity and as an extent further escalate fuelwood costs (see above). Studies in many parts of SSA have found that fuelwood scarcity might increase fuelwood costs (Scheid et al. 2019; Egeru et al. 2014; Sola et al. 2016), leading to decreasing profit margins for small-scale processors but also incentives for changing fuelwood procurement areas (Daurella and Foster 2009; Kammen and

Kirubi 2008; Lee et al. 2009). The second reason revolves around cultural factors and traditional taste preferences. On the one hand, according to many respondents, the salt produced through traditional techniques contains many impurities due to fuelwood combustion. However, traditional salt producers reported that due to the lower smoke emissions, mangrove fuelwood combustion tends to provide higher quality salt compared to salt produced with fuelwood from upland forests. Many studies have highlighted how cultural and taste-related reasons can affect the type of fuel used for household cooking and small-scale agricultural processing in some SSA contexts (Atanassov 2010; Kaburi and Medley 2011; Akintan et al. 2018).

Finally, the increasing urban demand drives to a large extent fuelwood extraction from mangrove forests. For example, most of the poles and chopped wood extracted from the mangrove forest are marketed to construction operators, food processors and fuelwood retailers in surrounding cities (Sects. 7.3.1–7.3.2). Although this was not part of this analysis, it is highly possible that the rising urban demand and the logging capabilities of external operators might outweigh the wood extraction abilities of local communities and forest regeneration capacity. Such effects have been identified in many different parts of SSA (Munslow, 2013; Mwampamba 2007; Chambwera 2004) and often cannot be sustained by local forests (Iiyama et al. 2014; Arnold et al. 2006).

7.4.2 Policy Implications and Recommendations

Throughout this chapter, we have outlined the strong linkages between multiple policy objectives related to energy use, ecosystem degradation and local livelihoods in coastal Guinea. These linkages suggest the strong connection between many SDGs, namely, SDG1 (No poverty), SDG2 (Zero hunger), SDG 7 (Affordable and clean energy), SDG 9 (Industry, innovation and infrastructure) and SDG15 (Life on land) (Sect. 7.1). Below we identify some policy and practice interventions that could possibly reduce negative trade-offs between the aforementioned policy objectives and underlying SDGs.

The negative effects associated with household energy consumption and wood extraction could be overcome with the promotion and dissemination of modern cookstoves and fuels such as kerosene, liquefied petroleum gas (LPG), biogas and electricity. This can help limit the reliance of local communities on traditional biomass for cooking, having many environmental and socioeconomic co-benefits that span multiple SDGs (Karanja and Gasparatos 2019; Rosenthal et al. 2018). However, as many studies across SSA have shown, it is a different thing to promote and disseminate clean cooking options, and a different thing to ensure their sustained adoption and use (Jürisoo et al. 2018; Shankar et al. 2014) (see also Chap. 2 Vol. 1; Chap. 5 Vol. 2). To achieve this, it would require a series of different interventions that address contextual factors such as fuel and stove affordability (Schlag and Zuzarte 2008; Shankar et al. 2014; ESMAP 2015; Karanja and Gasparatos 2019), local cooking traditions and socio-cultural issues (Akintan et al. 2018), negative perceptions of modern cooking technologies (Tamire et al. 2018), and fuel availability and accessibility (Dalaba et al. 2018) (see also Chap. 2 Vol. 1; Chap. 5 Vol. 2).

The negative impact associated with traditional salt production could be avoided through the promotion and adoption of improved and semi-improved salt production techniques (i.e. GS, MS) (Sect. 7.2.1). According to Balde et al. (2014c), improved salt production techniques could become a viable alternative to the traditional salt production techniques (i.e. TSP) that consume significant amounts of fuelwood from mangrove and upland forests (Sects. 7.3.1–7.3.2). This could reduce mangrove and upland forest over-exploitation and at the same time increase profit margins by reducing reliance to the increasingly expensive fuelwood from upland forests (Sect. 7.3.1). The adoption and sustained use of such techniques can also improve local livelihoods, considering that the salt producers using improved techniques tend to (a) have more household assets and better housing, (b) achieve higher profits and (c) work fewer hours investing the saved time to secondary income generation activities such as mangrove rice production (Balde et al. 2013b).

Similarly, some of the negative impacts of current rice production practices can be reduced through the implementation and proper maintenance of improved irrigation systems. Stabilizing (or even increasing) rice yields through such systems would reduce the need for encroaching new mangrove forest areas through slash-and-burn to expand paddy areas. However, many farmers lack the resources, funding and capacity to maintain properly such irrigation systems (Balde et al. 2013a, d).

Increasing the adoption of improved salt production and rice production techniques would require a series of interventions. First, it would be imperative to raise the awareness of local communities about the benefits of these practices on local livelihoods and mangrove conservation. Currently, local and international NGOs provide such services that reach, however, few local communities due to funding constraints (Balde et al. 2013b). The gradual involvement of other actors, and especially of government agencies, would be a necessary step for reaching more local communities.

7.5 Conclusions

This chapter explored the interaction of common livelihood options in the coastal region of Guinea with mangrove and upland forests. In particular, it explored the energy procurement and consumption patterns of salt producers, wood loggers and rice farmers, especially as it relates to fuelwood. The results suggest that all groups depend heavily on fuelwood from upland and mangrove forests and tend to use traditional stoves for cooking and livelihood activities.

However, the growing mangrove degradation associated with such practices led to the restriction of mangrove forest logging in the Balessourou and Bentya districts by the local government in the former case and the Darabo union in the latter case. However, partly due to these restrictions, the costs of upland fuelwood have been increasing for all groups. This is especially the case for traditional salt producers who use extensive amounts of fuelwood for salt production. Such escalating costs might influence many households to revert back to mangrove logging as a means of reducing energy procurement expenditures.

Thus, it is important to both raise awareness about the negative environmental and health impacts of prevailing fuelwood procurement and use practices as well as to provide appropriate technologies that reduce fuelwood consumption. For example, the dissemination of modern cooking fuels and stoves and requirement for improved salt production could have multiple socioeconomic and environmental benefits. Furthermore, the expansion and maintenance of improved irrigation systems could reduce mangrove conversion to rice paddies. However, such actions would require the coordinated action of many different stakeholders such as the local and national government, NGOs and farmers' organizations.

Acknowledgements Alice Karanja was supported by a Monbukagakusho scholarship offered by the Japanese Ministry of Education, Culture, Sports, Science, and Technology (MEXT) through the Graduate Program in Sustainability Science-Global Leadership Initiative (GPSS-GLI), at the University of Tokyo. Marcin Jarzebski developed Fig. 7.2.

References

AFDB (2010) Rural Electrification Project (PER) in the Republic of Guinea. African Development Fund. http://www.afdb.org/fileadmin/uploads/afdb/Documents/Project-and-Operations/Guinea_-_AR_-_Rural_Electrification_Project_pdf

Akintan O, Jewitt S, Clifford M (2018) Culture, tradition, and taboo: understanding the social shaping of fuel choices and cooking practices in Nigeria. Energy Res Soc Sci 40:14–22

Arnold JEM, Köhlin G, Persson R (2006) Woodfuels, livelihoods, and policy interventions: changing Perspectives. World Dev 34:596–611. https://doi.org/10.1016/j.worlddev.2005.08.008

Atanassov B (2010) Socio-cultural dimensions in household cooking energy choice: implications for energy transition in Catembe, Mozambique

Bailis R, Pennise D, Ezzati M, Kammen DM, Kituyi E (2004) Impacts of greenhouse gas and particulate emissions from woodfuel production and end-use in sub-Saharan Africa. Renewable and Appropriate Energy Laboratory, Berkeley, CA

Bailis R, Smith KR, Edwards R (2007) Kitchen performance test (KPT). Household Energy and Health Programme, Shell Foundation, London, UK

Bailis R, Drigo R, Ghilardi A, Masera O (2015) The carbon footprint of traditional woodfuels. Nat Clim Chang 5(3):266–272

Balde BS, Kobayashi H, Matsumura I, Nohmi M, Esham M, Tolno E (2013a) Present status and determinants of mangrove rice production in Dubreka prefecture in Guinea. Jpn J Farm Magement 51(3):97–102

Balde BS, Kobayashi H, Nohmi M, Ishida A, Matsumura I, Esham M, Tolno E (2013b) Socio-Economic Analysis of small-scale salt production techniques in the coastal area of Guinea: an alternative for improving livelihood status and mangrove forest management. Int J Res Eng IT Social Sci 3(9):1–23. http://www.indusedu.org/IJREISS/September2013(pdf)/1.pdf

Balde BS, Kobayashi H, Matsumura I, Esham M, Alam A, Tolno E (2014a) Land use changes and their determinants in the coastal area of Guinea: a study based on spatial and field survey. J Jpn Agric Syst Society 30(3):65–75

Balde BS, Kobayashi H, Ishida A, Nohmi M, Esham M, Matsumura I, Tolno E (2014b) Effects of rural livelihood activities on income inequality and poverty reduction in the Guinean coastal area. J Agric Sci 6(6):113–130. https://doi.org/10.5539/jas.v6n6p113

Balde BS, Kobayashi H, Nohmi M, Ishida A, Matsumura I, Esham M, Tolno E (2014c) A stochastic frontier approach for measuring technical efficiency of small-scale salt production in Guinea. Am J Appl Sci 11(8):1310–1320. https://doi.org/10.3844/ajassp.2014.1310.1320

Balde BS, Kobayashi H, Nohmi M, Ishida A, Esham M, Tolno E (2014d) An analysis of technical efficiency of mangrove rice production in the Guinean coastal area. J Agric Sci 6(8):179–196. https://doi.org/10.5539/jas.v6n8p179

Balmford A, Bruner A, Cooper P, Costanza R, Farber S, Green RE, Jenkins M, Jefferiss P, Jessamy V, Madden J, Munro K, Myers N, Naeem S, Paavola J, Rayment M, Rosendo S, Roughgarden J, Trumper K, Turner RK (2002) Economic reasons for conserving wild nature. Science 297:950–953

Bolognesi M, Vrieling A, Rembold F, Gadain H (2015) Rapid mapping and impact estimation of illegal charcoal production in southern Somalia based on WorldView-1 imagery. Energy Sustain Dev 25:40–49

Broadhead J, Bahdon J, Whiteman A (2001) Woodfuel consumption modelling and results. Annex 2 in Past trends and future prospects for the utilization of wood for energy. Working Paper No: GFPOS/WP/05, Global Forest Products Outlook Study. Food and Agriculture Organization of the United Nations (FAO), Rome

Cerutti PO, Sola P, Chenevoy A, Iiyama M, Yila J, Zhou W et al (2015) The socioeconomic and environmental impacts of wood energy value chains in Sub-Saharan Africa: a systematic map protocol. Environ Evidence 4(1):12

Chambwera M (2004) Economic analysis of urban fuelwood demand—the case of Harare in Zimbabwe. Wageningen Universiteit

Dahdouh-Guebas F, Mathenge C, Kairo JG, Koedam N (2000) Utilization of mangrove wood products around Mida creek (Kenya) amongst subsistence and commercial users. Econ Bot 54 (4):513–527

Dalaba M, Alirigia R, Mesenbring E, Coffey E, Brown Z, Hannigan M et al (2018) Liquified petroleum gas (LPG) supply and demand for cooking in Northern Ghana. EcoHealth 15:716–728. https://doi.org/10.1007/s10393-018-1351-4

Daurella DC, Foster V (2009) What can we learn from household surveys on inequalities in cooking fuels in sub-Saharan Africa. World Bank, Washington, DC, USA

Diallo A (1993) The mangrove swamps of Guinea. In: Diop ES, Field CD, Vannucci M (eds) Proceedings of a workshop on conservation and sustainable utilization of mangrove forests in Latin America and Africa Regions. Dakar, 20–22 January 1993. ITTO/ISME Project PD114/90 (F), p 6–7

Drigo R (2005) WISDOM, East Africa. Woodfuel Integrated Supply/Demand Overview Mapping (WISDOM) Methodology. Spatial Woodfuel Production and Consumption Analysis of Selected African Countries

Drigo R, Nzabanita V (2011) Spatial analysis of woodfuel production and consumption in Rwanda applying the Woodfuel Integrated Supply/Demand Overview Mapping methodology (WISDOM). FAO-Forestry Department-Wood Energy

Duke NC, Meynecke J-O, Dittmann S, Ellison AM, Anger K, Berger U, Cannicci S, Diele K, Ewel KC, Field CD, Koedam N, Lee SY, Marchand C, Nordhaus I, Dahdouh-Guebas F (2007) A world without mangroves? Science 317:41–42

Egeru A, Kateregga E, Majaliwa GJM (2014) Coping with firewood scarcity in Soroti District of Eastern Uganda. Open J For 4:70–74. https://doi.org/10.4236/ojf.2014.41011

Ellison JC (1997) Mangrove ecosystems of the western and the gulf provinces of Papua New Guinea, a review. Sci New Guinea 23(1):3–16

ESMAP (1994) Guinea household energy strategy. Report No. 163/94. Energy Sector Management Assistance Programme. http://www-wds.worldbank.org/servlet/WDSContentServer/WDSP/IB/1999/08/15/000009265_3961007215450/Rendered/PDF/multi_page.pdf

ESMAP (2015) The State of the Global Clean and Improved Cooking Sector. ESMAP Technical Paper No. 007/15. Authored by: Putti, Venkata Ramana; Tsan, Michael; Mehta, Sumi; Kammila, Srilata. Washington, DC: World Bank; 2015. https://openknowledge.worldbank.org/handle/10986/21878. License: CC BY 3.0 IGO

Ewel K, Twilley R, Ong J (1998) Different kinds of mangrove forests provide different goods and services. Glob Ecol Biogeogr Lett 7:83–94

Ezzati M, Kammen DM (2002) Evaluating the health benefits of transitions in household energy technologies in Kenya. Energy Policy 30:815–826. https://doi.org/10.1016/S0301-4215(01)00125-2

FAO (2001) Forests out of bounds: impacts and effectiveness of logging bans in natural forests in Asia-Pacific. Edited by Durst, Patrick B., T. R. Waggener, Thomas Enters, and Tan Laycheng. Food and Agriculture Organization of the United Nations, Rome

FAO (2007) State of the World's Forests 2007. Food and Agriculture Organization of the United Nations, Rome

FAO (2009) The relevance of mangrove forests to African fisheries, wildlife and water resources. Edited by Bojang, F., & Ndeso-Atanga, A. Food and Agriculture Organization of the United Nations, Rome

FAO (2017a) Incentivizing sustainable wood energy in sub-Saharan africa: a way forward for policy-makers. Food and Agriculture Organization of the United Nations, Rome. http://www.fao.org/3/a-i6815e.pdf

FAO (2017b) The charcoal transition: greening the charcoal value chain to mitigate climate change and improve local livelihoods. Food and Agriculture organization of the United Nations, Rome

FAO (n.d.) The role of wood energy in Africa. Food and Agriculture organization of the United Nations, Rome. http://www.fao.org/3/x2740e/x2740e02.htm#TopOfPage

Feka NZ, Ajonina GN (2011) Drivers causing decline of mangrove in West-Central Africa: a review. Int J Bio Sci, Ecosyst Ser Manag 7(3):217–230

Feka NZ, Manzano MG (2008) The implications of wood exploitation for fish smoking on mangrove ecosystem conservation in the South West Province, Cameroon. Trop Conserv Sci 1:222–241

GACC (2014) 100 million by 2020: the global alliance for clean cookstoves is expected to reach its phase I goal ahead of schedule. Global Alliance of Clean Cookstoves (GACC), Washington, DC

GoK (2013) An alysis of demand and supply of wood products in Kenya. Government of Kenya (GoK), ministry of environment and mineral resources, Nairobi. http://www.kenyaforestservice.org/documents/redd/Analysis of Demand and Supply of Wood Products in Kenya.pdf

IEA (2014) Africa energy outlook. A focus on the energy prospects in sub-Saharan Africa. World Energy Outlook Special Report, International Energy Agency Publication. https://doi.org/https://www.iea.org/publications/freepublications/publication/africa-energy-outlook.html

IEA (2017) Energy access outlook: from poverty to prosperity. Paris: International Energy Agency (IEA). https://webstore.iea.org/weo-2017-special-report-energy-access-outlook

IEA, IRENA, UNSD, WB, WHO (2019) Tracking SDG 7: the energy progress report 2019, Washington DC. https://www.worldbank.org/en/topic/energy/publication/tracking-sdg7-the-energy-progress-report-2019

Iiyama M, Neufeldt H, Dobie P, Njenga M, Ndegwa G, Jamnadass R (2014) The potential of agroforestry in the provision of sustainable woodfuel in sub-Saharan Africa. Curr Opin Environ Sustain 6:138–147. https://doi.org/10.1016/j.cosust.2013.12.003

IMF (2013a) Guinea: poverty reduction strategy paper PRS III (2013–2015). IMF country report No. 13/191. International Monetary Fund, Washington DC. https://www.imf.org/external/pubs/ft/scr/2013/cr13191.pdf

IMF (2013b) Guinea: poverty reduction strategy paper. PRS III (2013–2015). International Monetary Fund

IPBES (2018) Summary for policymakers of the regional assessment report on biodiversity and ecosystem services for Africa of the Intergovernmental Science-Policy Platform on Biodiversity and Ecosystem Services. In: Archer E, Dziba LE, Mulongoy KJ, Maoela MA, Walters M,

Biggs R, Cormier-Salem M-C, DeClerck F, Diaw MC, Dunham AE, Failler P, Gordon C, Harhash KA, Kasisi R, Kizito F, Nyingi WD, Oguge N, Osman-Elasha B, Stringer LC, de Morais LT, Assogbadjo A, Egoh BN, Halmy MW, Heubach K, Mensah A, Pereira L, Sitas N (eds) IPBES secretariat, Bonn, Germany. 49p

IRENA (2018) Sustainable rural bioenergy solutions in Sub-Saharan Africa: a collection of good practices. International Renewable Energy Agency, Abu Dhabi

Jasaw G, Saito O, Gasparatos A, Shoyama K, Boafo YA, Takeuchi K (2017) Ecosystem services trade-offs from high fuelwood use for traditional shea butter processing in semi-arid Ghana. Ecosyst Serv 27:127–138

Juo AS, Franzluebbers K, Backman CR (2003) Tropical soils: properties and management for sustainable agriculture. Oxford University Press on Demand

Jürisoo M, Lambe F, Osborne M (2018) Beyond buying: the application of service design methodology to understand adoption of clean cookstoves in Kenya and Zambia. Energy Res Soc Sci 39:164–176

Kaburi SM, Medley KE (2011) Community perspectives on fuelwood resources in East Africa. Mt Res Dev 31(4):315–325

Kammen DM, Kirubi C (2008) Poverty, energy, and resource use in developing countries. Ann Ny Acad Sci 1136:348–357. https://doi.org/10.1196/annals.1425.030

Karanja A, Gasparatos A (2019) Adoption and impacts of clean bioenergy cookstoves in Kenya. Renew Sust Energ Rev 102:285–306

Lee D, Neves B, Wiebe K, Lipper L, Zurek M (2009) Rural poverty and natural resources: improving access and sustainable management. Food and Agriculture Organization of the United Nations (FAO), Rome, Italy. http://www.fao.org/3/a-ak422e.pdf

Lukumbuzya K, Sianga C (2017) Overview of the timber trade in East and Southern Africa. TRAFFIC, Cambridge, UK

Maurice, J., Le Crom, M., Bouyer, O., Pesce, F., Cielo, P., Gachanja, M., & Bull, L. (2017). Improving efficiency in forestry operations and forest product processing in Kenya: a viable REDD+ policy and measure

Mwampamba TH (2007) Has the woodfuel crisis returned? Urban charcoal consumption in Tanzania and its implications to present and future forest availability. Energy Policy 35 (8):4221–4234

Naylor R, Drew M (1998) Valuing mangrove resources in Kosrae, Micronesia. Environ Dev Econ 3:471–490

Openshaw K (2010) Biomass energy: employment generation and its contribution to poverty alleviation. Biomass Bioenergy 34:365–378. https://doi.org/10.1016/j.biombioe.2009.11.008

REEEP (2012) Biomass POTENTIAL in Guinea. Renewable Energy, Energy Efficiency Partnership. https://www.reeep.org/guinea-2012

Republic of Guinea (2002) National strategy and action plan for biological diversity. CBD strategy and action plan–guinea part 1, English version. http://www.biodiv.org/doc/world/gn/gn-nbsap-01-p1-en.doc

Republic of Guinea (2012) Enquete Legere Pour L'evaluation De La Pauvrete (ELEP). Ministere Du Plan Institut National De La Statistique. https://www.undp.org/content/dam/guinea/docs/publications/poverty/rapport-final-evaluation-pauvrete-2012.pdf

RoG (2004) Living in the Coast people and livelihoods. Program Development Office for Integrated Coastal Zone Management Plan (PDO-ICZMP). Republic of Guinea (RoG). http://www.warpo.gov.bd/rep/liv/live1.PDF

RoG (2009) National strategy for the development of rice growing. Ministry of Agriculture and Livestock, Republic of Guinea

RoG (2015) Intended nationally determined contribution (INDC) under the United Nations Framework Convention on Climate Change (UNFCCC). Republic of Guinea (RoG)

Rosenthal J, Quinn A, Grieshop AP, Pillarisetti A, Glass RI (2018) Clean cooking and the SDGs: integrated analytical approaches to guide energy interventions for health and environment goals. Energy Sustain Dev 42:152–159

Rue O, Fontana A (1995) La zone littorale guineenne unite spatiale et coherence economique. In: Fontana A, Sow M, Rue O, Bangoura K, (Editeurs Scientifiques) 1995. Quel avenir pour la zone littorale guineenne? Actes du Seminaire sur la Programmation Du Projet (Analyse des Contraintes de Gestion et d'Amenagement de la Zone Littorale Guineenne)

Samoura K, Diallo L (2003) Environmental issues associated withthe main sectors of energy production in Guinea. AJEAM-RAGEE 5(1):28–38

Samoura K, Bouvier AL, Waaub JP (2007) Strategic environmental assessment for planning mangrove ecosystems in guinea. Knowledge Technol Policy 19(4):77–93

Scheid A, Hafner JM, Hoffmann HK, Kächele H, Uckert GB, Sieber S, Rybak C (2019) Adapting to fuelwood scarcity: the Farmers'perspective. Front Sustain Food Syst 3:28

Schlag N, Zuzarte F (2008) Market barriers to clean cooking fuels in Sub-Saharan Africa: a review of literature. An SEI (Stockholm Environment Institute) Working Paper 8, March 2008

SE4ALL (2012) Sustainable energy for all, a global action agenda: pathways for concerted action toward sustainable energy for all. p 1–32. https://www.seforall.org/sites/default/files/gather-content/SEFA-Action-Agenda-Final.pdf

Sedano F, Silva JA, Machoco R, Meque CH, Sitoe A, Ribeiro N et al (2016) The impact of charcoal production on forest degradation: a case study in Tete, Mozambique. Environ Res Lett 11 (9):094020

Shankar A, Johnson M, Kay E, Pannu R, Beltramo T, Derby E et al (2014) Maximizing the benefits of improved cookstoves: moving from acquisition to correct and consistent use. Global Health Sci Pract 2(3):268–274

Shankar AV, Onyura M, Alderman J (2015) Agency-based empowerment training enhances sales capacity of female energy entrepreneurs in Kenya. J Health commun 20(Suppl 1):67–75. https://doi.org/10.1080/10810730.2014.1002959

Smith HE, Hudson MD, Schreckenberg K (2017) (2016). Livelihood diversification: the role of charcoal production in southern Malawi. Energy Sustain Dev 36:22–36. https://doi.org/10.1016/j.esd.2016.10.001

Sola P, Ochieng C, Yila J, Iiyama M (2016) Links between energy access and food security in sub Saharan Africa: an exploratory review. Food Secur 8:635–642. https://doi.org/10.1007/s12571-016-0570-1

Tamire M, Addissie A, Skovbjerg S, Andersson R, Lärstad M (2018) Socio-cultural reasons and community perceptions regarding indoor cooking using biomass fuel and traditional stoves in Rural Ethiopia: a qualitative study. Int J Environ Res Public Health 15(9). https://doi.org/10.3390/ijerph15092035

UN (2015) Transforming our world: the 2030 agenda for sustainable development. United Nations General Assembly 70 Session. https://doi.org/10.1007/s13398-014-0173-7.2

UNEP-WCMC (2007) Mangroves of Western and Central Africa. United Nations Environment Programme (UNEP) Regional Seas Programme/UNEP—World Conservation Monitoring Centre, Cambridge, UK

USAID (2012) Guinea environmental threats and opportunities assessment. United States Agency for International Development

Valiela I, Bowen JL, York JK (2001) Mangrove forests: one of the world's threatened major tropical environments. BioSci 51:807–815

WHO (2014) Indoor air quality guidelines: household fuel combustion. World Health Organisation (WHO), Geneva. http://www.who.int/airpollution/guidelines/household-fuel-combustion/en/

WHO (2016) Burning opportunity: clean household energy for health, sustainable development, and wellbeing of women and children. World Health Organization (WHO), Geneva. http://www.who.int/airpollution/publications/burning-opportunities/en/

Wilkie ML, Fortuna S (2003) Status and trends in mangrove area extent worldwide. Working Paper FRA 63, Forest Resources Division, Forestry Department, UN-Food and Agriculture Organization, 292 pp

Woollen E, Ryan CM, Baumert S, Vollmer F, Grundy I, Fisher J et al (2016) Charcoal production in the Mopane woodlands of Mozambique: what are the trade-offs with other ecosystem services? Philos Trans R Soc Lond B Biol Sci 371(1703):20150315

World Bank (2011) Wood-based biomass energy development for Sub-Saharan Africa: issues and approaches. World Bank, Washington, DC

World Bank (2017) Global tracking framework 2017: progress towards sustainable energy. World Bank, Washington, DC. https://www.seforall.org/sites/default/files/GTF%20Executive%20Summary%202017.pdf

Chapter 8
Strategic Partnerships Between Universities and Non-academic Institutions for Sustainability and Innovation: Insights from the University of Ghana

Adelina Maria Mensah and Christopher Gordon

8.1 Introduction

Sustainability science is an emerging academic field that holds great promise for translating evidence-based research into practical long-term solutions. This is especially at the local level, where unique developmental paradigms and socioeconomic, cultural and environmental systems co-exist (e.g. Kates et al. 2001; Clark and Dickson 2003; Komiyama and Takeuchi 2006; Lang et al. 2012; Brandt et al. 2013; Miller 2014; Caniglia et al. 2017) (see Chaps. 4, 8 Vol. 2). In sustainability science, evidence is generated through inter- and transdisciplinary research approaches involving both academic and non-academic stakeholders (Komiyama and Takeuchi 2006; Lang et al. 2012; Caniglia et al. 2017) (Chaps. 4, 8 Vol. 2). However, different interpretations can affect the manner in which knowledge is generated, selected and understood for decision-making and implementation, especially where decisions regarding justice, social values and related moral and ethical choices are concerned (Hadorn et al. 2006; Juntti et al. 2010) (see Chap. 8 Vol. 2). The effective production and transfer of knowledge therefore requires a shift from simply producing information (or making information available to decision-makers) to proactively facilitating the sharing of different forms of knowledge as a means of improving the problem-solving capacity of stakeholders and decision-makers (Sheate and Partidário 2010).

A. M. Mensah (✉) · C. Gordon
Institute for Environment and Sanitation Studies, University of Ghana, Accra, Ghana
e-mail: ammensah@staff.ug.edu.gh; cgordon@staff.ug.edu.gh

A. Gasparatos et al. (eds.), *Sustainability Challenges in Sub-Saharan Africa I*, Science for Sustainable Societies, https://doi.org/10.1007/978-981-15-4458-3_8

Higher education institutions (HEIs)[1] have become increasingly important for enabling social transformations towards sustainability (Holmberg et al. 2008; Stephens and Graham 2010; Lambrechts et al. 2013; Lozano et al. 2013; Figueiro and Raufflet 2015; Verhulst and Lambrechts 2015) (see Chap. 8 Vol. 2). They are important microcosms, where researchers can engage between (and across) disciplines as well as with non-academic stakeholders using transdisciplinary, community-based and participatory research approaches (e.g. Polk and Knutsson 2008; Muhar et al. 2013; Lang et al. 2012). However, it is often challenging to implement effectively such research approaches, considering that academic researchers still try to better understand the practical processes that can catalyse this knowledge integration (Kaiser et al. 2016). Furthermore, there is scepticism about the reliability and validity of collaborative research with non-academic stakeholders, including, for example, the legitimacy of stakeholder claims, whether bargaining or deliberation is part of the stakeholder involvement process or the extent to which the underlying science and generated knowledge is autonomous (Lang et al. 2012; Yarime et al. 2012; Mielke et al. 2016). As sustainability science continues to mature, more effective processes for knowledge co-production and use are being explored (e.g. Michaels 2009; Dilling and Lemos 2011) (see Chap. 4 Vol. 2).

In developing countries, transdisciplinary research processes may need to follow different approaches that reflect the prevailing local socioeconomic and cultural contexts (van Breda and Swilling 2018). Furthermore, there is the need to achieve a better understanding of how collaborative research processes unfold in such countries with non-academic stakeholders such as industry, government and the general public (e.g. Zavale and Macamo 2016; Muriithi et al. 2018) (see Chap. 8 Vol. 2). This is particularly important for understanding whether (and how) universities are to make real contributions towards achieving sustainability. One key challenge to such processes of engagement in developing countries is the paucity of information regarding the effectiveness and performance of existing partnerships, and the available opportunities (Sá 2015; Gasparatos et al. 2017). Understanding how such transdisciplinary processes unfold can help develop successful collaborations and support international and national organizations in developing effective sustainability research strategies.

Many HEIs in SSA have declared regional commitments to address sustainability challenges. One such example is the Mainstreaming Environment and Sustainability in African (MESA) Universities Partnership. This partnership consists of institutions in 32 countries and helps universities practice sustainability. The UNESCO-drafted SSA Strategy for Education for Sustainable Development also aims at providing guidance for relevant government ministries (e.g. Ministries of Education and Training) and practitioners across the continent. At the same time, international

[1]For the purpose of this chapter an HEI is defined as (a) a university or a university college, (b) an institution conducting post-secondary education, or (c) an institution designated as eligible to receive support from funds administered by a National Council for Tertiary Education (NCTE).

donors have a great interest in catalysing ambitious and high-value partnerships to transform the quality, relevance, inclusiveness and value of Higher Education (HE) (Chap. 5 Vol. 1). For example, the UK Department for International Development (DFID) supports a GBP 45 million fund through its Strategic Partnerships for Higher Education Innovation and Reform (SPHEIR) programme.

In this respect, HEIs in SSA have the potential to generate the knowledge and practical solutions to solve many of the prevailing sustainability challenges discussed throughout these two volumes. However, despite the current pledges to promoting sustainability through education and research, an online survey of 73 HEIs across SSA highlighted the very low number of related research and outreach projects and academic involvement in sustainable development initiatives (GUNi et al. 2011). The top two barriers include the insufficient funding levels (mainly due to low state support for higher education), followed by the lack of human resources due to the prevailing shortage of staff development opportunities (GUNi et al. 2011). Other key barriers include the lack of awareness and information and inadequate university research agendas on programmes and activities related to sustainability (GUNi et al. 2011).

Despite its significant progress[2], Ghana still grapples with major sustainability challenges including a rapidly increasing population, persistent poverty, climate change, environmental degradation and urbanization (Thompson 2015) (see Chap. 6 Vol. 1). These sustainability challenges have led to ecosystem degradation and biodiversity loss, with important ramifications for critical sectors such as water, energy, food, health, shelter and security (e.g. Cobbinah et al. 2015a, b). Despite the multitude of policies, strategies and plans in place to tackle these sustainability challenges, there are still major challenges in their effective implementation and in achieving good performance (e.g. Domfeh et al. 2012; Fuseini and Kemp 2015). January 2016, Ghana adopted the 2030 Agenda for Sustainable Development, prioritizing issues such as climate change, economic equality, sustainable consumption, peace and justice. The Sustainable Development Goals (SDGs) have become a key element of the national development agenda seeking to address in a sustainable manner domestic challenges related to economic development including food security, health, sustainable energy, and natural resource management[3]. Although Ghana currently ranks 109th (out of 157 countries with a score of 59.9) in its overall SDG performance, it ranks sixth best in SSA (Sachs et al. 2017). Ghana aims at achieving

[2]Ghana achieved lower-middle income status in 2010, but still ranks 55th (out of 79 developing economies) in the Inclusive Development Index (IDI), exhibiting exceedingly high debt-to-GDP ratio, high poverty rate and insufficient protection of its natural capital (WEF 2017). A recurring theme within the national short- and medium-term development plans is to transform the economy to (a) achieve annual Gross Domestic Product (GDP) growth of at least 8 and 10% (non-oil- and oil-inclusive averages, respectively), and (b) sustain per capita income levels of at least USD 3,000 by 2020 (GoG-NDPC 2014). These figures would require substantial improvements compared to the 2016 levels of non-oil-inclusive GDP (4.9%) and per capita income (USD 1507) (GSS 2017).

[3]These goals have been integrated into the current Coordinated Programme of Economic and Social Development Policies (2017–2024) and the draft of the national Long-Term Development Plan.

zero poverty by 2030, but in order to attain this, there is a need for scientifically sound evidence and ground-breaking evidence-based solutions.

Currently, Ghana has 10 national public universities, 74 private universities, 5 polytechnics (one private and four public) and many other accredited institutions of tertiary education offering various degree programmes (NAB 2017). These HEIs are well positioned to contribute to the Ghanaian sustainable development agenda. They can play a major role in offering solutions to pertinent national sustainability challenges through related research and development (R&D), science and technology curricula, teaching and general engagement with society. Since the early 2000s, the government has shown renewed interest in developing science and technology policy frameworks to enhance economic development (Amankwah-Amoah 2016), including increasing the role of tertiary institutions in these efforts. The former government, for example, had pledged to have a public university in each of the ten regions of the country, convert polytechnics into technical universities and establish a University of Environment and Sustainable Development (UESD). So far, the Parliament has passed the Technical Universities Bill and the UESD Bill into law (NDC 2016).

Despite the ever-strengthened link between HEIs and sustainable development, the skillsets of graduates are generally inadequate to help achieve the required technological innovation (e.g. Amu and Offei-Ansah 2011; Bawakyillenuo et al. 2013; Jowi et al. 2013; Ababio 2016). Like many SSA universities, there are challenges in adequately integrating sustainability-related issues into curricula, research and other programmes (e.g. GUNi et al. 2011; Nyerere et al. 2014). Although the educational agenda and curricula related to sustainability should be delivered through teaching, learning and assessment, there are few sustainability-related policies and programmes to drive these. For example, out of the 43 taught courses in the curriculum of the Higher National Diploma in Purchasing and Supply Management programme offered by Polytechnics in Ghana, there were no sustainability focused courses and less than 4% of sub-topics were sustainability related (Etse and Ingley 2016).

HEIs in Ghana are also strategically placed to undertake research for new solutions to developmental issues to complement other national research institutions, such as the 13 institutes subsumed under the Council for Scientific and Industrial Research (CSIR). This has, however, not been as effective due to (a) archaic policies and institutional systems for science, technology and innovation that are not aligned to economic goals; (b) weak links and poor positive feedback between and among institutions (i.e. HEIs, research institutes and the private sector) and (c) few incentives and mechanisms to encourage communication and collaboration (UNCTAD 2011; Jowi et al. 2013).

In addition, although many university departments offer educational programmes that underline the core concept of sustainability science research (i.e., innovation and co-production of knowledge), there is limited documentation to track and evaluate them. In a review of six public Ghanaian universities, there were no explicit reports about sustainability performance (including research and curricula related to sustainability), although there was some general information was available in annual reports and on university websites (Hinson et al. 2015). This lack of documentation weakens the country's capacity to measure its progress and has possibly contributed to Ghana's failure to be ranked in the 2017 Global Innovation Index[4] (Cornell University et al. 2016, 2017, 2018) (Chap. 1 Vol. 1).

Using the University of Ghana (UG)[5] as a case study, this chapter explores the current mechanisms used to foster partnerships for sustainability research between the University and two key types of stakeholders, industry and government. Specifically, we aim to assess three key issues: (a) how the University carries out sustainability-related research with industry, (b) how the industry engages with UG researchers with regards to sustainability-related research areas and (c) how both of these processes support national development policies and structures. In particular, we employ an approach drawing from the Triple Helix model to understand the different structures and processes implemented by actors in academia, industry and government to facilitate collaborative research, innovation and research and development. We focus on (a) policies and regulatory frameworks; (b) institutions promoting research (c) research funding and collaborative research.

Section 8.2 outlines the main aspects of the Triple Helix model. Section 8.3 identifies the aforementioned structures and processes for the University of Ghana (Sect. 8.3.1), industry (Sect. 8.3.2) and government (Sect. 8.3.3). Section 8.4 puts these findings into perspective identifying critical points across the three analytical aspects (Sects. 8.4.1–8.4.4) and offers policy recommendations to enhance the development and effectiveness of partnerships between universities and non-academic institutions in Ghana and other parts of SSA (Sect. 8.4.5).

8.2 Methodology

This chapter adopts the Triple Helix model, developed by Etzkowitz and Leydesdorff (1995), to conceptualize the linkages and collaboration between different actors in research and development related to sustainability. The Triple Helix model assumes a collaborative relationship between academia, industry and government, where economic development is driven by the production and dissemination of socially relevant knowledge. This model reflects several scholars who increasingly perceive universities (and their collaboration with the government and industry) as central for technological innovation to create wealth and build a knowledge-based society (Soini et al. 2018).

[4]The Global Innovation Index measures the innovation performance of 127 economies for advances in agriculture and food value chains (Ghana ranked 107 out of 126 countries in 2018, and 102 out of 128 countries in 2016) (Cornell University et al. 2016, 2017, 2018).

[5]The University of Ghana is Ghana's oldest university. It was founded as the University College of the Gold Coast by Ordinance on 11 August 1948. Its mandate is to provide and promote university education, learning and research. Its current vision is "*to create a vibrant intellectual climate that stimulates relevant cutting-edge research and community engagement*" (UG 2017).

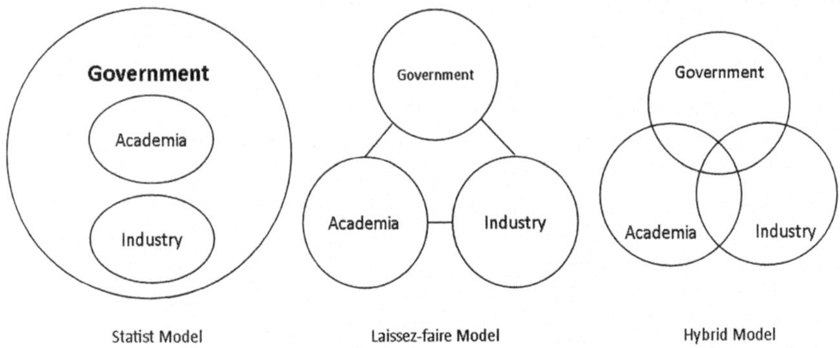

Fig. 8.1 Different conceptualizations of the academia–industry–government relationship in the Triple Helix model (Adapted from Etzkowitz and Leydesdorff 2000)

The Triple Helix model has evolved over time with various interpretations and critiques (Etzkowitz 2003; Etzkowitz and Leydesdorff 2000; Leydesdorff 2010) (Fig. 8.1). Different conceptualizations include: (a) the *statist* model, where government plans, controls and directs the relationship with academia and industry (with industry being the main agent of innovation and academia mainly responsible for teaching and academic research); (b) the *laissez-fair* model, where industry is the driving force of innovation, supported by academia and government who operate independently in their separate institutional arenas; and (c) the *hybrid* model, where, through a network, mutual collaboration is encouraged, and linkages are created among the three types of institutions (and with other organizations/disciplines), with innovations borne out of these interactions.

In the hybrid model, universities are perceived as having a stronger role in catalysing economic development through academic entrepreneurial activities that also resonate with industrial and economic objectives (Godin and Gingras 2000; Muller 2006; Etzkowitz and Leydesdorff 2000; Shinn 2002). There are numerous ontologies and reviews that characterize universities' collaborative processes and contributions towards sustainable development (e.g. Trencher et al. 2014; Draghicia et al. 2015; van der Hel and Biermann 2017; Soini et al. 2018; Paletta et al. 2019), yet there are significant knowledge gaps for such processes for universities in developing countries. According to Sarpong et al. (2017), many developing countries still operate within the *statist* or *laissez-faire* models (Fig. 8.1), as country-specific evolutionary pathways between models depend on the traditional activities and practices of institutions.

Through a synthesis of the literature and secondary data, this chapter describes the individual spheres of the Triple Helix model, that is, academia (UG) (Sect. 8.3.1), industry (Sect. 8.3.2) and the government of Ghana (Sect. 8.3.3). In particular, for each of these types of institutions we synthesize information from peer-reviewed articles and grey literature about the current (a) policies and regulatory frameworks, (b) institutions promoting research, and (c) research funding and opportunities for collaboration, among these three institutions.

8.3 Results

8.3.1 University of Ghana

8.3.1.1 Policies and Regulatory Frameworks

The National Council for Tertiary Education (NCTE) coordinates the activities of the UG as an institution of tertiary education in the country, alongside other specialized supporting agencies of the Ministry of Education (Table 8.1). The three main national legislative and regulatory instruments related to education are the (a) Ghana Education Service (GES) Act of 1995 (Act 506), (b) the Education Act of 2008 (Act 778) and (c) the Education Strategic Plan (ESP) 2010–2020.

As with other universities in the country, the UG is empowered to set its own priorities and agenda for academic programmes, educational curricula and research activities. The UG Basic Laws document and Strategic Plan (2014–2024) (in addition to numerous policies such as those on academic quality, code of conduct, sexual harassment, ethics), outline the various guidelines towards

Table 8.1 Key national institutions responsible for HEIs in Ghana

National Institution	Role
National Council for Tertiary Education (NCTE)	– Advise the Minister of Education on the development of institutions of tertiary education. – Formulate relevant policies.
National Accreditation Board (NAB)	– Accredit public and private institutions with regard to the contents and standards of their programmes.
National Board for Professional and Technician Examinations (NABPTEX)	– Formulate and administer examinations, evaluation, assessment and certification schemes for professional bodies, non-university tertiary institutions and private institutions.
Ghana Academy of Arts and Sciences (GAAS)	– Promote the pursuit, advancement and dissemination of knowledge in all branches of the sciences and the humanities.
National Service Scheme (NSS)	– Provide newly qualified graduates the opportunity to obtain practical exposure on the job (both in the public and private sectors), as part of their civic responsibility to the State.
Students Loan Trust Fund (SLTF)	– Provide facilities to enhance tertiary education to support students. – Provide funds to support any activities and programmes for relevant courses.
Council for Technical and Vocational Education and Training (COTVET)	– Co-ordinate and oversee all aspects of technical and vocational education and training in the country. – Formulate policies for skills development across the broad spectrum of pre-tertiary and tertiary education, in the formal, informal and non-formal sectors.

achieving its overall objectives. The type and nature of academic programmes, however, have to abide by the requirements and procedures of the National Accreditation Board (NAB), the National Council for Tertiary Education (NCTE) and the Governing Councils of the Institutions. To strengthen the partnership between education and industry, industrial liaison desks have been jointly set up by the NCTE and the Council for Technical and Vocational Education and Training (COTVET).

8.3.1.2 Institutions Promoting Research

At the UG, a collegiate system comprising four colleges was introduced in the 2014/2015 academic year: (a) Basic and Applied Science, (b) Education, (c) Health Sciences and (d) Humanities. This collegiate system provides a decentralized approach for schools and institutes to carry out independent collaborative research with local and international partners. Through its different departments, research institutes and centres, faculty members are involved in policy-relevant research that supports national development, often in collaboration with other universities and national and international institutions. The various university academic units have had long-standing formal and informal relationships with the private sector. However, there are only few documented cases where researchers from diverse disciplines work together to deliver sustainability-related knowledge, innovation and technology through transdisciplinary research processes (e.g. Box 8.1).

Box 8.1: Example of a National Transdisciplinary Research Platform Including UG Faculty Members

TERSUS Ghana is a private Environmental Sanitation Think-Tank, comprising of a team of senior researchers with different expertise from various national and international universities. The Board is currently chaired by Prof. Chris Gordon, from the Institute for Environment and Sanitation Studies at the University of Ghana. TERSUS has the mandate of providing adequate basic sanitation consultancy services in the country. The think-tank brings together partners from the private sector, government and academia that provide quality research, policy recommendations, and analysis on a full range of public policy issues pertinent to the waste management sector. The main objectives of TERSUS Ghana are to serve as a common national platform for advocacy on environmental sanitation, and engage with government to enhance its commitment towards a dedicated national sanitation goal.

To support such collaborative research efforts, the UG has established two formal institutional structures. The first is the Office of Research, Innovation and Development (ORID), which was established in 2010. Its mandate is to promote, facilitate and coordinate cutting-edge research within the university. This unit provides several services related to (a) research support, (b) grant management, (c) capacity building, (d) intellectual property and technology transfer, (e) research dissemination and (f) external funds. The Technology Development and Transfer Centre (TDTC), housed within ORID, facilitates effective partnership between the university and the industry for developing and transferring innovative demand-driven technologies to meet industry needs. The second is the Institute of Applied Science and Technology (IAST) (established in 2012), to build strategic partnerships between academia and industry for mutual benefit. The institute serves as a conduit between industry and society to allow the access to relevant academic knowledge, technical skills and solutions that can address their needs and challenges. IAST also provides a platform for academia to demonstrate research concepts before their full application and commercialization. Both institutions provide a space for systematic engagement with various stakeholders outside academia, which is usually announced through information on their websites. The ORID Annual Reports also provide an overview of the different activities with stakeholders. Despite the possibilities to generate multiple beneficial outcomes, there has not been any comprehensive evaluation of the extent to which national priorities and stakeholder concerns influence research and knowledge production in the UG (as well as the nature of the relationships between stakeholders and academic researchers).

UG does not have a specific policy on sustainability; however, its strategic plan aims to ensure that general sustainability principles are achieved through academic teaching, research programmes and campus-based activities. Examples of sustainability-related initiatives include the (a) Centre for Climate Change and Sustainability (created in 2017 with support from the Open Society Foundation) and a MPhil in Sustainability Science at the Institute for Environment and Sanitation Studies (started in 2018) that support UG efforts to promote inter- and transdisciplinary teaching, learning and research in sustainability science; (b) UG's central library, the Balme Library, displays information on one SDG every month in its Reference Hall to promote awareness to the general university community and (c) the Vice Chancellor's Green Project, a campus-based programme established in 2017 that seeks to green the university by ensuring sustainability for on-campus transport, water supply, sanitation and landscaping, among others.

To evaluate how UG initiatives and projects align to global SDGs, Gordon (2018) mapped the university's existing priorities and research units' activities with the individual goals and targets of the SDGs. Overall, the assessment showed that the UGs focus areas were SDGs 4 (Quality education), 8 (Decent work and economic growth), 9 (Industry, innovation and infrastructure) and 10 (Reduced inequalities) (Fig. 8.2). Other more specific analyses such as major themes of doctoral theses (2006–2016) showed priorities in SDGs 2 (End hunger), 3 (Good health and wellbeing) and 6 (Water and sanitation for all). Across all research institutions, staff publications prioritized SDGs 3 (Good health and wellbeing), 5 (Gender equality) and 11 (Safe and sustainable cities) for the same period.

Fig. 8.2 Key contributions of the University of Ghana towards achieving SDGs (Source: Gordon 2018)

8.3.1.3 Research Funding and Collaborative Research

The government is the main funder of the UG. However, according to the UG Strategic Plan (2014–2024), government support has decreased from 90% to less than 50% of the actual operating expenditure within the last decade (Table 8.2). Other funding comes from internally generated funds (mainly from academic fees), research grants and other sources.

Research fundraising has become a crucial avenue to raise funds for the UG research activities and outputs. ORID facilitates research fundraising by offering research support services, ranging from identifying research funding opportunities to processing grant applications (Table 8.3). In 2013, ORID received support from the Association of African Universities (AAU) to establish mechanisms for ensuring the effective coordination of activities as diverse as fundraising, monitoring and evaluation and managing donor relations. For IAST, despite the identification of various priority research areas, research funding for the period 2014–2017 was obtained from only a few industries, mainly the Japan International Research Center for Agricultural Sciences (JIRCAS) (USD 10,000), Bank of Ghana (USD 10,000) and Fuji Oil (USD 10,000) (pers. comm. IAST 2017).

The research funding managed by ORID and IAST has progressively increased in various key thematic areas (Table 8.4), in line with UGs vision to become a research intensive institution in four prioritized areas for international research collaboration: (a) malaria research, (b) climate change adaptation, (c) food production and processing and (d) development policy and poverty monitoring and evaluation. There is no funding that is being specifically allocated for sustainable development research. However, the Centre for Climate and Sustainability Studies (C3SS) (launched in March 2017) provides a Climate Change Resources Centre for

Table 8.2 Sources of funding for the University of Ghana

	Total annual expenditure (%)			
Funding source	2013	2014	2015	2016
Government subvention	48.09	45.69	44.30	42.74
Academic fees and other student charges	25.40	25.84	24.52	26.18
Research grants[a]	16.03	16.60	19.93	20.03
Other sources[b]	10.48	11.86	11.25	11.04

Source: University of Ghana Financial statements
[a]Research grants are mostly from external donors such as the Bill and Melinda Gates Foundation, Carnegie Foundation, Danish International Development Agency (DANIDA), European Union, Leverhulme Trust, National Science Foundation, Wellcome Trust and WIPO Re:Search, among others
[b]Includes bursaries and financial aid, interest income and other income

Table 8.3 Total research funding for the Academic Years 2014–2016 (in 1000 USD)

	2014–2015	2015–2016
Humanities	17,585	13,998
Health sciences	16,202	11,870
Basic and Applied Sciences	18,734	6,222
Education	4	—
Total	52,525	32,091

Note: Estimates are based on available ORID reports. The collegiate system took effect in August 2014

information on climate change and sustainability issues, in addition to the Masters programmes (MSc./MPhil.) in Climate Change and Sustainable Development (CCSD). The UG provides approximately USD 22,800 (GHS 120,000)[6] per year towards the running costs of the centre, in addition to government subventions for staff time costs.

8.3.2 Industry

8.3.2.1 Policies and Regulatory Frameworks

The Ghanaian industrial sector mainly consists of privately owned micro (55%) and small (39%) firms (Ackah et al. 2016). Over 90% of these firms operate in the manufacturing sector, which also accounts for the largest job creation among the industrial sub-sectors. The Ghana Industrial Policy (GIP) (2011) is based on national development agendas and was formulated to enhance productivity, efficiency and growth in the sector (especially the manufacturing sub-sector). The GIP acknowledges the limited existing research capacity and application of science, technology and innovation in the industrial sector. It prescribes the need to strengthen linkages

[6]In late 2018, 1 GHS = 0.19 US$

Table 8.4 Key cross-cutting research areas funded through ORID and IAST

Thematic area	ORID	IAST
Age and health	✓	—
Agribusiness and material technology	—	✓
Climate change	✓	—
Energy and climate change mitigation	✓	✓
Food processing	—	✓
Food security and safety	✓	✓
Gender and women's studies	✓	—
Health	✓	—
Health and traditional medicine development	✓	✓
Neglected tropical diseases	✓	—
Impacts of dams	✓	—
Infrastructure development	✓	✓
Maternal health	✓	—
Mental health	✓	—
Mining	✓	✓
Natural resource management and sustainable exploitation	✓	✓
Non-communicable diseases	✓	—
Policy	✓	—
Policy research	✓	—
Water and sanitation	✓	✓

Source: Annual ORID reports; Pers. comm, IAST 2017

with other key stakeholders such as universities, research and technology transfer institutions and between industry itself. However, these prescriptions are very general, with no clear guidance or indicative targets and timelines for achieving this (Bawakyillenuo et al. 2013).

8.3.2.2 Institutions Promoting Research

The Ministry of Trade and Industry (MOTI) is the main government agency mandated with the overall responsibility for the formulation, development, implementation, monitoring and evaluation of Ghana's internal and external trade. MOTI provides policy direction for Ghanaian exports and industrial trade and monitors and implements national private sector development programmes and activities. Several of these policies, programmes and special projects are implemented through various specialized agencies (Table 8.5). Additionally, several institutions provide support to the industrial sector, including the Association of Ghana Industries (AGI), Ghana National Chamber of Commerce and Industry (GNCCI), Private Enterprise Foundation (PEF) and the Federation of Association of Ghanaian Exporters (FAGE).

Table 8.5 Key agencies related to the industrial sector in Ghana

National Institution	Role
Export Development and Investment Fund (EDIF)	– Provide financial resources for the development and promotion of the export trade in Ghana.
Ghana Export Promotion Council (GEPC)	– Facilitate, develop and promote Ghanaian exports.
Ghana Free Zones Board (GFZB)	– Promote investment through the provision of a conducive business environment and attractive incentives.
Ghana Regional Appropriate Technology Industrial Service (GRATIS)	– Develop, promote and disseminate marketable technologies and skills for the growth of industry (particularly, micro-, small- and medium-scale enterprises) in Ghana and the West African sub-region.
Ghana Standards Authority (GSA)	– Promulgate standards, promote standardization and undertake conformity assessments to ensure that the produced products, goods and services (whether for local consumption or for export) are safe, reliable and of high quality.
National Board for Small Scale Industries (NBSSI)	– Promote and develop micro- and small-scale enterprises.

8.3.2.3 Research Funding and Collaborative Research

There is little documentation describing how industry funds innovation-related research in Ghana. Industry had, in 2016, the second largest contribution to GDP (24.2%) after the services sector (56.9%) (GSS 2017). Growth in the industrial sector has fluctuated over the past decade in response to various events. For example, in 2008, industry's contribution increased due to large infrastructural developments, while it declined in 2009 due to unreliable power supply and increasing fuel prices (Table 8.6). The increasing contribution from 2011 onwards can be credited to the extraction of crude oil that started in 2010 and increased the mining/quarrying sub-sector inputs from about 2% to over 8% since 2011. Capital availability has been the biggest challenge to industrial growth over the past four decades. This is both in terms of access to capital and the high cost of borrowing due to interest rates, which sometimes exceeded 42% during that period. The declining value of the Ghanaian currency[7] has also limited the amount of foreign direct investment (FDIs) in Ghanaian industries.

UG has successfully collaborated with many domestic industries, mainly through the transfer of applied science and technology know-how that was introduced into their manufacturing processes (e.g. Box 8.2). Additional research areas have been recommended in order to forge a stronger collaboration between the UG and the industry immediately and in the short- and long term (Table 8.7).

[7]The Ghana Cedi is the only legal unit of currency nationally since 2008, after replacing the Cedi. Since its introduction, its value has gradually depreciated to one-fifth of its original value by the end of 2018.

Table 8.6 Gross Domestic Product (GDP) contribution by economic sector (in percent)

	2006	2007	2008	2009	2010	2011	2012	2013	2014	2015	2016
Agriculture	30.4	29.1	31.0	31.8	29.8	25.3	22.9	22.4	21.5	20.3	18.9
Industry	20.8	20.7	20.4	19.0	19.1	25.6	28.0	27.8	26.6	25.1	24.2
Mining	2.8	2.8	2.4	2.1	2.3	8.4	9.5	9.4	8.0	5.3	4.2
Oil and gas	0.0	0.0	0.0	0.0	0.4	6.7	7.7	8.2	7.2	4.1	2.1
Manufacturing	10.2	9.1	7.9	6.9	6.8	6.9	5.8	5.3	4.9	4.8	4.6
Electricity	0.8	0.6	0.5	0.5	0.6	0.5	0.5	0.4	0.4	0.9	1.1
Water/sewerage	1.3	1.0	0.8	0.7	0.8	0.8	0.7	0.6	0.5	0.6	0.5
Construction	5.7	7.2	8.7	8.8	8.5	8.9	11.5	12.0	12.7	13.5	13.7
Services	48.8	50.2	48.6	49.2	51.1	49.1	49.1	49.8	51.9	54.6	56.9

Source: (GSS 2017)

Box 8.2: Examples of Collaboration Between UG and Industry

The UG Department of Nutrition and Food Science has many collaborations with industry, which have either been initiated by industry, or by the department through faculty research, extension of knowledge, product development and commercialisation of products. One example is its collaboration with the Cocoa Processing Company Ltd., with support from the British Council's African Knowledge Transfer Partnership Programme, to introduce a low calorie sugar free chocolate (ASPIRE) launched on 10 February 2010. The department played a key role in the process, from conceptualisation through product formulation, package development and marketing plan/product launch. Additional benefits from this collaboration were (a) a better understanding of how to fit academic curricula to industry needs, (b) stronger ties with industry through guest lectures from industry experts, (c) use of industry facilities by academic staff and students, (d) student internships, and (e) sponsorship for students' projects.

The College of Agriculture and Consumer Sciences (CACS) (consisting of the School of Agriculture, the School of Veterinary Medicine, and the Institute of Agriculture Research) have also had numerous linkages with various local and international industries. In one collaboration with the Forum for Agricultural Research in Africa (FARA) and Concern Universal, a British NGO based in Ghana, CACS linked up with Bioplastic Ghana Ltd., to develop biodegradable plastic bags for the control of post-harvest grain losses. The Hermetic Triple Layer Sack Project also involved extension agents from the Ministry of Food and Agriculture and farmers in two maize growing communities in the Upper West and Brong Ahafo regions.

Source: IAST (2012)

Despite manufacturing being the key industrial sub-sector in Ghana, the level of engagement with UG is, however, still quite inadequate. Many of the research findings and recommendations generated within academia have not been implemented by industry on account of missing interlinkages and lack of commitment (IAST 2012). Other challenges for achieving a more effective collaboration between industry and academia in Ghana include:

- Secrecy and mismatch between research orientation of industry and academia
- Difficulties in negotiating partnerships
- Preference towards engaging private consultants for research work
- High research charges of academics
- Lack of research and development considerations in business plans
- Local industries not willing to invest in capacity building
- Import of cheap products

Table 8.7 Priority research areas for collaborative research between industry and academia

Research area	Immediate	Short term	Long term
Agricultural Engineering	—	✓	✓
Agriculture (breeding)	—	✓	✓
Best practices	✓	—	—
Bio-remediation, Composting	—	✓	—
Chemical industry	—	✓	✓
Clay/ceramics industry	—	✓	✓
Customer surveys and feedback	✓	—	✓
Data aggregation and management	✓	✓	✓
Financial management and projections	✓	—	✓
Food safety	—	✓	—
Food processing	—	✓	✓
Food supplements	—	✓	✓
Hazards analysis and risk assessment	✓	—	—
Health and safety	✓	—	✓
Herbal medicines	—	✓	✓
Intellectual property	—	✓	✓
Insulation and tiling	—	✓	—
Internship for ideas	✓	—	✓
Market demand	✓	—	—
Nutraceuticals (e.g. plant extracts, pigments)	—	—	—
Packaging (quality of and biodegradable options)	—	✓	✓
Pharmaceutical industry	—	✓	✓
Preservation of traditional foods	—	✓	—
Process audit	✓	—	—
Process design	—	—	✓
Product innovation	—	✓	✓
Quality assurance and product innovation	—	✓	✓
Residue analysis in foods	—	✓	—
Small and medium-scale accounting procedures	✓	—	—
Software development	—	✓	✓
Standardization (SOP)	—	✓	✓
Technology transfer	—	✓	✓
Training	✓	—	✓
Trend analysis	✓	—	—

Source: IAST (2012)

- Poor adherence to local content and local participation laws[8]
- Absence of a legal framework to ensure that intellectual property rights (IPR) are not forfeited by investors
- Multi-national companies' use of imported research results for their application in local factories

It has been proposed that collaboration between academia and industry should be driven by appropriate policies in order to be mutually efficient and beneficial. IAST employs various strategies to achieve this, mainly through organized events such as exhibitions, symposia, workshops and seminars to enhance engagement and show-case technologies as well as support project-based research fundraising activities. Outputs from these activities have indicated various opportunities that can be explored to improve the engagement between academia and industry (Table 8.8).

8.3.3 Government Institutions

8.3.3.1 Policies and Regulatory Frameworks

The Government of Ghana prioritizes the further development of Research and Development (R&D) or Science, Technology and Innovation (STI) initiatives for addressing its SDG targets (Box 8.3). In 2017, the country finally adopted an STI policy that could guide actors in the various economic sectors and prioritize investments. The revised National Science, Technology and Innovation Policy (2017–2020) highlights innovation in a new framework of actions and programmes to apply science and technology towards achieving social and economic objectives. The policy also provides sector-specific programmes and activities for agriculture, health, education, environment, energy, trade, industry, natural resources, human settlements and communications.

However, it is still unclear how tertiary institutions can contribute to achieving these development plans or addressing the needs of industry, although their relevance has always been acknowledged in national plans. In fact, it has been suggested that there is a need to develop and implement a comprehensive tertiary education policy framework that is situated within the long-term development plan of the country (Bawakyillenuo et al. 2013).

[8]In the extractive industry, the local content law applies to a quantum or percentage of materials, personnel, financing, goods and services that must be locally produced. Local participation laws stipulate that foreign investors must partner with local companies at recommended share percentages.

Table 8.8 Opportunities for improving engagement between academia and industry

Aspects	Opportunities
Collaboration mechanisms	• Policies should include a platform for stakeholder engagement • Invite industry practitioners as guest lecturers • Create avenues for academics to utilize industry facilities • Develop internships for students, sponsorship of student projects and increase relevance of student projects
Communication/ Dissemination	• Facilitate regular interactions for sharing the benefits of new discoveries and innovations • Develop user-friendly databases of academic personnel, indicating researchers and their research profiles • Develop Research and Development (R&D) units within companies, which will serve as a common platform for engagement with academia • Develop products for marketing academic research findings
Employment	• Generate new jobs through R&D activities, especially at the community level • Develop incubator systems to facilitate research and training for graduate students
Funding	• Raise sustainable funding for research/R&D through tax levies (e.g. corporate tax) • Develop joint funding proposals • Solicit for more public–private partnerships • Commercialize R&D outputs through independent companies • Enact cost-sharing arrangements on projects of national importance that would also benefit industrial partners
Incentives	• Develop Memoranda of Understanding (MOUs) to reduce service charges to below commercial rates, as a means of encouraging industry to link up with academia • Base collaboration between industry and academia on appropriate policies
Intellectual property system	• Offer appropriate and competitive payment for academic research services • Implement non-disclosure and benefit-sharing agreements for academia–industry partnerships • Sign MOUs for specified periods of time • Publicly recognize researchers who provide information through various media
Marketing	• Enhance the publicity and marketing of academic research findings, particularly through reports that are relevant and could provide solutions to the needs of industry • Develop academic research findings that are of commercial value and interest
Quality of outputs	• Set up quality standards for local products in terms of packaging, product quality and durability • Frame/base relevant academic research on the needs of industry

Source: *Based on* IAST (2012)

Box 8.3: Role of Science and Technology in Ghanaian Development Policy

The Coordinated Programme of Economic and Social Development Policies (2017–2024) is Ghana's current development policy. This policy places premium importance on mainstreaming science and technology towards achieving its agenda for job generation, creating prosperity and equal opportunity for all. In the short- to medium-term, one key intervention highlighted for implementing flagship projects and initiatives is "*Leveraging on Science, Technology and Innovation*". This will support the achievements of flagship projects and initiatives such as the "*Planting for Food and Jobs*" campaign to stimulate food production and generate income, the "*One-Village-One Dam*" initiative to ensure farming throughout the year, especially in the semi-arid northern regions, and the "*One District, One Factory*" initiative to establish at least one industrial enterprise in each of the 216 Ghanaian districts through public-private partnerships. To achieve these, the policy proposes to set up technology and innovation incubation centres on the campuses of public research institutions, as a means of providing basic facilities that the technology innovators need, to scale up and pilot their innovations to commercial levels.

8.3.3.2 Institutions Promoting Research

The Ministry of Science, Technology and Innovation (MESTI) is the main coordinating ministry for the formulation of policies regarding the application of scientific research and technology. On the other hand, the Ministry of Education (MoE) is responsible for STI-related education policies. Departments and agencies under MESTI include the Council for Scientific and Industrial Research (CSIR), the Ghana Atomic Energy Commission (GAEC), the Environmental Protection Agency (EPA) and the Town and Country Planning Department, whose collective primary responsibilities relate to applied R&D in Ghana (MESTI 2017a).

The Science, Technology and Innovation Directorate, one of MESTI's six directorates, is responsible for formulating STI policies, and communicating, coordinating and monitoring programmes and activities of relevant agencies and departments within the Ministry, at various levels of implementation. The new STI policy proposed an apex statutory body, the Presidential Advisory Council on Science, Technology and Innovation (PACSTI), that links the ministry to the President, as a means of ensuring the effective coordination and harmonization of the new policy and other national STI programmes. In addition, MESTI launched in 2018 the Ghana Innovation and Research Commercialization Centre (GIRC-Centre). The aim of this body is to facilitate the translation of research findings into commercial products and services to achieve the objectives of the current national development plan.

8.3.3.3 Research Funding and Collaborative Research

In Ghana, R&D is financed by the Ministry of Education, the GETFund[9], institutional internally generated funds and multilateral and bilateral donors (Table 8.9) (MoE 2016). The Export Development and Agricultural Investment Fund (EDIF) through the Ministry of Trade and Industry also provides R&D grants for the agro-processing industry to facilitate improved products and financial returns to increase the competitiveness of the local industries (i.e. SMEs) on the local and international markets.

R&D funding has averaged around 0.38% of the Ghanaian GDP in the last decade, placing the country 74th globally (out of 113 countries) and 12th in SSA (OECD 2017). The STI policy, however, makes provisions for setting up a National Science, Technology and Innovation Fund to raise R&D investments to 1% of GDP, in the short- to medium term, as recommended by the African Union.

Government expenditure for education amounted to USD 2.3 billion (GHS 8.7 billion) in 2015 (an increase of 24.6% from 2014 levels), which was equivalent to 6.16% of the national GDP (Fig. 8.3). Of this, tertiary education accounted for 23% (GHS 1.8 billion), which was higher than spending for senior high, junior high and basic schools. Comparatively, other countries in the ECOWAS or the West Africa sub-region spend on average 4.3% of their national GDP for tertiary education, making Ghana a front-runner in the region.

To ensure that in future the budget's national financial priorities are aligned with SDG priorities, the Ministry of Finance published the SDG Budget Baseline Report in 2018. This report is the first of an annual series of SDG budget reports and provides a methodology and framework to develop a more advanced and comprehensive tracking system. The 2018 report cross-mapped the objectives and costs in the national 2018 budget, with SDGs, targets and indicators. In the case of quality education (SDG 4) approximately USD 266,000 (GHS 1.4 million) was allocated, sourced from the government's other funds (60%), internally generated funds (30%) and donor partners (10%), respectively (MoF 2018). Comparatively, the commitments to transformative investments in economic and social infrastructure as an avenue towards job creation, poverty reduction, sustainable industrialization and innovation was estimated at USD 427.5 million (GHS 2.25 billion).

[9]The Ghana Education Trust Fund (GETFund) is a public sector agency set up by an Act of Parliament in 2000, managed by a Board of Trustees and a Secretariat. The main function of GETFund is to support the government's financing of public education institutions (from pre-tertiary to tertiary).

Table 8.9 Sources of R&D funding in Ghana (in 1000 GHS)

Sources of funds	2013	2014	2015
Public sector (IGF)	15,298	20,378	12,791
Public sector (Government)	52,811	73,592	89,508
Private sector firms	31	–	–
Private not-for profit organizations	31	27	28
Foreign donor agencies	171,303	205,462	388,994
Total (1000 GHS)	239,474	299,458	491,321
Total (1000 USD)	101,904	92,999	128,956

Source: Modified from (MESTI 2017b)

Note: Exchange rate of USD to GHS exchange rate 31 Dec. 2017 (1: 2.35 for 2013; 1: 3.22 for 2014; 1: 3.81 for 2015)

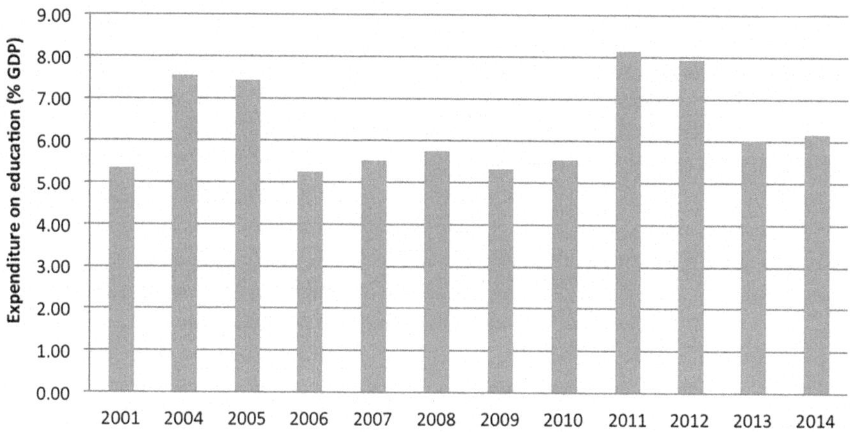

Fig. 8.3 Government expenditure on education as a fraction of the Gross Domestic Product (GDP) (*Source*: Global Economy 2017)

8.4 Discussion

As outlined throughout this chapter there have been some efforts to foster collaborative and innovative research towards sustainability in Ghana. Section 8.3 outlined thoroughly the different supportive structures in academia, private sector and government in terms of (a) policies and regulatory frameworks (Sects. 8.3.1.1, 8.3.2.1, 8.3.3.1); (b) institutions promoting research (Sects. 8.3.1.2, 8.3.2.2, 8.3.3.2) and (c) research funding and collaborative research (Sects. 8.3.1.3, 8.3.2.3, 8.3.3.3). Sections 8.4.1–8.4.4 identify some of the major challenges across these three domains in Ghana (and other parts of SSA), as well as some of the possible solutions and key policy recommendations to overcome them (Sect. 8.4.5).

8.4.1 Policies and Regulatory Frameworks

Ghana's new STI policy has provided general and sectoral guidelines towards achieving both regional (e.g. Africa Agenda 2063) and national sustainable development priorities. However, it has been criticized for viewing innovation only as an off-shoot of science and technology (Oduro-Marfo 2015). Scholars have suggested that national innovation policies and criteria could also promote processes and innovations that are not necessarily only science based (Oduro-Marfo 2015) (see Chap. 10 Vol. 1; Chaps. 3-4 Vol. 2). Such an example in the case of Ghana is the National Friday Wear programme (a traditional dress code on Fridays) that has created domestic demand for local prints and fabrics and has boosted the dying textile manufacturing industry. Similarly, the celebration of the National Chocolate Day has also had positive implications for cocoa-based industries in Ghana.

The STI policy can still have a far-reaching effect on both how academic research can help solve current sustainability challenges as well as how Sustainability Science research will be conducted in Ghana. However, the effectiveness and success of the STI policy would require strong actions to address the (a) limitations related to its implementation (e.g. lack of accurate budget estimates for actions) and the (b) broad factors that have curtailed STI performance in developing countries, Ghana included (e.g. constraints related to tertiary education institutions, intellectual property, innovation, productivity and competitiveness) (UNECA 2016) (Chap. 1 Vol. 1). As discussed below, improving the performance of HEIs and enhancing innovation would be critical for the effective implementation of the STI policy.

First, although there has been recognition of the importance of HEIs for sustainable development in numerous national policies, there is a lack of specific prescriptions on how exactly to achieve this. With an average of 6.75% tertiary school enrollment in Ghana, minimum of 0.69% in 1971 and maximum of 16.01% in 2017 (Global Economy 2018), higher education programmes need to be restructured in a way to attract more students. More importantly these programmes should aim at fostering key competencies and capacity for critical thinking that can help solve sustainability challenges (e.g., Weik et al. 2011; Filho et al. 2018). This could be attained through specific policy interventions, such as a tertiary education policy on STI, or increased spending on high quality tertiary education.

Second, SSA countries (including Ghana) are generally poorly performing in global innovation indices[10] (Chap.1 Vol. 1), with only seven SSA countries ranking in the top 100 in 2018 (Cornell University et al. 2018) (Chap. 1 Vol. 1). In order to improve their ranks, SSA countries need to prioritize investment and data collection for reporting (see Chaps. 1, 5 Vol. 1). Innovation is also linked to competitiveness (i.e. the set of institutions, policies and factors that determine national productivity). In terms of competitiveness Ghana performs rather poorly for business

[10]It should be noted that IP-related indices have been criticised on the grounds of having been developed to account for more Western-oriented modes of output, and inability to account for how creativity actually occurs in SSA settings (de Beer et al. 2014).

sophistication and innovation, ranking 111th (out of 137 countries) in the 2017–2018 Global Competitiveness Index (WEF 2018). Some of the identified factors include the (lack of) access to financing, high tax rates, high corruption, inadequate supply of infrastructure, high inflation and poor foreign currency regulations, among others (see Chaps. 1, 5 Vol. 1).

To enhance innovation in the Ghanaian economy, universities will also need to re-conceptualize their roles on how they contribute to innovation factors that drive sustainable development. In many developed countries, universities have been supported through broader national initiatives seeking to commercialize technology through patenting, licencing and spin-offs (e.g. through science parks, business incubators, public seed capital funds and bridging institutions that link universities to industry) (Mathews and Hu 2007; EENEE 2014). Consistent with such efforts, the UG (and HEIs in general in Ghana) will need to develop indicators for the evaluation and periodic monitoring of technology transfer mechanisms. Ideally, this should offer a good basis to formulate more appropriate supporting policies.

8.4.2 Institutional Structures Promoting Research

At the national level, the effectiveness of innovation systems for spurring long-term sustainable economic growth depends on the quality of formal institutions (e.g. laws, regulatory bodies) as well as on the linkages and knowledge flows between the various actors (e.g. universities, private sector, governments, and other institutions) (Watkins et al. 2014). However, this is very complicated in practice, especially in SSA countries such as Ghana. Apart from the clear need to develop and implement an integrative approach to boost innovative activities and technology development (as proposed by the triple helix model), scholars have acknowledged the underlying methodological and practical complexities for realistically mapping out the activities, connections and architecture of the institutional landscape. Sarpong et al. (2017), for example, illustrate how the three organizing practices of "research capacities", "accountability" and "collective entrepreneurship" related to the "sayings" and "doings" of the various actors, shape the extent to which countries (un) successfully transition to a triple helix model. Thus, steering a country successfully towards a triple helix model would also require mutual collaboration between all relevant actors at the micro-level (Sarpong et al. 2017).

As discussed throughout this chapter, universities are increasingly required to transform from predominately centres of learning, to sources of socially relevant knowledge through collaborative research with non-academic institutions (see Chap. 8 Vol. 2). Co-innovation for sustainability therefore becomes a consequence of these partnerships, rather than a specific recommendation from one institution. However, in order to achieve the objectives of these institutions (and in the process contribute meaningfully towards generating sustainable solutions for society and the economy), the formal/informal institutional arrangements and rules/regulations have to be modified to support these interactions (Yarime et al. 2012). Within universities, this can include changes in (a) internal rules/regulations and structures to integrate the various disciplines required to conduct sustainability science research (e.g. through faculty programmes, interdepartmental collaboration, fellowships, exchange programmes) as well as (b) external rules/regulations (e.g. policies on intellectual property and technology transfer, information-sharing, multi-stakeholder engagement, and outreach programmes) (Talwar et al. 2011; Lang et al. 2012; Yarime et al. 2012; Ávila et al. 2017). These, more hybrid organizational models of governance, could possibly enable universities to cope with the complex networks and linkages between research-based innovation and economic development (Jongbloed 2015).

However, concrete guidelines on how to effectively restructure institutional dynamics are currently lacking. Possibly, an important first step towards improved collaborative research would be to begin periodic evaluations of these relationships to address weaknesses in institutional designs and working practices that have hindered effective stakeholder interaction and learning in the past (Saad 2004).

8.4.3 Research Funding Strategies

As already mentioned in this chapter, funding remains a key factor that can dictate the success of innovation strategies and effective collaboration and research for sustainability (see Chap. 1, 5 Vol. 1). In line with national and regional initiatives, the government of Ghana aims to establish a National Science, Technology and Innovation Fund (MESTI 2017a). The target is to allocate the recommended minimum of 1% of GDP annually to support science and technology activities across various institutions. This allocation may, however, not adequately reflect the national development priorities. Actually, various combinations of different measures under the national STI policy could rather be budgeted separately (UNECA 2016).

Conversely, although the financial support towards HEIs has so far been generally higher from the government compared to the private sector (Sect. 8.3.1), the increasing constraints on public spending may reduce future public financing of education (Sect. 8.3.3). In this context, HEIs are being pressured to diversify their research funding options as well as enhance collaboration with industry to attract funding. The government already encourages engagement with the private sector through various Public–Private Partnerships (PPPs) schemes (MOFEP 2011). This can translate into one or more variations of the archetypical university–industry relationships, such as the "entrepreneurial" university (Gulbrandsen et al. 2011; Yarime et al. 2012). In this model, industry benefits from access to research and expertise, while universities obtain additional funds to purchase equipment and conduct research. Technology transfer offices become the main avenues to maximize knowledge transfer and reconcile the potentially conflicting interests of stakeholders, while R&D contracts manage these possibly win–win collaborations (Berbegal-Mirabent and Ribeiro-Soriano 2015; Fernandes et al. 2015). The development of appropriate analytical tools and the systematic monitoring of key performance indicators can help measure the success of such partnerships in order to adjust or improve their outputs (e.g., Fernandes et al. 2017).

There are, however, some concerns about this gradual shift in funding sources. For example, there is generally a high complementarity between government and private sector funding to universities (Sá 2015) such that reducing public funding can negatively affect university–industry collaboration (and their external fundraising capacity) (Muscio et al. 2013). In addition, scientific originality may be compromised as researchers become more business oriented and prioritize research projects along “fundable” commercial interests (e.g. Besley et al. 2017; Larsen 2011). Indeed, higher budget shares from industry are sometimes associated with declining publication output (in terms of quality and quantity) in science and engineering departments but can have a positive impact on the number of patent citations (Hottenrott and Thorwarth 2011). This implies an increasing quality of applied research, which can be a desirable outcome in some applied research fields. However, it can be counterproductive in fields with high societal relevance such as Sustainability Science. Overall, through this closer interaction between universities and the private sector can be welcome in terms of transdisciplinarity and research co-design, it may limit real advances in knowledge and out-of-the-box thinking. In this sense, there is a real need for appropriate internal operating policies and incentive structures to encourage academics to strike the right balance between basic and applied research (Berbegal-Mirabent and Ribeiro-Soriano 2015).

8.4.4 Opportunities for Collaborative Research

The formation and operation of effective partnerships for sustainability research require that the institutional priorities of each of the partners are addressed in a mutually satisfying and equitable manner (Chap. 8 Vol. 2). In providing the science and evidence to guide sustainability decision-making, the UG, for example, has to meet its traditional obligations towards education, learning, research and knowledge generation. However, at the same time it also has to acknowledge the priority of industrial partners towards the production of goods and services for economic benefit as well as the values and politics of the government. Currently, the potential mutual benefits of collaborative research do not manifest strongly, as the UG faces scepticism and a lack of confidence from the industrial sector. Furthermore, the modalities through which the government uses research findings are still unclear.

However, there are opportunities for improving these collaborations, both from the side of academia and industry. From the side of academia, a fundamental first step towards strengthening these partnerships is an understanding of how researchers themselves perceive the relationship between science and society (and their role for transformative change). According to van der Hel (2018), sustainability researchers tend to engage with the values and politics of societal change in different ways by (a) focusing on the production of knowledge and are neutral/impartial to societal impacts, (b) perceiving themselves as political actors engaged in shaping and changing society, (c) being hesitant in becoming agents of change for sustainability or (d) stressing the complex web of actors, interests and perspectives related to sustainability but emphasizing humility with respect to their societal role. That said, researchers often still feel removed from decision-making processes for sustainable development. Scholars have argued that additional capacity and skills are needed to effectively integrate scientific knowledge and quality within social practices and principles and to address challenges that arise (especially) when scientific credibility becomes incompatible with stakeholders' values and beliefs (Van Kerkhoff and Lebel 2006; Yarime et al. 2012; Pohl et al. 2010) (Chap. 8 Vol. 2).

From the side of the industry, the increasing implementation of corporate social responsibility (CSR) policies (see Chap. 4 Vol. 1) provides universities with another opportunity to redefine how they can engage meaningfully with companies, especially in the context of the SDGs. Currently, these relationships are usually short term, project based and unsustainable after the initial funding finishes.

Overall, the SDGs provide universities with a unique opportunity to re-interpret institutional strategies and re-determine the structures and mechanisms required for strengthening partnerships at the local level (El-Jardali et al., 2018). For example, the Sustainable Development Goals Philanthropy Platform launched in Ghana in July 2015 involves multi-stakeholder partnerships and engagement processes in the thematic areas of water, education and enabling environments. The platform aims at helping both local and global institutions align their priorities in the context of SDG implementation efforts within the country and broker development frameworks (see Chap. 5 Vol. 1). Universities can support such processes by adopting fit-for-purpose research governance mechanisms, orienting internal decision-making processes, allocating resources effectively, redesigning organizational structures and providing an incentive system for research, reporting activities and accountability (Paletta et al. 2019).

8.4.5 Policy Implications and Recommendations

Ghana's Coordinated Programme of Economic and Social Development Policies (2017–2024) highlights five key strategic anchors to drive growth and development: (a) revitalizing the economy; (b) transforming agriculture and industry; (c) strengthening social protection and inclusion; (d) revamping economic and social infrastructure and (e) reforming public service delivery institutions. The government has recognized the importance of research to achieving these, emphasizing that

> An appropriate environment will be created to promote formation of strong partnerships with research institutions, academia and industry, to ensure that research outputs are turned into industrial applications. This will ultimately lead to a reduction in imports, increase in exports, and serve as a catalyst for job creation and economic growth (GoG 2017: 86).

Academic institutions (as exemplified by UG) can provide evidence-based knowledge and advanced solutions for a broad range of economic sectors, such as agriculture, water and sanitation, health and energy, among others. The SDGs provide an important framework for guiding the formulation, implementation and monitoring of research activities as they relate to development.

Enhancing the collaboration between academia and stakeholders from industry and government can contribute towards meeting national targets related to SDG4 (Quality Education), SDG 8 (Decent Work and Economic Growth), SDG 9 (Industry, Innovation and Infrastructure) as well as SDG 12 (Sustainable Production and Consumption) and SDG17 (Partnerships for the Goals) (Chap. 1 Vol. 1). This showcases that such partnerships are important avenues for catalysing progress across multiple SDG (and individual targets), thus adding tremendous value in national efforts for achieving sustainable development. Government can play a key role in setting up systems and structures that can advance such partnerships, further supporting academic institutions in Ghana (and possibly elsewhere in SSA) to foster innovation and technological development for sustainability.

The government can create supportive contexts to facilitate the fruitful international engagement and strong funding mechanisms for research and innovation (Pogodaeva et al. 2015). For example, through the Africa Centers of Excellence Project[11], West and Central African governments receive funds as loans from the World Bank to find solutions to regional development challenges. Experts are trained in various research fields to deliver high quality training and applied research.

The government can also act as a facilitator by ensuring the viability of these partnerships through specific policies that reflect the institutional mandates and priorities of each partner. The UK Government, for example, announced in late 2015 the GBP 1.5 billion Global Challenges Research Fund (GCRF)[12] to support cutting-edge research that addresses the challenges faced by developing countries. GCRF is creating complimentary programmes alongside other partners to promote challenge-led disciplinary and interdisciplinary research, including the participation of researchers who may not previously considered the applicability of their work to development issues. These collaborations can be reinforced through formal and informal partnering agreements that include equitable intellectual property and commercial rights, in order to maintain openness, transparency and flexibility for all. This ensures that research benefits are equitably shared among all partners in relation to intellectual property, data sharing and ownership.

Finally, as the positive outcomes of such collaborations will be felt directly or indirectly across society, there is a need for policies that support or enhance participatory and coordination processes between the different types of stakeholders. The application of Strategic Environmental Assessment (SEA) to Ghana's Poverty Reduction Strategy (GPRS) in 2004, for example, allowed a wide range of

[11]For more information, see https://ace.aau.org/about/

[12]For more information, see https://www.ukri.org/research/global-challenges-research-fund/

stakeholders to participate in the reformulation of the GPRS policy statement to integrate national policy goals and practical delivery of sustainable development (Boakye-Agyei 2007). These processes therefore incorporate the key values of individuals (and the broader society), including traditions, culture and religion that are collectively critical in shaping economic development and social coexistence. National platforms that share research findings and applications will also increase public awareness and encourage public participation in innovation processes for development.

8.5 Conclusions

There is an increasing call on SSA universities to strengthen partnerships with industry and government, to co-develop innovative solutions to meet national sustainable development goals and optimize economic growth. However, despite the strong national policy focus on sustainability in Ghana, numerous institutional factors influence how such partnerships are formed and how the generated evidence, knowledge, technology and research outputs are applied. This chapter provides an in-depth and illustrative understanding of key facets on the relationship between the University of Ghana with external actors from industry and government. In particular, we identified and critically discussed the different structures and processes implemented by these actors as a means of facilitating collaborative research, innovation and research and development in the context of sustainable development.

Our findings suggest that although significant progress has been made recently through the establishment of supportive national policies and flexible institutional structures, the actual collaboration and co-production of knowledge is still weak. For one, universities need to improve dialogue with policy makers and substantiate their capacity as credible sources of technology/knowledge generation and transfer for industrial uptake. This can be achieved by (a) increasing internal capabilities (e.g. capabilities of individual researchers), (b) developing structures to produce and formally transfer technologies and (c) monitoring the performance of R&D-funded contracts. Researchers engaging in sustainability research should clearly outline knowledge co-development processes that enable the articulation and integration of multiple perspectives, values and methodologies.

When it comes to other actors, the industrial sector would need to adopt a more proactive approach in providing inputs for research priorities, assessing research progress, and co-funding research that is relevant for industrial uptake. Government should focus on developing policies that could effectively support collaborative processes and structures, such as providing credibility and standards for hybrid organizations and a supportive environment to all institutions through minimal bureaucracy. Finally, the statutory and non-statutory functions of local traditional systems are also critical for local economic and social conditions in many parts of SSA. They can be used to demystify research theories and solutions, thus encouraging community ownership and involvement in transdisciplinary research.

References

Ababio BT (2016) Actions to improve current state of student employability skills development in Ghana's public universities: A case study of Geography departments. Int J Educ Eval 2 (9):65–82

Ackah C, Adjasi C, Turkson F (2016) Industrial Policy in Ghana. In: Newman C, Page J, Rand J, Shimles A, Söderbom M, Tarp F (eds) Manufacturing transformation: comparative studies of industrial development in Africa and emerging Asia. Oxford University Press, Oxford

Amankwah-Amoah J (2016) The evolution of science, technology and innovation policies: A review of the Ghanaian experience. Technol Forecast Soc Change 110:134–142

Amu MEK, Offei-Ansah C (2011) Linking tertiary institutions to industries: Evidence from the Vocational and Technical Education Department of the University of Cape Coast. Int J Vocat Tech Educ 2(5):53–60

Ávila LV, Filho WL, Brandli L, Macgregor CJ, Molthan-Hill P, Özuyar PG, Moreira RM (2017) Barriers to innovation and sustainability at universities around the world. J Clean Prod 164:1268–1278

Bawakyillenuo S, Akoto IO, Ahiadeke C, Aryeetey EB-D, Agbe EK (2013) Tertiary education and industrial development in Ghana, A Working Paper. International Growth Centre, London

de Beer, J., Armstrong, C., Oguamanam, C., Schonwetter, T. (2014) Innovation and intellectual property: collaborative dynamics in Africa, De Beer, J., Armstrong, C., Oguamanam, C., Schonwetter, T. Innovation and intellectual property: collaborative dynamics in Africa., UCT Press, Cape Town. ISBN: 978-1-919895-99-4. Available at SSRN. https://ssrn.com/abstract=2572909

Berbegal-Mirabent J, García JLS, Ribeiro-Soriano DE (2015) University-industry partnerships for the provision of R&D services. J Bus Res 68:1407–1413

Besley JC, McCright AM, Zahry NR, Elliott KC, Kaminski NE, Martin JD (2017) Perceived conflict of interest in health science partnerships. PLoS One 12(4):e0175643

Boakye-Agyei K (2007) Stakeholder participation for Strategic Environmental Assessments (SEA). The case of Ghana's Growth and Poverty Reduction Strategy (GPRS). Consultant Report for World Bank, November 2007. Available at http://documents.worldbank.org/curated/en/162271468340804618/pdf/690790ESW0whit0holder0Participation.pdf

Brandt P, Ernst A, Gralla F, Luederitz C, Land DJ, Newig J, Reinert F, Abson DJ, von Wehrden H (2013) A review of transdisciplinary research in sustainability science. Ecol Econ 92:1–15

van Breda J, Swilling M (2018) The guiding logics and principles for designing emergent transdisciplinary research processes: learning experiences and reflections from a transdisciplinary urban case study in Enkanani informal settlement, South Africa. Sustain Sci. https://doi.org/10.1007/s11625-018-0606-x

Caniglia G, Schäpke N, Lang DJ, Abson DJ, Luederitz C, Wiek A, Laubichler MD, Gralla F, von Wehrden H (2017) Experiments and evidence in sustainability science: A typology. J Clean Prod 169:39–47

Clark WC, Dickson NM (2003) Sustainability science: the emerging research program. Proc Natl Acad Sci 100(14):8059–8061

Cobbinah PB, Erdiaw-Kwasie MO, Amoateng NS, P. (2015a) African's urbanisation: Implications for sustainable development. Cities 47:62–72

Cobbinah PB, Erdiaw-Kwasie MO, Amoateng NS, P. (2015b) Rethinking sustainable development within the framework of poverty and urbanisation in developing countries. Environ Dev 13:18–32

Cornell University, INSEAD, and WIPO (2018) The Global Innovation Index 2018: Energizing the World with Innovation. Ithaca, Fontainebleau, and Geneva

Cornell University, INSEAD, and WIPO (2016) The Global Innovation Index 2016: Winning with Global Innovation, Ithaca, Fontainebleau, and Geneva.

Cornell University, INSEAD, and WIPO (2017) The Global Innovation Index 2017: Innovation Feeding the World, Ithaca, Fontainebleau, and Geneva.

Dilling L, Lemos MC (2011) Creating usable science: Opportunities and constraints for climate knowledge use and their implications for science policy. Glob Environ Chang 21:680–689

Domfeh KA, Ahenkan A, Bawole JN (2012) Is sustainable development achievable in Ghana? An analysis of Ghana's development policy achievements and challenges. Int J Environ Sustain Dev 11(3):304–317

Draghicia A, Baban C-F, Gogan M-L, Ivascu L-V (2015) A knowledge management approach for the university-industry collaboration in open innovation. Proc Econ Finan 23:23–32

El-Jardali F, Ataya N, Fadlallah R (2018) Changing roles of universities in the era of SDGs: rising up to the global challenge through institutionalising partnerships with governments and communities. Health Res Policy Syst 16:38. https://doi.org/10.1186/s12961-018-0318-9

Etse D, Ingley C (2016) Higher education curriculum for sustainability. Int J Sustain Higher Educ 17(2):269–280

Etzkowitz H (2003) Innovation in innovation: the triple helix of university–industry–government relations. Soc Sci Inform 42(3):293–337

Etzkowitz H, Leydesdorff L (1995) The triple helix-university-industry-government relations: A laboratory for knowledge based economic development. East Rev 14(1):14–19

Etzkowitz H, Leydesdorff L (2000) The dynamics of innovation: from National Systems and "Mode 2" to a Triple Helix of university–industry–government relations. Res Pol 29 (2):109–123

European Expert Network on Economic of Education (EENEE) (2014) The contribution of universities to innovation, (regional) growth and employment, EENEE Analytical Report No. 18. Prepared for the European Commission.

El-Jardali F, Ataya N, Fadlallah R (2018) Changing roles of universities in the era of SDGs: rising up to the global challenge through institutionalising partnerships with governments and communities. Health Research Policy and Systems 16 (1)

Fernandes G, Pinto EB, Machado RJ, Araújo M, Pontes A (2015) A program and project management approach for collaborative University-Industry R&D funded contracts. Proc Comput Sci 64:1065–1074

Fernandes G, Pinto EB, Araújo M, Magalhães P, Machado RJ (2017) A method for measuring the success of collaborative University-Industry R&D funded contracts. Proc Comput Sci 212:451–460

Figueiro PS, Raufflet E (2015) Sustainability in higher education: a systematic review with focus on management education. J Clean Prod 106:22–33

Filho WL, Raath S, Lazzarini B, Vargas VR, de Souza L, Anholon R, Quelhas OLG, Haddad R, Klavins M, Orlovic VL (2018) The role of transformation in learning and education for sustainability. J Clean Prod 199:286–295

Fuseini I, Kemp J (2015) A review of spatial planning in Ghana's socio-economic development trajectory: a sustainable development perspective. Land Use Policy 47:309–320

Gasparatos A, Takeuchi K, Elmqvist T, Fukushi K, Nagao M, Swanepoel F, Swilling M, Trotter D, von Blottnitz H (2017) Sustainability science for meeting Africa's challenges: setting the stage. Sustain Sci 12(5):635–640

Global Economy (2017) Ghana: education spending, percent of GDP. Retrieved from https://www.theglobaleconomy.com/Ghana/Education_spending/

Global Economy (2018) Ghana: tertiary school enrollment. Retrieved from https://www.theglobaleconomy.com/Ghana/Tertiary_school_enrollment/

Godin B, Gingras Y (2000) The place of universities in the system of knowledge production. Res Pol 29(2):273–278

GoG-NDPC (Government of Ghana—National Development Planning Commission) (2014) Ghana Shared Growth and Development Agenda (GSGDA II), 2014-2017, vol. 1: Policy framework. Medium term National Development Framework, Accra

Gordon C (2018) Charity begins at home: The University and the delivery of the Sustainable Development Goals, Inter-College Lecture Series, College of Basic and Applied Sciences

University of Ghana, April 12 2018. Available at https://www.youtube.com/watch?v=ZiVhVcHM11U
Government of Ghana (GoG) (2017) The coordinated programme of economic and social development policies 2017-2024: an agenda for jobs-creating prosperity and equal opportunity for all. In: Presented by Nana Addo Dankwa Akufo-Addo, President of the Republic to the 7th Parliament of the 4th Republic, 20 October 2017. Accra, Ghana
GSS (Ghana Statistical Service) (2017) Provisional 2016 Annual Gross Domestic Product, April 2017 Edition, Accra
Gulbrandsen M, Mowery D, Feldman M (2011) Introduction to the special section: heterogeneity and university–industry relations. Res Pol 40(1):1–5
GUNi, IAU and AAU (2011) The promotion of sustainable development by higher education institutions in sub-Saharan Africa, Survey Report, Available online. https://www.iau-hesd.net/sites/default/files/documents/promotion_sd_by_heis_sub_saharan_africa.pdf
Hadorn GH, Bradley D, Pohl C, Rist S, Wiesmann U (2006) Implications of transdisciplinarity for sustainability research. Ecol Econ 60(1):119–128
van der Hel S, Biermann F (2017) The authority of science in sustainability governance, A structured comparison of six science institutions engaged with the sustainable development goals. Environ Sci Pol 77:211–220
Hinson R, Gyabea A, Ibrahim M (2015) Sustainability reporting among Ghanaian universities. Communication 41(1):22–42
Holmberg J, Svanstrom M, Peet DJ, Mulder K, Ferrer-Balas D, Segalas J (2008) Embedding sustainability in higher education through interaction with lecturers: case studies from three European technical universities. Eur J Eng Educ 33(3):271–282
Hottenrott H, Thorwarth S (2011) Industry funding of university research and scientific productivity. KYKLOS 64(4):543–555
IAST (Institute of Applied Science and Technology) (2012) Report on interactive forum between industry and University of Ghana faculty researchers, Organized by the IAST in collaboration with the UG-Carnegie Next Generation of Academics in Africa Project funded by Carnegie Corporation of New York, 30 April 2012. Centre for African Wetlands, Legon, Ghana
Jongbloed B (2015) Universities as hybrid organisations. Int Stud Manag Organ 45(3):207–225
Jowi JO, Obamba M, Sehoole C, Barifaijo M, Oanda O, Alabi G (2013) Governance of higher education, research and innovation in Ghana, Kenya and Uganda, Programme on Innovation, Higher Education and Research for Development (IHERD) hosted by OECD, funded by the Swedish International Development Cooperation Agency (SIDA). www.oecd.org/sti/Governance%20of%20higher%20education%20research%20and%20innovation%20in%20Ghana%20Kenya%20and%20Uganda.pdf
Juntti M, Russel D, Turnpenny K (2010) Evidence, politics and power in public policy for the environment. Environ Sci Policy 12(3):207–215
Kaiser DB, Köhler T, Weith T (2016) Knowledge management in sustainability research projects: Concepts, effective models, and examples in a multi-stakeholder environment. Appl Environ Educ Commun 15(1):4–17
Kates R, Clark WC, Hall JM, Jaeger C, Lowe I, McCarthy JJ, Gallopin GC (2001) Sustainability science. Science 292:641–642
Komiyama H, Takeuchi K (2006) Sustainability science: building a new discipline. Sustain Sci 1 (1):1–6
Lambrechts W, Mula I, Ceulemans K, Molderez I, Gaeremynck V (2013) The integration of competencies for sustainable development in higher education: an analysis of bachelor programs in management. J Clean Prod 48:65–73
Lang DJ, Wiek A, Bergmann M, Stauffacher M, Martens P, Moll P, Thomas CJ (2012) Transdisciplinary research in sustainability science: practice, principles, and challenges. Sustain Sci 7 (1):25–43
Larsen MT (2011) The implications of academic enterprise for public science: An overview of the empirical evidence. Res Pol 40(1):6–19

Leydesdorff L (2010) The knowledge based economy and the triple helix model. Annu Rev Inf Sci Technol 44:367–417

Lozano R, Lukman R, Lozano FJ, Huisingh D, Lambrechts W (2013) Declarations for sustainability in higher education: becoming better leaders, through addressing the university system. J Clean Prod 48:10–19

Mathews JA, Hu M-C (2007) Enhancing the role of universities in building national innovation capacity in Asia: The case of Taiwan. World Dev 35(6):1005–1020

Michaels S (2009) Matching knowledge brokering strategies to environmental policy problems and settings. Environ Sci Policy 12(7):994–1011

Mielke J, Vermaßen H, Ellenbeck S, Milan BF, Jaeger C (2016) Stakeholder involvement in sustainability science—a critical review. Energy Res Soc Sci 17:71–81

Miller TR (2014) Reconstructing sustainability science. knowledge and action for a sustainable future. Science in Society Series, Earthscan, Routledge

Ministry of Education (MoE) (2016) Education sector performance report. Government of Ghana, Accra. www.moe.gov.gh/assets/media/docs/ESPR2016_Final_Version_Final.pdf

Ministry of Environment, Science, Technology and Innovation (2017a) National science, technology and innovation policy (2017-2020), May 2017. Accra, MESTI

Ministry of Environment, Science, Technology and Innovation (2017b) Overview of Ghana's Science, Technology and Innovation System, presented by Adelaide A. Asante, at a consultative workshop organised by MESTI and hosted by the Council for Scientific and Industrial Research. Available at http://mesti.gov.gh/wp-content/uploads/2017/07/Overview-of-Ghana%E2%80%99s-STI-system-new-2.pptx

Ministry of Finance (2018) Ghana's SDG Budget baseline report. Available at https://www.mofep.gov.gh/sites/default/files/news/Ghana's-SDG-Budget-Baseline-Report-Aug-09-18.pdf

Ministry of Finance and Economic Planning (MOFEP) (2011) National Policy on Public Private Partnerships: Private participation in Infrastructure and Services for Better Public Services Delivery, Government of Ghana. Available at https://www.mofep.gov.gh/sites/default/files/reports/economic/ppp_policy.pdf

Muhar A, Visser J, van Breda J (2013) Experiences from establishing structures inter- and transdisciplinary doctoral programs in sustainability: a comparison of two cases in South Africa and Austria. J Clean Prod 61:122–129

Muller P (2006) Exploring the knowledge filter: how entrepreneurship and university industry relationship drive economic growth. Res Pol 35(20):1499–1508

Muriithi P, Horner D, Pemberton L, Wao H (2018) Factors influencing research collaborations in Kenyan universities. Res Pol 47:88–97

Muscio A, Quaglione D, Vallanti G (2013) Does government funding complement or substitute private research funding to universities? Res Pol 42:63–75

National Accreditation Board (NAB) (2017). http://www.nab.gov.gh/

National Democratic Congress (NDC) (2016). Changing lives transforming Ghana: 2016 manifesto. www.codeoghana.org/assets/downloadables/2016-Manifesto.pdf

Nyerere J, Mfune O, Fuh D, Sulemana N, Mutisya E, Yiran G, Fadairo O, Ameyaw J, Odingo A (2014) The role of higher education in building a sustainable African society. Afr J Online 4 (3):1–16

Oduro-Marfo S (2015) Toward a national innovation strategy: a critique of Ghana's science, technology and innovation policy. Innov J 20(3):5. https://www.innovation.cc/case-studies/20_3_5_oduro-marfo_national-strategy.pdf

OECD (2017) Gross domestic spending on R&D (indicator). https://doi.org/10.1787/d8b068b4-en. Accessed on 25 October 2017

Paletta A, Fava F, Ubertini F, Bastioli C, Gregori G, La Camera F, Douvan AR (2019) Universities, industries and sustainable development: Outcomes of the 2017 G7 Environment Ministerial Meeting. Sustain Prod Consump 9:1–19. https://doi.org/10.1016/j.spc.2019.02.008

Pogodaeva T, Zhaparova D, Efremova I (2015) Changing role of the university in innovation development: New challenges for Russian regions. Pro Soc Behav Sci 214:359–367

Pohl C, Rist S, Zimmermann A, Fry P, Gurung GS, Schneider F, Speranza CI, Kiteme B, Boillat S, Serrano E, Hadorn GH, Wiesmann U (2010) Researchers' roles in knowledge co-production: experience from sustainability research in Kenya, Switzerland, Bolivia and Nepal. Sci Publ Pol 37(4):267–281

Polk M, Knutsson P (2008) Participation, value rationality and mutual learning in transdisciplinary knowledge production for sustainable development. Environ Educ Res 14(6):643–653

Sá CM (2015) Perspective of Industry's engagement with African Universities. For the African Association of Universities. http://www.heart-resources.org/doc_lib/perspective-of-industrys-engagement-with-african-universities/

Saad M (2004) Issues and challenges arising from the application of innovation strategies based on the triple helix culture: experience of the incubation system in Algeria. Int J Technol Manage Sustain Dev 3(1):17–34

Sachs J, Schmidt-Traub G, Kroll C, Durand-Delacre D, Teksoz K (2017) SDG Index and Dashboards Report 2017. Bertelsmann Stiftung and Sustainable Development Solutions Network (SDSN), New York

Sarpong D, AbdRazak A, Alexander E, Meissner D (2017) Organising practices of university, industry and government that facilitate (or impede) the transition to a hybrid triple helix model of innovation. Technol Forecast Soc Change 123:142–152

Sheate WR, Partidário MR (2010) Strategic approaches and assessment techniques—potential for knowledge brokerage towards sustainability. Environ Impact Assess Rev 30:278–288

Shinn T (2002) The triple helix and new production of knowledge pre-packaged thinking on science and technology. Soc Stud Sci 32(4):599–614

Soini K, Jurgilevich A, Pietikäinen J, Korhonen-Kurki K (2018) Universities responding to the call for sustainability: A typology of sustainability centres. J Clean Prod 170:1423–1432

Stephens JC, Graham AC (2010) Toward an empirical research agenda for sustainability in higher education: exploring the transition management framework. J Clean Prod 18(7):611–618

Talwar S, Wiek A, Robinson J (2011) User engagement in sustainability research. Sci Publ Pol 38 (5):379–390

Thompson NM (2015) Black Star Rising: A 40-Year Journey towards a Just. Free and Prosperous Society, The National Development Planning Commission (NDPC), Accra International Conference Centre, Ghana

Trencher G, Yarime M, McCormick KB, Doll CNH, Kraines SB (2014) Beyond the third mission: Exploring the emerging university function of co-creation for sustainability. Sci Publ Pol 41:151–179

United Nations Conference on Trade and Development (UNCTAD) (2011) Science, technology and innovation policy review. Ghana, United Nations, Switzerland

United Nations Economic Commission for Africa (UNECA) (2016) Chapter 5: Africa's science, technology and innovation policies—national, regional and continental. In: Assessing Regional Integration in Africa VII—Innovation, Competitiveness and Regional Integration. doi: https://doi.org/10.18356/73293cf1-en

University of Ghana (UG) (2017). UG's 3rd research report launched. https://orid.ug.edu.gh/news/ug%E2%80%99s-3rd-research-report-launched

van der Hel S (2018) Science for change: a survey on the normative and political dimensions of global sustainability research. Global Environ Chang 52:248–258

Van Kerkhoff L, Lebel L (2006) Linking knowledge and action for sustainable development. Annu Rev Env Resour 31:445–477

Verhulst E, Lambrechts W (2015) Fostering the incorporation of sustainable development in higher education. Lessons learned from a change management perspective. J Clean Prod 106:189–204

Watkins, A., Papaioannou, T., Kale, D., and Mugwagwa, J. (2014). National innovation systems, developing countries, and the role of intermediaries: a critical review of the literature. In: 15th International Conference of the International Joseph A. Schumpeter Society (ISS), 27-30 Jul 2014, Jena, Germany

Weik A, Withycombe L, Redman CL (2011) Key competencies in sustainability: a reference framework for academic program development. Sustain Sci 6:203–218

World Economic Forum (2017) Insight report: the inclusive growth and development report 2017, January 2017. WEF, Geneva

World Economic Forum (2018) The global competitiveness report 2017–2018: full data edition. WEF, Geneva

Yarime M, Trencher G, Mino T, Scholz RW, Olsson L, Ness B, Frantzeskaki N, Rotmans J (2012) Establishing sustainability science in higher education institutions: towards and integration of academic development, institutionalization, and stakeholder collaborations. Sustain Sci 7 (1):101–113

Zavale NC, Macamo E (2016) How and what knowledge do universities and academics transfer to industry in African low-income countries? Evidence from the stage of university-industry linkages in Mozambique. Int J Educ Dev 49:247–261

Part III
Central Africa

Chapter 9
Long-Term Vegetation Change in Central Africa: The Need for an Integrated Management Framework for Forests and Savannas

Julie C. Aleman and Adeline Fayolle

9.1 Introduction

Tree-dominated forests and grass-dominated savannas represent the two main tropical biomes covering the overwhelming majority of sub-Saharan Africa (SSA) (White 1983). Forests and savannas have a different, even antagonistic, ecological functioning (Staver et al. 2011a), but they both provide critical services to local populations (see Chaps. 7 and 10, Vol. 1; Chaps. 5 and 6, Vol. 2). Tropical forests are dominated by trees, forming closed canopies and complex vertical structures, and contain in their understory C3 grasses that are more adapted to humidity and shadow. Such forests are encountered in areas of high annual rainfall and limited seasonality (Malhi et al. 2009), and are very sensitive to disturbances. In contrast, trees and C4 grasses in tropical savannas coexist, are more adapted to aridity, are shade intolerant, and are found in areas that are drier and have higher seasonality (Ratnam et al. 2011). Savannas rely on frequent disturbances due to fires and/or mega-herbivores that maintain an open canopy and species diversity (Bond et al. 2005; Sankaran et al. 2005; Staver and Bond 2014).

The occurrence of tropical forests and savannas is, however, not rigidly determined by climatic conditions. Recent analyses of remotely sensed tree cover at the global (Hirota et al. 2011; Staver et al. 2011b) and at the regional (Favier et al. 2012) scales, supported by theoretical work (Staver et al. 2011a; Staver and Levin 2012),

J. C. Aleman (✉)
Gembloux Agro-Bio Tech, University of Liège, Gembloux, Belgium

Department of Geography, University of Montréal, Montreal, QC, Canada

A. Fayolle
Gembloux Agro-Bio Tech, University of Liège, Gembloux, Belgium
e-mail: Adeline.Fayolle@uliege.be

A. Gasparatos et al. (eds.), *Sustainability Challenges in Sub-Saharan Africa I*, Science for Sustainable Societies, https://doi.org/10.1007/978-981-15-4458-3_9

have demonstrated that for intermediate rainfall regimes,[1] forests and savannas can both occur in the same regions. Under these climatic conditions they represent alternative stable states maintained by a positive feedback between fire and canopy cover (Hirota et al. 2011; Staver et al. 2011b; Favier et al. 2012). These two ecosystems are unique, highly biodiverse, and can provide multiple ecosystem services that are directly related to their unique functioning (see Chaps. 7 and 10, Vol. 1; Chaps. 5 and 6, Vol. 2).

Tropical forests harbor more than 60% of all known species (Laurance and Useche 2009), are characterized by unique food webs and high endemism and diversity (Malhi et al. 2014). Tropical forests are also critical for the global climate system (Spracklen et al. 2018) as they represent one of the largest carbon sinks, which constitutes 40–50% of terrestrial carbon stocks (Pan et al. 2011; Achard et al. 2014).

The Congo basin is the largest forest block in SSA in terms of size, and is the world's second largest behind the Amazon. However, tropical forests in SSA have been heavily used for a very long time, and can hardly be considered as pristine (Chap. 10, Vol. 1), see also Willis et al. (2004) and Roberts et al. (2017) for evidence of human activity across the tropics, and Morin-Rivat et al. (2017) across central Africa). At the same time, tropical forests are facing the new pressures of the Anthropocene (Malhi et al. 2014), such as accelerated deforestation and forest degradation due to slash-and-burn subsistence agriculture, and more recently conversion for cash crops (e.g., cocoa, hevea) production (Sonwa et al. 2011), among many other pressures (IPBES 2018). Land use change, that is, the conversion of forest land into another land use, is the most important threat for tropical forests worldwide (Sala et al. 2000; Malhi et al. 2014; Kehoe et al. 2017), and in particular for SSA (Aleman et al. 2016, 2017). Deforestation occurs particularly around big cities, and along major road axes and rivers (Verhegghen et al. 2012), and is expected to continue along such patterns in the near future (Laurance and Arrea 2017).

The logging industry is dominated by industrial scale operations, also representing a major driver of forest fragmentation across central SSA (Laporte et al. 2007). Timber production is of extreme importance for the national economies in the region, with production forests spanning 180 million ha (De Wasseige et al. 2012) (Chap. 1, Vol. 1). However, timber production does not necessarily induce land use change (Pan et al. 2013). Indeed, temporal changes in forest cover due to forest exploitation do not imply a conversion to another land use due to the quick recovery of tropical forests after disturbance (Poorter et al. 2016) or silvicultural interventions (Gourlet-Fleury et al. 2013). Moreover, forest management in central SSA is highly selective with only few trees per hectare extracted over 25–30 year rotations (Ruiz Pérez et al. 2005). This allows for the maintenance of a permanent forest cover after logging, and offers the possibility to manage forests for timber

[1]Intermediate rainfall regimes are 1000–2000 mm for SSA (Staver et al. 2011a) and 1000–2500 mm for the global tropics (Staver et al. 2011b).

extraction and carbon storage (Nasi et al. 2012). Yet, hunting is a non-negligible side-effect, as it nearly always accompanies logging in tropical forests of SSA, having dramatic consequences on animal populations and diversity (Poulsen et al. 2011). Defaunation is indeed occurring at an alarming rate in central SSA (Nasi et al. 2011; Abernethy et al. 2016), potentially having substantial impacts on ecosystem functioning, and specifically tree regeneration and recruitment as demonstrated in other parts of the world (for example, see Terborgh et al. (2008) for southeastern Peru).

Tropical savannas cover 20% of the global land surface, account for 30% of terrestrial net primary productivity, and sustain 20% of the human population, while hosting most of the remaining megafauna (Scholes and Archer 1997; Veldman et al. 2015a; Bond 2016). Tropical savannas cover the overwhelming majority of SSA, while provide important services to local populations (IPBES 2018). Furthermore, savannas and other grassy ecosystems harbor important biodiversity, including the largest herbivores on Earth (Parr et al. 2014; Bond 2016).

Savannas are commonly used in SSA for agriculture, pastoralism, and fuelwood extraction, catering for the livelihoods of many local communities (Scholes and Archer 1997; IPBES 2018) (see also Chaps. 3 and 5, Vol. 2). Recent land use change has been threatening savanna ecosystems across SSA, with many savannas identified as potentially interesting areas for large-scale conversion to industrial/cash crops (Searchinger et al. 2015) (Chap. 3, Vol. 1), biofuel feedstock production (Alexandratos and Bruinsma 2012) (Chap. 2, Vol. 1; Chap. 5, Vol. 2), and forest plantations for forest restoration and CO_2 sequestration (Laestadius et al. 2012). Contrary to tropical forests, savannas rely on chronic disturbances for maintaining an open and diverse ecosystem (as detailed above). In fact, savannas require antagonist management strategies such as recurrent or prescribed fires in high rainfall areas (Lehmann et al. 2014; Osborne et al. 2018), which are not taken into account by the forest scientific community (Veldman 2016).

Mesic savannas are located in areas where forest can also occur, and are especially targeted for the types of land conversion outlined above (Searchinger et al. 2015; Aleman et al. 2016). These types of savannas are indeed often mistaken as "degraded" landscapes, or even as secondary successional stages following deforestation, which are of little value for biodiversity conservation, see arguments in Bond and Parr (2010), Parr et al. (2014), and Veldman et al. (2015b). This is mainly due to the fact that there is no recognition regarding their status as an alternative stable state (Bond and Zaloumis 2016). This is clearly illustrated in the ongoing debate on how to define forests.[2] A recent study that reassessed the extent of forests in dryland biomes worldwide (Bastin et al. 2017), mistakenly considered most

[2]The FAO defines forests as areas of more than 0.5 ha that have tree cover >10% (Pan et al. 2013). Other authors have proposed higher thresholds such as tree cover >65% (e.g., Hirota et al. 2011; Staver et al. 2011a).

savannas as dry forests[3] (Bond and Zaloumis 2016; Griffith et al. 2017). The above suggest that the FAO definition of forest is problematic, as most mesic savannas would be consequently considered as forests (more specifically degraded forests), and could thus be targeted for reforestation (Laestadius et al. 2012).

For example, the Miombo woodlands that cover much of southern Africa has been identified as a particular ecoregion (Olson et al. 2001), and is considered as forest by Bastin et al. (2017) or other type of woodland by Achard et al. (2014). However, Miombo woodlands in which trees are widely spaced and do not form a closed canopy cover are associated to the savanna flora (Linder 2014). They are also associated to the savanna biome as they have a grassy understory and experience regular fires (Ratnam et al. 2011; Dexter et al. 2015; Pennington et al. 2018). The Miombo woodlands host a high floristic richness and specificity, and are dominated by species of the *Brachystegia*, *Julbernardia*, and *Isoberlinia* genera (White 1983). They also provide many different types of natural resources and ecosystem services such as timber products, fuelwood, and a huge variety of non-timber forest products including beeswax and honey, mushrooms, edible caterpillars, wild fruits, and livestock grazing (Lawton 1982; Malaisse 2010) (see Chap. 5, Vol. 2). With millions of people living in (and depending on) the Miombo woodlands (Bradley and McNamara 1993; Dewees 1994), these landscapes are heavily affected by human activity and are under threat from increasing fragmentation and deforestation (Chazdon 2008).

The above suggests that forests and savannas are both two very important tropical biomes in SSA and face the new threats posed by the Anthropocene. Land use change (e.g., conversion to agriculture in forest areas, and industrial plantations in savanna areas) and mismanagement (e.g., extensive logging and woodfuel harvesting, uncontrolled fire regimes) threaten biodiversity and ecosystem services from these biomes. Furthermore, they can compromise the critical role that tropical forests and savannas play for local livelihoods and global climate regulation.

In this respect the loss of tropical forests and savannas and the need for their effective management have become major sustainability challenges in SSA, and especially in central Africa. In particular, there is a need for their integrated management, taking into account that they can be alternative stable states in some areas (i.e., under intermediate rainfall) (Staver et al. 2011a; Favier et al. 2012), and that savannas are not simply degraded forests only useful for reforestation and afforestation projects. In this context, it is important to identify the appropriate management responses to the drivers of vegetation change in forests and savannas of central Africa.

The aim of this chapter is to provide a comprehensive analysis of the past, present, and future vegetation changes in central African forests and savannas. We focus particularly on historical and future changes related to climate change and land

[3]Tropical dry forests are an important biome in the Neotropics, the third tropical biome (Dexter et al. 2018). However in SSA tropical dry forests are supposed to be restricted only to a very small area in East Africa (Pennington et al. 2018).

use change. We start by reviewing the vegetation changes since the Last Glacial Maximum (approximately 21,000 years ago), and assessing the current distribution of major vegetation types in relation to different drivers, mainly related to climate and fire regimes. We then project future vegetation cover based on rainfall variability and agricultural activity, using different scenarios. Section 9.2 outlines the main historical milestones shaping vegetation change in central Africa, and the main data collection and analysis approaches. Section 9.3 highlights the main past (Sect. 9.3.1), current (Sects. 9.3.2), and future (Sect. 9.3.3) vegetation changes, and their drivers. Section 9.4 synthesizes the main findings and discusses some of the main policy implications for restoring ecosystems, expanding protected areas, and designing sustainable forest management approaches in the region.

9.2 Methodology

9.2.1 *Central Africa and Its History*

9.2.1.1 Study Area

This chapter focuses on vegetation changes in central Africa, an area that covers the Congo Basin, and adjacent savannas and woodlands (including the Miombo). For practical reasons, the study area is defined by the boundaries of Cameroon, the Central African Republic (CAR), the Democratic Republic of Congo (DRC), Equatorial Guinea, Gabon, and the Republic of Congo, and spans more than 4 million km^2. According to the Tropical Rainfall Measuring Mission (TRMM, Nicholson et al. 2003), the environmental conditions vary widely across this vast study area. Mean annual rainfall ranges between 440 and 3220 mm, from the driest sites in North Cameroon to the wettest sites in the Coast of Cameroon.

The distribution of forests and savannas in this area is the legacy of a long history of climate changes and human impacts. Table 9.1 summarizes the main periods of climate and vegetation changes since 21,000 cal yr BP (i.e., calibrated years before present, the present being 1950 by convention).

9.2.1.2 Paleo-History

Paleo-environmental reconstructions[4] suggest that central Africa has experienced a succession of dry and humid periods, with tropical forests contracting or expanding in response (Maley 1989, 1996; Vincens et al. 1999).[5] The Last Glacial Maximum

[4]Paleoenvironmental reconstructions entail the reconstruction of past vegetation, climate and disturbances using bio-proxies (e.g., pollen, phytoliths, diatoms, charcoal particles) and the bio-geochemical analysis of natural archives (e.g., sediments or paleosols).

[5]In comparison the Amazon remained much more stable during the same period (Anhuf et al. 2006).

Table 9.1 Main periods of climate and vegetation changes in central Africa

Period	Climate and vegetation
Last glacial maximum (~21,000 cal yr BP)	Low rainfall and temperature. Open vegetation, Afromontane species, the Congo Basin is supposedly reduced to few forest refuges.
Younger Dryas (from ~12,900 to ~11,700 cal yr BP)	Intensely dry period. Open vegetation.
African humid period (from 11,700 to 4000 cal yr BP)	Higher rainfall than currently, increase in temperature. Largest forest extent of the Congo Basin.
Third millennium rainforest crisis (~4000 to 1200 cal yr BP)	Rainfall modification, droughts. Contraction and fragmentation of the Congo Basin.
Since 1200 cal yr BP	Increase in rainfall. Forest transgression.

Note: cal yr BP stands for calibrated years before present, the present being 1950 by convention

(~21,000 cal yr BP) was a period of very low rainfall and temperature, while the subsequent Younger Dryas period (from ~12,900 to ~11,700 cal yr BP) was a short, but intensely dry, period. During these two periods, forests in central Africa were supposedly reduced to small patches (Maley 1989, 1996). Rainfall started increasing at the beginning of the Holocene (~11,700 cal yr BP), reaching higher rainfall levels than currently observed, with this period commonly referred to as the African Humid Period (de Menocal et al. 2000; Shanahan et al. 2015). During this period, tropical forests were more widespread across central Africa than currently (Vincens et al. 2010, 1998; Lebamba et al. 2016).

The African Humid Period ended abruptly at ~4000 cal yr BP, and was followed by the "third millennium rainforest crisis." This period was characterized by low rainfall and major droughts that lasted until 1200 cal yr BP (Vincens et al. 1999), resulting in major perturbations across the Congo forest block. The relative roles of climate and people during the third millennium crisis have been hotly debated in the literature. Some scholars favor the climate hypothesis (Maley et al. 2012; Neumann et al. 2012b; Lézine et al. 2013; Giresse et al. 2018), and others suggest that the migration of Bantu people from the border of Cameroon and Nigeria also contributed to large-scale forest disturbances (Bayon et al. 2012; Garcin et al. 2018). The Bantu people were farmers and metallurgists (Bostoen et al. 2015) who used slash-and-burn farming techniques and required large quantities of wood for metal processing. They tended to farm pearl millet (Neumann et al. 2012a) and raise cattle (Grollemund et al. 2015) within the tropical forest of western and central Africa. Even if they were not responsible for this large-scale event, they may have caused more localized perturbations (Garcin et al. 2018), through canopy opening and wood collection (Van Gemerden et al. 2003; Neumann et al. 2012a). Additionally, charcoal has been found in lakes, wetlands and soils, in the deepest parts of the forest, suggesting localized forest burning (Hubau et al. 2013; Tovar et al. 2014; Biwolé et al. 2015; Morin-Rivat et al. 2016). There was an increasing occurrence of charcoals, as seasonality increased drastically in ~2500 cal yr BP (Hubau et al.

2015). Moreover, people may have played a role in maintaining the newly formed savannas in areas located in the periphery of the Congo forest.

Rainfall started increasing again after 1200 cal yr BP, leading to forest expansion, a trend still observed today in some areas of Cameroon (Youta Happi 1998) and CAR (Guillet et al. 2001). However, the more recent human activity is also important for explaining the current distribution and composition of forests and savannas in central Africa (Willis et al. 2013; Morin-Rivat et al. 2017).

9.2.1.3 Recent History

Human occupation patterns were strongly modified following the colonization of the central African region in the 1900s, which had impacts on current forest composition (Van Gemerden et al. 2003; Engone Obiang et al. 2014; Morin-Rivat et al. 2017). Indeed, even though local communities used to live deeper into the forest before the colonial period, they were then forced to gather in the European trading centers and villages along the main roads. This demographic change is now reflected in forest composition, as the populations of tree species that dominate the canopy are now aging (Engone Obiang et al. 2014; Morin-Rivat et al. 2017).

Following European colonization, deforestation and forest degradation also started to increase. Indeed, the industrialization of the region (starting in ~1920) was associated with the development of extensive oil palm, coffee, and rubber plantations (Van Reybrouck 2012) (see Chap. 3, Vol. 1). This process is anecdotally recorded in lake sediments, where forest degradation has been associated with the increasing prevalence of fire within forest areas (Aleman et al. 2013). This supports the notion that plantations and logging activities, coupled with the development of roads, has been a major cause of forest degradation in the region.

However, despite these recent changes (and their impacts on the vegetation), the tropical forests of central Africa have experienced relatively low deforestation rates since the 1900s (Aleman et al. 2018), and even since the 1980s (Pan et al. 2011; Achard et al. 2014), compared to the other tropical regions (Sodhi et al. 2010; ter Steege et al. 2015). The socioeconomic and political instability that has plagued most countries in the region since their independence from colonial powers have prevented the large-scale expansion of agricultural activities and deforestation. More recently, forest degradation, and specifically forest fragmentation (Malhi et al. 2014) due to road development (Laurance et al. 2017) and logging (Laporte et al. 2007) have been identified as a major threat for the biodiversity (Poulsen et al. 2011; Abernethy et al. 2016; Ziegler et al. 2016, as also reported in the IPBES report for Africa). If the broad region eventually stabilizes, a surge of land conversion may be expected, particularly linked to the expanding oil palm sector (IPBES 2018).

The recent history of land use change also affected vegetation structure and composition within savannas. First, fire was seen as a destructive force during the colonial period that was responsible for the deforestation of extensive areas. Indeed, most mesic savannas are located in areas where the rainfall range allows forests and savannas to represent alternative stable states (i.e., 1000–2000 mm), and these savannas contain gallery forests and forest patches. Mesic savannas were then seen as relatively new and formed through anthropogenic activity (Aubréville 1939, 1947; Bond and Zaloumis 2016). During this period, many fire suppression policies were developed in an attempt to protect tropical forests (see Bond and Zaloumis 2016). Fire practices were heavily modified, triggering changes in fire regimes, and ultimately leading to woody encroachment in savannas (Aubréville 1947).

More recently, Sahelian pastoralists from west and central Africa started migrating in mesic savannas in the 1960s, in areas where Mbororo farmers dominated historically the agricultural landscape (Bassett and Boutrais 2000; Ankogui-Mpoko 2003). This has led to important land conflicts between pastoralists and farmers in the CAR (Ankogui-Mpoko 2003) and Cameroon (Ouikon 2003), and ultimately changed the vegetation structure (Bassett and Boutrais 2000). In particular, changes in agricultural practices modified the fire regime (i.e., from large fires for hunting, to small and early fires for cattle herding), which, coupled with overgrazing, resulted in woody encroachment (Bassett and Boutrais 2000).[6] Pasturelands became less productive "forcing" cattle herders to migrate southwards (Ankogui-Mpoko 2003).

9.2.2 Data Collection and Analysis

9.2.2.1 Reconstruction of Past Vegetation Changes

In this chapter, we conduct a meta-analysis of published studies to better understand and illustrate past changes on vegetation cover, and their determinants (e.g., fires, human presence, and climate) since the Last Glacial Maximum (~21,000 cal yr BP) (see Sect. 9.3.1). We specifically focus on vegetation reconstruction studies from 14 paleo-ecological sites (Table 9.2), for which palynological information is available.

Pollen preserved in lake sediments can be used to infer past vegetation composition and structure (Vincens et al. 1999). For example, we used data downloaded from the African Pollen Database[7] to explain composition changes at Lake Barombi Mbo. Each of the different pollen sites was recorded, and then plotted on a map of central Africa. Dated charred material (e.g., oil palm endocarps, charcoals) can be used as a proxy for local past fire events (Vleminckx et al. 2014). We plotted this information on the same map described above to illustrate the location of past fire events, even those located very deeply within the forest block.

We also reviewed the evidence of past human presence and activity in central Africa, using all of the dated information on human presence assembled by Oslisly

[6]Woody encroachment in savannas is currently widespread throughout SSA, and even in other parts of the world (Stevens et al. 2017). While local encroachment has been directly linked to land use change, global climate change, and specifically CO_2 fertilization, has also been identified as a possible driver (Buitenwerf et al. 2012).

[7]For more information refer to: http://fpd.sedoo.fr/fpd/

Table 9.2 List of palynological sites used to explain vegetation change

#	Site	Latitude	Longitude	References
1	Mbalang	7.32	13.73	Vincens et al. (2010)
2	Tizong	7.25	13.58	Lebamba et al. (2016)
3	Bambili	5.94	10.24	Assi-Kaudjhis (2012)
4	Barombi Mbo	4.67	9.40	Giresse et al. (1994)
5	Ossa	3.80	10.75	Reynaud-Farrera et al. (1996)
6	Nyabessan	2.67	10.67	Ngomanda et al. (2009)
7	Maridor	−0.17	9.35	Ngomanda et al. (2007)
8	Nguene	−0.20	10.47	Ngomanda et al. (2009)
9	Mopo Bai	2.23	16.26	Brncic et al. (2007, 2009)
10	Goualougo	2.16	16.51	Brncic et al. (2007, 2009)
11	Sinnda	−3.83	12.80	Vincens et al. (1994)
12	Kitina	−4.25	11.98	Elenga et al. (1996)
13	Coraf	−4.75	11.85	Elenga et al. (2001)
14	Tilla	10.39	12.13	Salzmann (2000)

et al. (2013) and complemented by Morin-Rivat et al. (2014). The combined dataset consists of 585 ^{14}C-dated records from archaeological sites covering the period 5000 to 100 cal yr BP. Each record is calibrated using the IntCal13 curve and a probability density is obtained through a Bayesian modeling procedure in the "Bchron" package (R Core Team 2015). For each date, we used the median of the probability density, and bin the number of dates every 50 years.

Finally, we illustrated climate change effects during this period, by reporting sea surface temperature (SST) reconstructed from a core in the Guinea Gulf (Weldeab et al. 2007).

9.2.2.2 Assessment of Current Vegetation Cover and Change

We used remotely sensed information about vegetation types (Verhegghen et al. 2012) at the regional scale with a pixel resolution of 900 m, to understand current vegetation types and recent vegetation cover changes across the study region (i.e., between 2000 and 2015) (see Sect. 9.3.2). We aggregated the different vegetation types into the following eight classes: (1) tropical forest; (2) montane forest; (3) swamp forest; (4) mangrove; (5) savanna (containing the different savanna types of the original map); (6) forest–savanna mosaic; (7) Miombo woodland; and (8) areas impacted by humans. The areas affected by humans contain all such classes from the original map, including agricultural areas, human settlements, and roads, among others. To delineate the main biogeographic zones (i.e., phytochoria) in central Africa, we used the biogeographic regions from White (1983) and Linder et al. (2012).

Altitude and topography are important determinants of vegetation distribution across the Tropics, including in central Africa. A very important distinction in the study area is between mountain and lowland forests (Letouzey and Aubréville 1968; White 1983). Within the lowland forest category, swamp forests and mangroves are distinguished from *terra firme* forests (Letouzey and Aubréville 1968), and can be remotely sensed (Verhegghen et al. 2012). Even though such forests occupy vast areas in the Congo basin, and hold huge amount of carbon (Dargie et al. 2017), we focused here only on *terra firme* vegetation, excluding mangrove and swamps from our analysis.

We further studied how tree cover and tree cover changes are distributed across forests and savannas (grouping classes 5–7 outlined above). This is because tree cover is a key variable for both forest and savanna functioning, enabling the differentiation of forests and savannas at a broader scale (Staver et al. 2011a; Aleman et al. 2016; Aleman and Staver 2018). Tree cover within savannas has been shown to be highly variable (Sankaran et al. 2005) and largely determined by annual rainfall (Bucini and Hanan 2007). However, even though annual rainfall indeed constraints maximum tree cover, disturbances (especially fire and herbivory), complex interactions between disturbances (Hanan et al. 2008; Staver and Bond 2014) and soil composition (Sankaran et al. 2008) can also reduce tree cover from its climatic potential in a less predictable way (Sankaran et al. 2005, 2008; Staver 2018).

We used remotely sensed information on tree cover, initially available at 30-m resolution, and aggregated at the resolution of the vegetation types (900 m) (Hansen et al. 2013). Tree cover losses between 2000 and 2015 are also available from the same dataset (Hansen et al. 2013) and were used to identify the forest and savanna areas that have been heavily modified during that period.

9.2.2.3 Identification of Natural and Anthropogenic Drivers of Vegetation Change

A wide array of drivers is currently influencing vegetation structure (Aleman et al. 2017) and composition in forests (Fayolle et al. 2012, 2014b) and savannas (Fayolle et al. 2019). We focused here on annual rainfall and fire as drivers of vegetation change (Sect. 9.3.3), as they are important in determining forest and savanna as alternative stable states in the climate zone where both forests and savannas co-occur (Staver et al. 2011a).

We used for annual rainfall the 3B43 Monthly gridded rainfall product from TRMM, available at 0.25° resolution (Nicholson et al. 2003). For fires, we used the burned area monthly product from the MODerate-resolution Imaging Spectroradiometer (MODIS) sensor available at 500-m resolution (Giglio et al. 2010). Both datasets are resampled at the resolution of the vegetation types (900 m) (Sect. 9.2.2.2), using a nearest neighbor procedure and the package "raster" in the R environment (R Core Team 2015). We computed annual rainfall between 2002 and 2015, and the number of times an area represented by a pixel was burned between 2002 and 2015 as an index of fire frequency.

As mentioned in Sect. 9.2.1, ecosystems in central Africa are facing an increasingly high pressure from human activity, but are also threatened by ongoing climate change (see also Chap. 10, Vol. 1). Through a literature review, we identified the main threats that forests and savannas are experiencing in central Africa. Additionally, we extracted spatial information on land allocated for biodiversity conservation (i.e., protected areas) and to economic exploitation, specifically logging concessions. We used protected area maps[8] for the six countries of central Africa mentioned (Sect. 9.2.1) and data from the World Resources Institute to produce a map of the areas under logging concessions and identified as having forest restoration opportunities.

9.2.2.4 Prediction of Future Vegetation Change

We estimated the area of current forest and savanna that may be impacted by future rainfall and agricultural activity (Sect. 9.3.3). In particular, we used the projected annual rainfall and cropland area for 2070 in central Africa derived from the two most contrasted scenarios of future global change: the Representative Concentration Pathways (RCP) 2.6 and 8.5. These RCPs represent different trajectories for greenhouse gases (GHG) concentrations and radiative forcing for the year 2100 (Van Vuuren et al. 2011). RCPs contain different assumptions about socioeconomic trajectories (e.g., demography, economy, land use).

RCP 8.5 is the most pessimistic scenario (Van Vuuren et al. 2011). In this scenario, radiative forcing and GHG concentration continue to increase at the current rate, while cropland and pasture land also expand significantly due to large population increase (Hurtt et al. 2011). Conversely, RCP 2.6 represents the most optimistic scenario in terms of GHG emissions, as they are expected to eventually decline due to declining fossil fuel consumption and aggressive climate policies implemented to reduce climate change. A crucial element of RCP 2.6 is the use of bioenergy, and carbon capture and storage technologies, which are expected to result in negative emissions (see Chap. 2, Vol. 1). However, the extensive bioenergy expansion is expected to cause a large increase in the cropland dedicated to biofuel production (Hurtt et al. 2011).

Future climate projections for the year 2070 (average 2061–2080) were taken from all general circulation models (GCM) for the two RCPs, which are available downscaled and calibrated against Worldclim 1.4 as baseline climate (Hijmans et al. 2009), using relative change for rainfall. Future land use projections for cropland area for the year 2070 are averaged between 2061 and 2080 for the two RCPs, in order to be consistent with the climate data.

We compared areas that are currently forests and savannas, and that will be impacted by future changes in annual rainfall (>100 mm year^{-1}) and by changes in the fraction of pixels affected by cropland (>0.25). This analysis is conducted for forests that are currently under logging concession and for forests/savannas areas that are currently protected areas. The rationale is that protected areas can potentially buffer against climate and/or land use change (Loarie et al. 2009; Aleman et al. 2016).

---

[8]For more information refer to: https://www.protectedplanet.net (accessed August 2017).

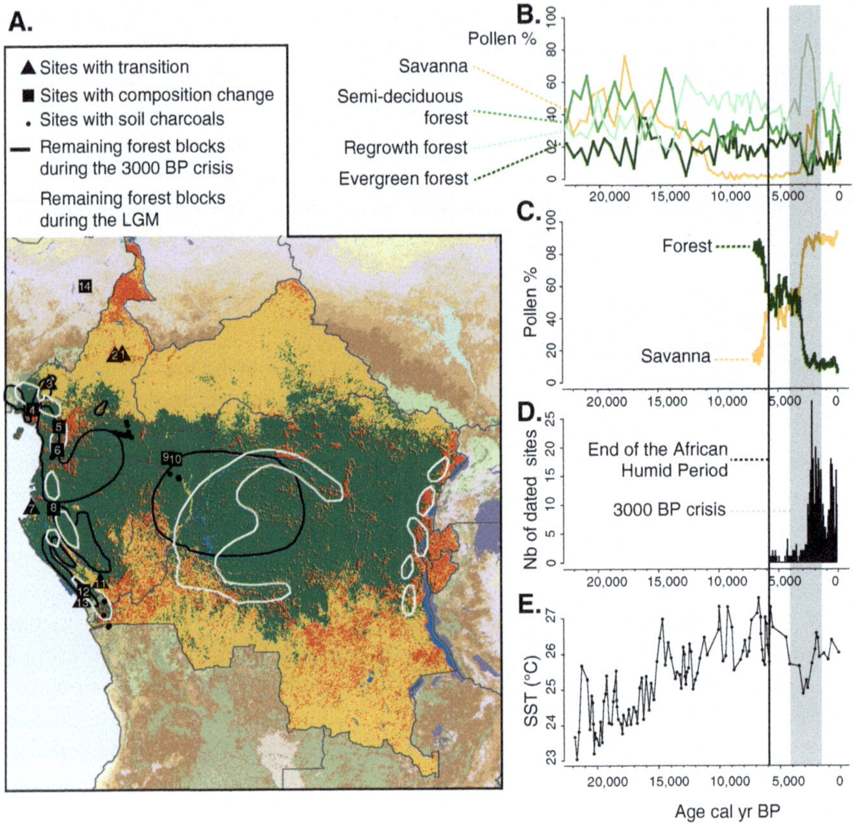

Fig. 9.1 Location of 14 pollen sites and other proxies in central Africa (**a**), pollen assemblages for Lake Barombi-Mbo (**b**, #4 in Table 9.2) and Lake Mbalang (**c**, #1 in Table 9.2), archeological sites (**d**) and sea surface temperature (**e**). Note: In (**b**)–(**e**), the black line corresponds to the end of the African Humid Period, and the beginning of the dryer conditions that led to the 3000 cal yr BP crisis (gray highlight)

9.3 Results

9.3.1 Past Vegetation Changes and Its Drivers

9.3.1.1 Trends from the Last Glacial Maximum to the African Humid Period

Figure 9.1a illustrates the hypothesized forest extent during the Last Glacial Maximum (in light gray) from Maley (1989), and during the 3000 cal yr BP crisis (in black) from Maley (2002). The vegetation map is a simplified version of the map included in Verhegghen et al. (2012) that contains forests, savannas, and human-impacted ecosystems in central Africa, using as a background the land-cover map of Mayaux et al. (2004) for Africa.

After the Last Glacial Maximum (21,000 cal yr BP), tropical forests were reduced to only few refugia in central Africa. This is suggested both through the direct paleo-data (Dechamps et al. 1988; Maley 1989) and indirect information from endemism centers (Maley 1989). After the Younger Dryas, which was a short but intensely dry period (from ~12,900 to ~11,700 cal yr BP, see Table 9.1, Sect. 9.2.1), rainfall levels started increasing at the early stages of the Holocene (Fig. 9.1e). During that period, rainfall levels were even higher than current levels, with this period called the African Humid Period (de Menocal et al. 2000; Shanahan et al. 2015). As a result, tropical forests were more extensive across central Africa, even reaching the Adamawa Plateau in Cameroon (sites 1–2, Fig. 9.1a (Vincens et al. 2010; Lebamba et al. 2016)) and the Niari Valley in the Republic of Congo (site 11, Fig. 9.1a (Vincens et al. 1998)).

Figure 9.1b–e presents information about the different proxies and determinants outlined in Sect. 9.2.2.1. Figure 9.1b, c highlights two examples of changes in pollen assemblages, (a) a site that had undergone major changes in vegetation composition during the LGM and the 3000 cal yr BP crisis (Fig. 9.1b): Lake Barombi Mbo from Giresse et al. (1994) (site 4, Fig. 9.1a) and (b) a site that had experienced a transition from forest to savanna during the 3000 cal yr BP crisis (Fig. 9.1c): Lake Mbalang from Vincens et al. (2010) (site 1, Fig. 9.1a).

Figure 9.1d presents the frequency of dated archaeological evidence in sites binned every 50 years. This is used to illustrate the changes in human impact across the Congo basin using available data since 5000 cal yr BP (Oslisly et al. 2013; Morin-Rivat et al. 2014). The period indicated in red corresponds to a period of population increase across the Congo basin.

The long-term dynamics of sea surface temperature (SST) illustrates changes in climate, and specifically rainfall (Fig. 9.1e) (Weldeab et al. 2007). It is worth noting the period spanning the 3000 cal yr BP crisis, which is a period characterized by abrupt changes in SST as indicated in the gray area (Fig. 9.1e).

9.3.1.2 Trends During the Late Holocene

The African Humid Period ended abruptly ~4000 cal yr BP (Fig. 9.1e). The subsequent period was characterized by low rainfall and major droughts that lasted until 1200 cal yr BP (Vincens et al. 1999). This period is called the "third millennium rainforest crisis" and is divided into two major phases (Maley 2002).

The first phase started shortly after 4000 cal yr BP, and is associated with an abrupt decrease in rainfall (see increase in SST in Fig. 9.1e). This trend affected the peripheral areas of the Congo basin (see sites 1–2 and the Niari Valley in Fig. 9.1a, c), and was responsible for the opening of the coastal savannas in central Africa (see site 7, Fig. 9.1a) (Elenga et al. 1992; Ngomanda et al. 2009) and the Dahomey Gap in west Africa (Salzmann and Hoelzmann 2005, not seen in Fig. 9.1a). During the same period, savanna vegetation was also heavily modified, as suggested by the pollen record of Lake Tilla (site 14, Fig. 9.1a), with a gradual (during the African Humid Period termination) and abrupt (during 3000 cal yr BP crisis) floristic shifts from Guinean to Sudano-Guinean savanna (Salzmann 2000). As illustrated in Lake Mbalang (site 1, Fig. 9.1a) (Vincens et al. 2010), the first phase of this abrupt climatic change triggered a transition from forest to savanna taxa (Fig. 9.1c) in some sites.

The second phase was rather short and abrupt, lasting between 2500 and 2000 cal yr BP. The SST reconstructions (Fig. 9.1e) and geological limestone zones (Maley et al. 2012, 2018) suggest a strong climatic seasonality. During this phase (in gray, Fig. 9.1b–e), vegetation reconstructions from pollen data show an increasing abundance of pioneer and secondary forest trees, and even in grasses within forest areas (Vincens et al. 1999) (Fig. 9.1b). This suggests that forests were highly disturbed during this period, probably exhibiting mosaics of open patches and closed forests. Some authors even suggest the opening of a north–south savanna corridor in the Sangha River Interval (Maley and Willis 2010) that could have permitted the migration of Bantu-speaking people. However, the savanna corridor has been controversial (Bremond et al. 2017). The evidence of increased human population and activity in the forest zone is dated from ~2500 cal yr BP, with the first iron-age settlements (Fig. 9.1d) (Wotzka 2006). Human presence in the region generally shows a bimodal pattern (Fig. 9.1d, red highlight), with two phases of human population expansion, and an intermediate phase of depopulation, between ~1300 and ~700 cal yr BP, but of unknown origin (Morin-Rivat et al. 2014, 2017).

9.3.2 Current Vegetation Distribution

9.3.2.1 Vegetation Types

Currently, tropical forest (in green, Fig. 9.2a) is the most important vegetation type in central Africa, covering 42% of the study area, followed by savanna (yellow, 28.4%), Miombo woodland (brown, 7%), mosaics of tropical forest and mesic savanna (light green, 5%), and other types of forested vegetation, such as swamp and swamp forest (3.0%), montane forest (1.3%), and mangrove (1%). The area affected by human activity amounts to 13.3% of the study area (red, Fig. 9.2a).

Terra firme forests occupy a vast and continuous area across the Congo basin, forming mosaics with southern savannas, such as the Nyanga river area and the Batéké Plateau. Deforested areas (in red, Fig. 9.2a) are prevalent around major cities such as Douala and Yaoundé in Cameroon, Brazzaville, and Ouesso in Congo, Kinshasa, and Kisangani in DRC, and along major road networks and rivers that connect them (Laurance et al. 2017). Deforestation is also prevalent in the densely populated areas in the mountains of eastern DRC. Large savanna areas are located in the northern and southern edge of central Africa, and are associated with high human impacts, for example, in northern Cameroon, southern Congo, and DRC (Fig. 9.2a).

There is substantial floristic variation within these vegetation types. *Terra firme* forests can broadly be divided into (a) evergreen forests under wet and aseasonal

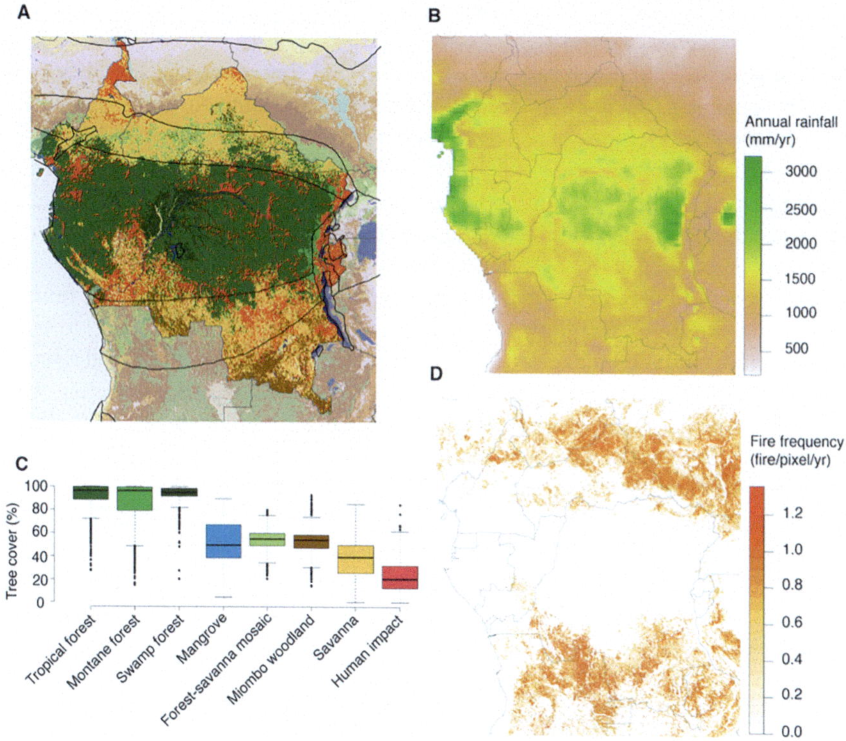

Fig. 9.2 Vegetation types (**a**), fire frequency (**b**), tree cover (**c**), and rainfall frequency (**d**) in central Africa. Note: Figure (**a**) is modified from Verhegghen et al. (2012). The background of the vegetation map (Fig. 9.1a) is based on the land-cover map of Mayaux et al. (2004) and the limits of the major phytochoria (black lines) according to White (1983). Mean annual rainfall (**b**) is expressed as millimeter per year, averaged over the period 2002–2015. Fire frequency (**d**) is expressed as the number of times that the area of each pixel is burned per year during the period 2002–2015

climates (see rainfall in Fig. 9.2b), (b) semi-deciduous forests under moist and seasonal climates (Fayolle et al. 2014a), (c) a transition type generally recognized in between the previous two types (De Namur 1990). Similarly, savannas can be divided into several types based on physiognomy, from woodlands to grass savannas (Aubréville 1957), or based on tree species composition (Fayolle et al. 2019; Osborne et al. 2018).

Tree cover can be used to distinguish the physiognomy of broad vegetation types, such as forest and savanna. We find that although highly variable, tree cover is (a) high in forested vegetation types such as tropical forests, montane forests, and swamp forests (>90%, Fig. 9.2c), (b) intermediate in mangroves, forest–savanna mosaics, and miombo woodlands (~50%, Fig. 9.2c), and (c) very low in savannas (<40%, Fig. 9.2c) and areas modified through human activity (~20%, Fig. 9.2c).

9.3.2.2 Determinants of Vegetation

The spatial distribution of mean annual rainfall and fire occurence highlights the different ecological functioning of forest and savanna in central Africa (Fig. 9.2b, d). The continuous forest block experiences only a few fire events each year (Figs. 9.2b and 9.3b), and encompasses wet (>2000 mm) and moist sites (1500–2000 mm) (Figs. 9.2d and 9.3a). In contrast, northern and southern savannas and woodlands experience frequent fires (Fig. 9.3d), and are generally encountered under drier conditions (Figs. 9.2d and 9.3c).

As expected, forests and savannas are located on the two extremes of a rainfall gradient, ranging from evergreen forests (high rainfall: >2500 mm $year^{-1}$) to open savannas (low rainfall: <500 mm $year^{-1}$). Tree cover increases with rainfall (Bucini and Hanan 2007), at least in the lower part of the rainfall gradient, interacting with the fire regime (Staver et al. 2011a). For intermediate rainfall (here, 1000–2500-mm $year^{-1}$) forests and savannas represent alternative stable states (Staver et al. 2011a, b), meaning that both biome types can co-occur (Fig. 9.3a–c, gray area). However, in central Africa, forests and savannas are spatially separated (Fig. 9.2a) (Aleman and Staver 2018). Forests in Congo and forest–savanna mosaics in the CAR occur under the same rainfall conditions, but one biome is dominated by trees

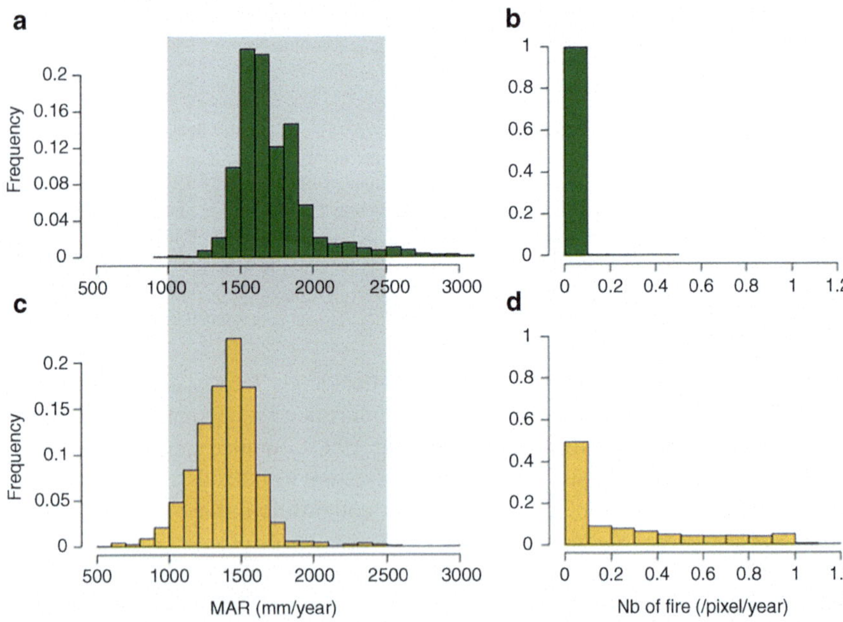

Fig. 9.3 Mean annual rainfall (in mm) (**a**, **c**) and fire occurrence between 2002 and 2015 (**b**, **d**). Note: Fire occurrence was estimated through the number of fires per pixel between 2002 and 2015. Green figures illustrate the mean annual rainfall and fire occurrence for tropical forest (**a**, **b**) and the yellow for savannas (**c**, **d**). The mean annual rainfall range for which forest and savanna co-occur is highlighted in gray (i.e., 1000–2500 mm $year^{-1}$)

(forest), while the other is dominated by grasses (savanna). The two alternative states are maintained through a positive feedback between fire and an open canopy, with fire being common in savannas (Fig. 9.3d), while nearly absent in forests (Fig. 9.3b). This seems to be the legacy of long-term vegetation distribution, and not some kind of founder effect.

Finally, even though rainfall and fire are important determinants of the distribution of tropical forests and savannas, other drivers can also become significant, including soils (Lloyd et al. 2015), and herbivores, specifically mammal browsers (Charles-Dominique et al. 2016). However, we have not addressed these effects in this chapter.

9.3.3 Changes in Vegetation

9.3.3.1 Ongoing Vegetation Changes

The remotely sensed data indicate strong ongoing vegetation changes, and specifically deforestation and forest degradation in the forests and woody encroachment in savannas (see Mitchard et al. 2011). At the regional scale, tree cover loss is far more prevalent in forests than in savannas (Fig. 9.4a). Overall, although deforestation is relatively low in central Africa (Pan et al. 2011; Achard et al. 2014; Aleman et al. 2018), tree cover loss can be extremely important locally, especially along the main roads and the surroundings of major cities, specifically in the DRC (Fig. 9.4a). This trend is due to the increasing fuelwood demand and need for agricultural lands, as population grows (Schure et al. 2014; see also Chaps. 7 and 10, Vol. 1). Logging is also responsible for forest degradation and fragmentation, as tree cover loss is much more prevalent in logging concessions than in protected areas (Fig. 9.4b). Approximately 29% of the forests in central Africa are currently granted to logging companies, ranging, however, from 13% in DRC to more than 60% of Gabon and Congo (Fig. 9.4b, Laporte et al. 2007). Although selective logging (Ruiz Pérez et al. 2005), specifically under Forest Stewardship Council (FSC) criteria, might not be very detrimental to forest biodiversity (Putz et al. 2012), the sustainability of conventional logging practices elsewhere in the region is certainly questionable, especially due to the progression of logging roads (Laporte et al. 2007).

In this chapter, we only analyzed tree cover loss between 2000 and 2015. However, there is a longer term trend of woody encroachment in savannas areas across SSA (Stevens et al. 2017), and also more regionally in central Africa (Youta Happi 1998; Mitchard et al. 2009). Soil carbon isotopes signature analysis has demonstrated forest encroachment into savanna areas over the past century in the CAR (Runge 2002), Gabon (Delègue et al. 2001), and Cameroon (Guillet et al. 2001). However, the causes of this widespread woody encroachment in savannas are hotly debated in the academic literature (Devine et al. 2017). Some scholars have proposed that this is a response to fire suppression policies during the last century

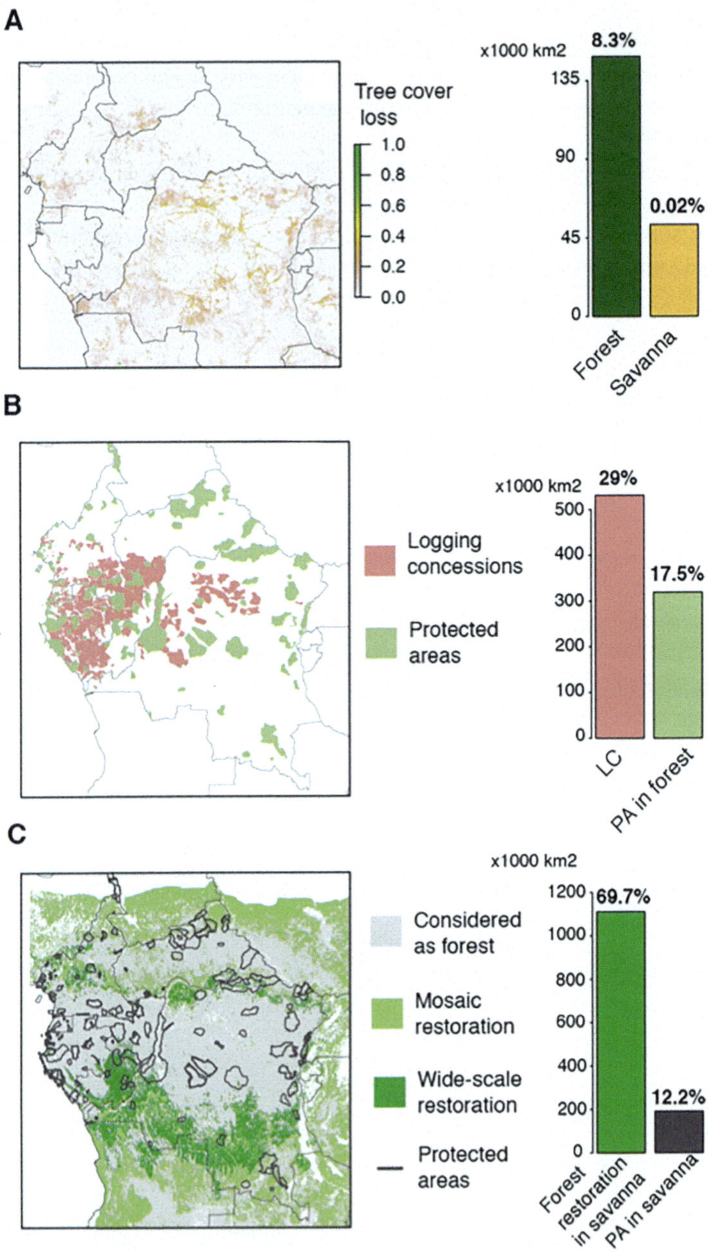

Fig. 9.4 Areas of tree cover loss between 2000 and 2015 in forests and savannas (**a**), in logging concessions and protected areas (**b**) and according to restoration potential (**c**). Note: Map in (**a**) shows tree cover loss at the pixel scale. Map in (**b**) delineates protected areas and areas granted to logging companies. Map in (**c**) highlights the location of forest restoration opportunities according to the WRI. Fractions in bar charts indicate (**a**) the percentage of pixels with tree cover loss above 10% for forest and savanna pixels, (**b**) for forest areas granted to logging concessions or falling within protected areas, and (**c**) for savanna areas protected or granted for reforestation

(Veldman et al. 2017; Bond and Zaloumis 2016), with more recent acceleration possibly being due to increasing CO_2 levels (Buitenwerf et al. 2012).

Restoration and reforestation efforts have been aiming to reverse deforestation and forest degradation in the region (Fig. 9.4c). However, such efforts are sometimes misplaced, with large savanna areas being targeted for tree plantation purposes (Fig. 9.4c). Approximately 30% of savanna areas are currently identified as forest (in gray, Fig. 9.4c), while 70% are targeted for restoration, including both savannas in mosaic and large extent of savannas. However, these areas have actually been savannas since the Last Glacial Maximum, and essentially represent a strongly stable state, considering that savannas and forests are alternative stable states in areas of intermediate rainfall (Staver et al. 2011a, b; Veldman et al. 2015a; Bond and Zaloumis 2016; Veldman 2016) (Sect. 9.1).

Protected areas are crucial for sustaining biodiversity and ecosystem processes and services, in both forest and savanna areas. They can also protect landscapes against misplaced reforestation projects and unsustainable logging practices. Currently, protected areas in central Africa cover approximately 17.5% of forests and 12.2% of savannas (Fig. 9.4b, c). This represents a relatively lower fraction of protected area coverage, compared to other parts of SSA such as East and South Africa (Aleman et al. 2016; IPBES 2018).

9.3.3.2 Predicted Vegetation Changes

We then explore the effects of future land use and climatic change in the area currently covered by forests and savannas in central Africa. Cropland expansion is expected to affect significantly savanna areas, considering that more than 40,000 and 150,000 km^2 of land will be converted for crop production by 2070 according to RCP 8.5 and RCP 2.6, respectively (Fig. 9.5a, b). The changes according to the most "optimistic" scenario in terms of emissions and radiative forcing (RCP 2.6) are projected to be more significant in savanna areas (14% converted area) compared to RCP 8.5 (2.8% converted area) (Fig. 9.5a). Forests are expected to be less affected, with only 0.5% and 0.2% of their area converted, respectively (Fig. 9.5a). This is mainly due to the fact that forests are much more protected under these scenarios than savannas (IPCC 2014). This large increase in cropland under RCP 2.6 reflects the large expansion of biofuel production (Van Vuuren et al. 2011) (see Chaps. 2–4, Vol. 1; Chap. 5, Vol. 2).

When it comes to climate change impacts, it is projected that rainfall patterns will change and the probability of drought and extreme events will increase. Rainfall

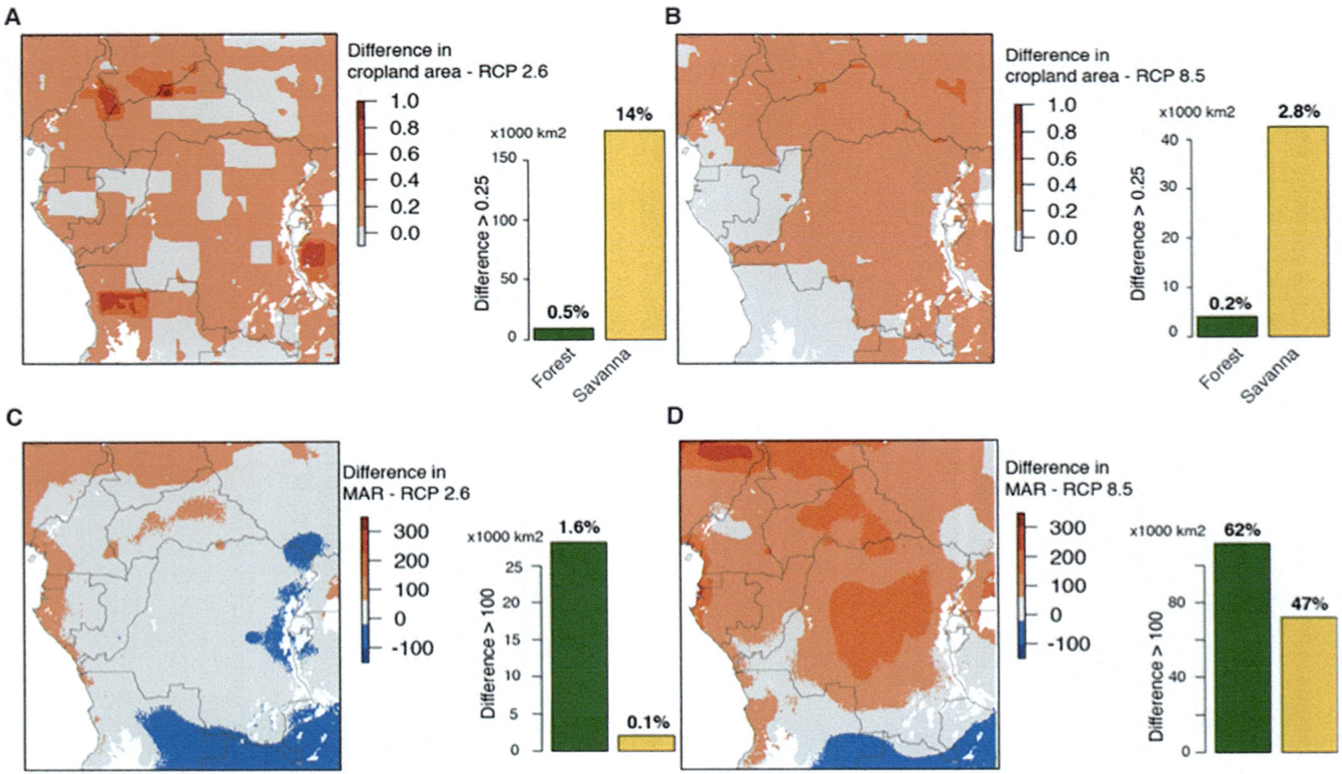

Fig. 9.5 Change in cropland area and rainfall according to RCP 2.6 (**a**, **c**) and RCP 8.5 (**b**, **d**). Note: Change in cropland area in each pixel is estimated as the difference between current and future (in 2070) cropland area in each pixel for the scenarios RCP 2.6 (**a**) and RCP 8.5 (**b**). Barplots represent the area for which this difference is expected to exceed 25% for forest and savanna areas in central Africa. Changes in rainfall (mean annual rainfall in mm $year^{-1}$) are estimated as the difference between current and future (in 2070) rainfall for the scenarios RCP 2.6 (**c**) and RCP 8.5 (**d**). Barplots represent the area for which this difference is expected to exceed 100 mm $year^{-1}$ for forests and savannas in central Africa

change within RCP 2.6 is expected to be less intense and extensive, with only 1.6% of forest areas (located at the western part of central Africa) expected to be affected by rainfall change (Fig. 9.5c). On the contrary, according to RCP 8.5, as much as 62% of forests and 47% of savannas will be impacted by a decrease (or increase) in annual rainfall, with a massive drying trend over central Africa, and a localized increase in rainfall in the extreme south (Fig. 9.5d).

9.4 Discussion

9.4.1 Changes in Central African Forests and Savannas

The evidence reviewed in this study highlights that since the Last Glacial Maximum (21,000 cal yr BP), the area covered by tropical forests experienced a succession of dry and humid periods, which resulted in forest expansion and contraction (Maley 1989, 1996; Vincens et al. 1999) (Sect. 9.3.1). In contrast, forests in Amazonia remained much more stable during that same period (Anhuf et al. 2006). Savannas were more widespread and changed both in terms of extent and composition following climate fluctuations. However, for some areas, such as Lake Bambili in Cameroon (Site 4, Fig. 9.1a), there is accumulated evidence that savanna vegetation has been maintained since the LGM, indicating a very stable state at this relatively high altitude (>2000 m above sea level).

When analyzing the current extent and distribution of forest and savanna according to the most recent vegetation types available for central Africa (Verhegghen et al. 2012), we confirm that forests dominate the Congo basin, are found under moist and wet climates, and experience only few fire events (Sect. 9.3.2). In contrast, savannas occupy drier areas in the northern and southern edge of central Africa, and correspond to the Sudanian and Zambezian Savanna Regions (White 1983). These savannas occur in areas of relatively high rainfall and experience frequent fires (Lehmann et al. 2014; Osborne et al. 2018). Forests and savannas can both occur at intermediate rainfall (1000–2500 mm $year^{-1}$) (Sect. 9.3.2). This seems to be the outcome of feedback mechanisms between canopy closure and fire, past climatic changes, and, to a lesser extent, human occupation, as shown at the continental (Sankaran et al. 2005; Aleman and Staver 2018) and the global (Staver et al. 2011b) scales.

As the ecological functioning of forests and savannas depend on factors related to both climate and land use, we then examined such ongoing changes in areas currently covered by forests and savannas (Sect. 9.3.3). Over the past decades, woody encroachment has been reported in savannas worldwide (Stevens et al. 2017), probably due to land use change and/or CO_2 fertilization effects. More recently (i.e., between 2000 and 2015), there has been extensive tree cover loss in the forest area (Hansen et al. 2013) and forest fragmentation due to industrial logging (Laporte et al. 2007). This has been associated to hunting (Poulsen et al. 2011; Abernethy et al. 2013) (Sect. 9.3.3).

Both ongoing climate change and land use change threaten the functioning and resilience of forests and savannas in central Africa. Increasing annual rainfall will certainly favor forests (Malhi et al. 2009) and could even improve crop yields in savanna areas (Sonwa et al. 2011; IPBES 2018). Even though we have not explored how seasonality and temperature are projected to change by 2070, other studies have predicted important shifts in composition (Sala et al. 2000) and vegetation structure (Aleman et al. 2016, 2017). In addition, even though forest in central Africa are carbon sinks that sequester large volumes of carbon annually (Lewis et al. 2009), massive emissions could be expected in specific years due to drought-related tree dieback (Allen et al. 2010; Bennett et al. 2015), as reported for the Amazon (Phillips et al. 2009). This phenomenon has, however, not yet been reported in central Africa.

Section 9.3.3 outlined how savannas are specifically targeted for reforestation and the production of biofuel and cash crops, largely because they are perceived to have lower conservation value than forests (see Alexandratos and Bruinsma 2012; Searchinger et al. 2015). Currently >30% of forests in central Africa have been granted to logging companies. Even though increasing protected area coverage may help prevent the further degradation of both forests and savannas (especially savannas that only have 12.2% of their area under protection), there is a widespread concern that protected areas are not effective in the region, primarily due to conflicts and poor governance (Chape et al. 2005) (Chap. 1, Vol. 1). This potentially compromises their effectiveness in maintaining forest and savanna resilience. Indeed, protected areas in tropical areas are increasingly threatened due to population increase (Laurance et al. 2014) and especially funding constraints (Bruner et al. 2004) (Chap. 1, Vol. 1). Illegal logging, grazing, and agriculture commonly occur inside poorly enforced protected areas (Laurance et al. 2012), especially in west and central Africa (Tranquilli et al. 2014). Agricultural expansion near protected areas tends to erode biodiversity and ecosystem services due to edge effects (Wittemyer et al. 2008; Laurance et al. 2014). However, if the integrity of protected areas can be maintained, for example, by involving local communities (Vodouhê et al. 2010), then central African forests and savannas may remain relatively resilient in the face of these threats, and future climate change (Chap. 10, Vol. 1).

9.4.2 Policy Implications and Recommendations

Based on the comprehensive analysis of the past, present, and future vegetation patterns in central African forests and savannas, we identify two main policy relevant findings: (a) mesic savannas are not degraded forests and (b) forests and savannas have experienced a long history of climate changes and human activity, which offers insights for their effective management.

Considering that forests and savannas in central Africa offer habitat for biodiversity and contribute manifold to the livelihoods of local communities (see

Chap. 10, Vol. 1), these findings carry important implications for meeting the sustainable development goals (SDGs) and especially SDG 15 "Life on land." Below, we discuss some important aspects related to reforestation (Sect. 9.4.2.1), designation of protected areas (Sect. 9.4.2.2), and sustainable forest management (Sect. 9.4.2.3).

9.4.2.1 Target Degraded Areas for Ecosystem Restoration

Our findings indicate that savannas have unique ecological functioning, based on chronic disturbances (Sankaran et al. 2005). These savannas exist for millennia (Salzmann and Hoelzmann 2005), but are currently threatened by recent land use change, especially crop expansion and reforestation projects (Sect. 9.3.3). This is often done because savannas are mistaken as degraded forests (Sect. 9.1), despite the fact that they are alternative stable states to forest and are maintained through a positive feedback between vegetation structure and fire events (Sect. 9.3.2).

Due to this mistaken view of mesic savannas being degraded forests, many studies have suggested to target these savannas for forest restoration and tree planting for carbon sequestration (Laestadius et al. 2012; WRI 2014; Chazdon and Laestadius 2016). Even though many studies have suggested that the ecological restoration of degraded savanna and forest landscapes can have substantial ecological benefits (Lamb et al. 2005; Buisson et al. 2019), it is important to focus restoration efforts only to those areas that have actually been degraded, and are not mistaken as such. This is particularly important in the context of SDG15, and especially Target 15.2, where forest restoration is seen as an important intervention.

To this end, there is a need to develop integrated restoration frameworks that both consider forest and savanna ecosystems, their respective ecological functioning, and their long-term location and history. This could help avoid misrepresenting savannas as degraded forest (Veldman et al. 2015a, c), and thus implementing restoration interventions that could have counterproductive conservation outcomes. This would require more retrospective studies to accurately reconstruct the past locations of forests and savannas (Aleman et al. 2018).

In those cases where reforestation and forest restoration are deemed appropriate, then it would be crucial to favor native species (Chazdon 2008; Doucet et al. 2016). Indeed, exotic tree species such acacia and eucalypt currently dominate timber plantations in central Africa, following a systematic replication of the same experimental design (Proces et al. 2017) (Chap. 10, Vol. 1).

9.4.2.2 Expand Protected Areas

Section 9.3.3 highlighted that protected areas in central Africa cover approximately 17.5% of tropical forests and 12.2% of savannas. According to the Aichi Biodiversity Target 11, 17% of terrestrial ecosystems need to be conserved through protected areas by 2020. In this respect, it can be argued that the protection of savanna areas falls well below established targets, possibly due to the fact that savanna areas can be mistaken for degraded forests (Sect. 9.4.2.1). In view of the renewed conservation efforts through SDG Target 15.1 and the currently negotiated post-2020 global biodiversity framework, it can be argued that both savannas and forests in central Africa may require better protected areas networks.

Boosting formal conservation efforts (through national parks and otherwise) within savanna and forest ecosystems may help mitigate the negative future outcomes of land use and climate changes (Parr et al. 2014; Veldman et al. 2015a). However, it is interesting to point that savanna ecosystems may also lend themselves to alternative conservation practices. For example, protected areas in savannas are often associated with lower plant diversity compared to communally managed areas (Dahlberg 2000; Nacoulma et al. 2011). This is because savannas are intrinsically associated with disturbances such as fires (Veldman et al. 2015c) (Sect. 9.3.2). In this sense, traditional land management might not degrade savanna habitats (Nacoulma et al. 2011) (see Chap. 10, Vol. 1; Chap. 3, Vol. 2).

In such contexts, there might be a need to reevaluate land use change planning practices to explicitly include the conservation and agricultural value of savanna ecosystems (Veldman et al. 2015c). Some scholars have even argued that effective biodiversity conservation will rely on associating traditional and communally managed areas, with reserves (Abel and Blaikie 1989; Nacoulma et al. 2011). Such approaches could potentially create important synergies between SDG15 (especially as it pertains to the sustainable use of biodiversity), SDG13 (Climate action), and SDGs directly related to human livelihoods such as SDG2 (zero hunger) and SDG1 (no poverty).

There would also be a need to develop appropriate management frameworks for grass-dominated biomes (Veldman et al. 2015a, c), and to better estimate the environmental value of such biomes. Historically, the vast majority of conservation efforts have focused on forests (Laurance et al. 2014; Tranquilli et al. 2014), but savannas are at equal, if not greater, risk of change due to land use change (Aleman et al. 2016). Mesic savannas and transition zones between forests and savannas in west and central Africa may be particularly susceptible, not the least due to the fact that they contain few protected areas (Sect. 9.3.3). Sustainable management plans for such ecosystems should include strong monitoring components that serve both biodiversity conservation and human development needs.

9.4.2.3 Promote Sustainable Forest Management Approaches

Finally, some of the new directions in biodiversity conservation and ecosystem management suggest to take into account the long-term history of ecosystems (Gillson 2015; Gillson and Marchant 2014; Barnosky et al. 2017). The peculiar long-term history of forests and savannas in central Africa, especially in relation to climate change and human activity (e.g., migration and activity of Bantu people), could inform the development of more sustainable logging practices.

Logging concessions in central Africa constitute a major land use, accounting for more than 60% of forest area in some countries, while at the same time being an engine of economic growth (Bayol et al. 2012; De Wasseige et al. 2012) (Chap. 1, Vol. 1). Even though current logging practices are not entirely sustainable, considering that the initial stock of timber accumulating for centuries cannot be recovered in 25–30 years of a timber rotation, some good logging practices such as selective logging following Forest Stewardship Council standards could limit negative biodiversity outcomes (Putz et al. 2012) and ensure at least 50% recovery of the timber resource for the targeted species (Bayol et al. 2012). With logging roads closing within only a few years (Kleinschroth et al. 2015) and biomass recovering within approximately 20 years following logging (Gourlet-Fleury et al. 2013), it is possible that production forests could provide multiple ecosystem services, at levels even comparable to "intact" or old-growth forests.

Nevertheless, logging can be a major driver of forest fragmentation, having negative biodiversity outcomes through habitat fragmentation and forest accessibility (Poulsen et al. 2013; Ziegler et al. 2016). Thus enhancing conservation actions and preventing poaching/hunting are key priorities in logging concessions. In addition, even though timber recovery rates are supposed to reach 50%, regeneration deficits have been observed for many timber species (Engone Obiang et al. 2014; Morin-Rivat et al. 2017). Thus prior to logging management inventories should be used to identify target species and how the regeneration is organized. Indeed, dissimilarities between canopy species and regeneration have been observed in different forest types (e.g., Swaine and Hall 1988 in Ghana along a rainfall gradient), which can be explained by forest successional dynamics. The conditions that promoted the regeneration of canopy species may no longer be encountered. As a result (and in order to be logged in the future) the species targeted for timber in central Africa will need to be favored, and specific silvicultural interventions will be required. Among them, forest enrichment (Doucet et al. 2009) and plantations (Doucet et al. 2016) have already provided good results at a relatively small scale.

The above suggest that designing nuanced sustainable management approaches based on the long-term understanding of forest and savanna dynamics could contribute substantially in meeting SDG Target 15.2 (sustainable forest management) and 15.7 (end poaching). More importantly such interventions can have important synergies with some of the targets related to SDG12 (responsible consumption and production) and SDG8 (decent work and economic growth).

9.5 Conclusions

In this chapter, we analyzed the distribution of the two main biomes in central Africa, tropical forests and savannas, as well as the underlying factors affecting their distribution. Interestingly, in intermediate rainfall regimes, forest and savanna represent alternative stable states maintained by a positive feedback between vegetation structure and fire events. Forest-savanna bistability spans a huge portion of central

Africa, with 36% of African savannas currently occurring where rainfall is sufficient for forest canopy closure. Consequently, these mesic savannas occurring in areas of annual rainfall between 1000 and 2500 mm have often been mistakenly considered as degraded forests.

However, forests and savannas in central Africa have experienced a long history of climate events and human activity that have both triggered important contractions and expansions in the Congo Basin, as well as modifications in forest floristic composition. During the past century, new threats have emerged for these biomes, as we have entered the Anthropocene. Apart from climate change, the alarming rates of land use change related to cropland expansion, logging concessions, and misplaced forest plantations may represent an even more substantial threat to forest and savanna structure, composition, and ultimately ecosystem services provision.

As a conclusion, we urge academics and practitioners to develop integrated management frameworks for both forest and savanna ecosystems, to avoid further ecosystem degradation. Furthermore, we point the need to restore the importance of savannas in biodiversity conservation efforts. We advice taking into account knowledge from the long-term history of these ecosystems to build more sustainable management plans that can guide ecosystem restoration, protected area expansion, and the design of sustainable forest management approaches in the region, and beyond.

References

Abel N, Blaikie P (1989) Land degradation, stocking rates and conservation policies in the communal rangelands of Botswana and Zimbabwe. Land Degrad Dev 1:101–123

Abernethy K, Coad L, Taylor G, Lee M, Maisels F (2013) Extent and ecological consequences of hunting in Central African rainforests in the twenty-first century. Philos Trans R Soc B 368:20120303

Abernethy K, Maisels F, White LJ (2016) Environmental issues in central Africa. Annu Rev Environ Resour 41:1–33

Achard F, Beuchle R, Mayaux P, Stibig HJ, Bodart C, Brink A, Carboni S, Desclée B, Donnay F, Eva HD (2014) Determination of tropical deforestation rates and related carbon losses from 1990 to 2010. Glob Chang Biol 20:2540–2554

Aleman JC, Staver AC (2018) Spatial patterns in the global distributions of savanna and forest. Glob Ecol Biogeogr 27:792–803

Aleman JC, Blarquez O, Bentaleb I, Bonté P, Brossier B, Carcaillet C, Gond V, Gourlet-Fleury S, Kpolita A, Lefèvre I, Oslisly R, Power MJ, Yongo O, Bremond L, Favier C (2013) Tracking land-cover changes with sedimentary charcoal in the Afrotropics. The Holocene 23:1853–1862

Aleman JC, Blarquez O, Staver CA (2016) Land-use change outweighs projected effects of changing rainfall on tree cover in sub-Saharan Africa. Glob Chang Biol 22:3013–3025

Aleman JC, Blarquez O, Gourlet-Fleury S, Bremond L, Favier C (2017) Tree cover in Central Africa: determinants and sensitivity under contrasted scenarios of global change. Sci Rep 7:41393

Aleman JC, Jarzyna MA, Staver AC (2018) Forest extent and deforestation in tropical Africa since 1900. Nat Ecol Evol 2:26

Alexandratos N, Bruinsma J (2012) World agriculture towards 2030/2050: the 2012 revision. ESA Working Paper 3

Allen CD, Macalady AK, Chenchouni H, Bachelet D, McDowell N, Vennetier M, Kitzberger T, Rigling A, Breshears DD, Hogg ET (2010) A global overview of drought and heat-induced tree mortality reveals emerging climate change risks for forests. For Ecol Manag 259:660–684

Anhuf D, Ledru M-P, Behling H (2006) Paleo-environmental change in Amazonian and African rainforest during the LGM. Palaeogeogr Palaeoclimatol Palaeoecol 239:510–527

Ankogui-Mpoko GF (2003) Gestion des espaces agropastoraux en zones de savanes centrafricaines: De l'arbitrage de l'Etat à une cogestion. In: Seiny Boukar L, Jamin JY, Floret C (eds) Savanes africaines: des espaces en mutation, des acteurs face à de nouveaux défis., Garoua, Cameroun. Prasac, N'Djamena, Tchad—Cirad, Montpellier

Assi-Kaudjhis C (2012) Vegetation evolution in the mountains of Cameroon during the last 20 000 years: pollen analysis of Lake Bambili sediments. In: Blanco JA (ed) Forest ecosystems—more than just trees. IntechOpen, Rijeka

Aubréville A (1939) Forêts reliques en Afrique Occidentale Française. Revue de botanique appliquée et d'agriculture coloniale 19:479–484

Aubréville AMA (1947) The disappearance of the tropical forests of Africa. Unasylva (FAO) 1:5–11

Aubréville A (1957) Nomenclature of African vegetation types at Yangambi. Bois For Trop 51:23–27

Barnosky AD, Hadly EA, Gonzalez P, Head J, Polly PD, Lawing AM, Eronen JT, Ackerly DD, Alex K, Biber E (2017) Merging paleobiology with conservation biology to guide the future of terrestrial ecosystems. Science 355:eaah4787

Bassett TJ, Boutrais J (2000) Cattle and trees in the West African savanna. In: Contesting forestry in West Africa. Ashgate Publishing Ltd., Aldershot, pp 242–263

Bastin J-F, Berrahmouni N, Grainger A, Maniatis D, Mollicone D, Moore R, Patriarca C, Picard N, Sparrow B, Abraham EM (2017) The extent of forest in dryland biomes. Science 356:635–638

Bayol N, Demarquez B, De Wasseige C, Eba'a Atyi R, Fisher J-F, Nasi R, Pasquier A, Rossi X, Steil M, Vivien C (2012) Forest management and the timber sector in Central Africa. In: The forests of the Congo Basin—state of the forest 2010, pp 43–61

Bayon G, Dennielou B, Etoubleau J, Ponzevera E, Toucanne S, Bermell S (2012) Intensifying weathering and land use in Iron Age Central Africa. Science 335:1219–1222

Bennett JM, Clarke RH, Horrocks GF, Thomson JR, Mac Nally R (2015) Climate drying amplifies the effects of land-use change and interspecific interactions on birds. Landsc Ecol 30:2031–2043

Biwolé AB, Morin-Rivat J, Fayolle A, Bitondo D, Dedry L, Dainou K, Hardy OJ, Doucet J-L (2015) New data on the recent history of the littoral forests of southern Cameroon: an insight into the role of historical human disturbances on the current forest composition. Plant Ecol Evol 148:19–28

Bond WJ (2016) Ancient grasslands at risk. Science 351:120–122

Bond WJ, Parr CL (2010) Beyond the forest edge: ecology, diversity and conservation of the grassy biomes. Biol Conserv 143:2395–2404

Bond W, Zaloumis NP (2016) The deforestation story: testing for anthropogenic origins of Africa's flammable grassy biomes. Philos Trans R Soc B 371:20150170

Bond W, Woodward F, Midgley G (2005) The global distribution of ecosystems in a world without fire. New Phytol 165:525–538

Bostoen K, Clist B, Doumenge C, Grollemund R, Hombert J-M, Muluwa JK, Maley J (2015) Middle to late holocene paleoclimatic change and the early Bantu expansion in the rain forests of Western Central Africa. Curr Anthropol 56:354–384

Bradley E, McNamara K (1993) Living with trees. Policies for forestry management in Zimbabwe. World Bank Group, Washington, DC

Bremond L, Bodin S, Bentaleb I, Favier C, Canal S (2017) Past tree cover of the Congo Basin recovered by phytoliths and δ 13 C along soil profiles. Quat Int 434:91–101

Brncic T, Willis K, Harris D, Washington R (2007) Culture or climate? The relative influences of past processes on the composition of the lowland Congo rainforest. Philos Trans R Soc B Biol Sci 362:229–242
Brncic T, Willis K, Harris D, Telfer M, Bailey R (2009) Fire and climate change impacts on lowland forest composition in northern Congo during the last 2580 years from palaeoecological analyses of a seasonally flooded swamp. The Holocene 19:79–89
Bruner AG, Gullison RE, Balmford A (2004) Financial costs and shortfalls of managing and expanding protected-area systems in developing countries. Bioscience 54:1119–1126
Bucini G, Hanan NP (2007) A continental-scale analysis of tree cover in African savannas. Glob Ecol Biogeogr 16:593–605
Buisson E, Le Stradic S, Silveira FAO, Durigan G, Overbeck GE, Fidelis A, Fernandes GW, Bond WJ, Hermann J-M, Mahy G, Alvarado ST, Zaloumis NP, Veldman JW (2019) Resilience and restoration of tropical and subtropical grasslands, savannas, and grassy woodlands. Biol Rev 94:590–609
Buitenwerf R, Bond W, Stevens N, Trollope W (2012) Increased tree densities in South African savannas: >50 years of data suggests CO2 as a driver. Glob Chang Biol 18:675–684
Chape S, Harrison J, Spalding M, Lysenko I (2005) Measuring the extent and effectiveness of protected areas as an indicator for meeting global biodiversity targets. Philos Trans R Soc B Biol Sci 360:443–455
Charles-Dominique T, Davies TJ, Hempson GP, Bezeng BS, Daru BH, Kabongo RM, Maurin O, Muasya AM, van der Bank M, Bond WJ (2016) Spiny plants, mammal browsers, and the origin of African savannas. Proc Natl Acad Sci 113:E5572–E5579
Chazdon RL (2008) Beyond deforestation: restoring forests and ecosystem services on degraded lands. Science 320:1458–1460
Chazdon RL, Laestadius L (2016) Forest and landscape restoration: toward a shared vision and vocabulary. Am J Bot 103:1869–1871
Dahlberg AC (2000) Vegetation diversity and change in relation to land use, soil and rainfall—a case study from North-East District, Botswana. J Arid Environ 44:19–40
Dargie GC, Lewis SL, Lawson IT, Mitchard ET, Page SE, Bocko YE, Ifo SA (2017) Age, extent and carbon storage of the central Congo Basin peatland complex. Nature 542:86–90
de Menocal P, Ortiz J, Guilderson T, Adkins J, Sarnthein M, Baker L, Yarusinsky M (2000) Abrupt onset and termination of the African Humid Period: rapid climate responses to gradual insolation forcing. Quat Sci Rev 19:347–361
De Namur C (1990) Aperçu sur la végétation de l'Afrique centrale atlantique. In: Paysages quaternaires de l'Afrique centrale atlantique. ORSTOM, Paris, pp 60–67
De Wasseige C, De Marcken P, Bayol N, Hiol-Hiol F, Mayaux P, Desclée B, Nasi R, Billand A, Defourny P, Atyi EA (2012) The forests of the Congo basin: state of the forest 2010. Publications Office of the European Union, Luxembourg
Dechamps R, Lanfranchi R, Le Cocq A, Schwartz D (1988) Reconstitution d'environnements quaternaires par l'étude de macrorestes végétaux.(Pays bateke, RP du Congo). Palaeogeogr Palaeoclimatol Palaeoecol 66:33–44
Delègue M, Fuhr M, Schwartz D, Mariotti A, Nasi R (2001) Recent origin of a large part of the forest cover in the Gabon coastal area based on stable carbon isotope data. Oecologia 129:106–113
Devine AP, McDonald RA, Quaife T, Maclean IM (2017) Determinants of woody encroachment and cover in African savannas. Oecologia 183:939–951
Dewees PA (1994) Social and economic aspects of Miombo woodland management in southern Africa: options and opportunities for research. CIFOR, Jakarta
Dexter K, Smart B, Baldauf C, Baker T, Balinga MB, Brienen R, Fauset S, Feldpausch T, Ferreira-Da Silva L, Ilunga Muledi J (2015) Floristics and biogeography of vegetation in seasonally dry tropical regions. Int For Rev 17:10–32

Dexter KG, Pennington RT, Oliveira-Filho AT, Bueno ML, de Miranda PLS, Neves DM (2018) Inserting tropical dry forests into the discussion on biome transitions in the tropics. Front Ecol Evol 6:104

Doucet J-L, Kouadio YL, Monticelli D, Lejeune P (2009) Enrichment of logging gaps with moabi (Baillonella toxisperma Pierre) in a Central African rain forest. For Ecol Manag 258:2407–2415

Doucet J-L, Daïnou K, Ligot G, Ouédraogo D-Y, Bourland N, Ward SE, Tekam P, Lagoute P, Fayolle A (2016) Enrichment of Central African logged forests with high-value tree species: testing a new approach to regenerating degraded forests. Int J Biodivers Sci, Ecosyst Ser Manag 12:83–95. https://doi.org/10.1080/21513732.2016.1168868

Elenga H, Schwartz D, Vincens A (1992) Changements climatiques et action anthropique sur le littoral congolais au cours de l'Holocène. Bull Soc Géol Fr 163:83–90

Elenga H, Schwartz D, Vincens A, Bertaux J, deNamur C, Martin L, Wirrmann D, Servant M (1996) Holocene pollen data from Kitina lake (Congo): palaeoclimatic and palaeobotanical changes in the Mayombe forest area. Comptes Rendus De L Academie Des Sciences Serie Ii Fascicule a-Sciences De La Terre Et Des Planetes 323:403–410

Elenga H, Vincens A, Schwartz D, Fabing A, Bertaux J, Wirrmann D, Martin L, Servant M (2001) Le marais estuarien de la Songolo (Sud Congo) a l'Holocene moyen et recent. Bull Soc Géol Fr 172:359–366

Engone Obiang NL, Ngomanda A, Hymas O, Chézeauxl É, Picard N (2014) Diagnosing the demographic balance of two light-demanding tree species populations in central Africa from their diameter distribution. For Ecol Manag 313:55–62

Favier C, Aleman J, Bremond L, Dubois MA, Freycon V, Yangakola JM (2012) Abrupt shifts in African savanna tree cover along a climatic gradient. Glob Ecol Biogeogr 21:787–797

Fayolle A, Engelbrecht B, Freycon V, Mortier F, Swaine M, Réjou-Méchain M, Doucet J-L, Fauvet N, Cornu G, Gourlet-Fleury S (2012) Geological substrates shape tree species and trait distributions in African moist forests. PLoS One 7:e42381

Fayolle A, Picard N, Doucet J-L, Swaine M, Bayol N, Bénédet F, Gourlet-Fleury S (2014a) A new insight in the structure, composition and functioning of central African moist forests. For Ecol Manag 329:195–205

Fayolle A, Swaine MD, Bastin JF, Bourland N, Comiskey JA, Dauby G, Doucet JL, Gillet JF, Gourlet-Fleury S, Hardy OJ (2014b) Patterns of tree species composition across tropical African forests. J Biogeogr 41:2320–2331

Fayolle A, Swaine MD, Aleman J, Azihou AF, Bauman D, te Beest M et al (2019) A sharp floristic discontinuity revealed by the biogeographic regionalization of African savannas. J Biogeogr 46:454–465

Garcin Y, Deschamps P, Ménot G, De Saulieu G, Schefuß E, Sebag D, Dupont LM, Oslisly R, Brademann B, Mbusnum KG (2018) Early anthropogenic impact on Western Central African rainforests 2,600 y ago. Proc Natl Acad Sci 115:3261–3266

Giglio L, Randerson JT, van der Werf GR, Kasibhatla PS, Collatz GJ, Morton DC, DeFries RS (2010) Assessing variability and long-term trends in burned area by merging multiple satellite fire products. Biogeosciences 7:1171–1186

Gillson L (2015) Biodiversity conservation and environmental change: using palaeoecology to manage dynamic landscapes in the anthropocene. OUP, Oxford

Gillson L, Marchant R (2014) From myopia to clarity: sharpening the focus of ecosystem management through the lens of palaeoecology. Trends Ecol Evol 29:317–325

Giresse P, Maley J, Brenac P (1994) Late quaternary palaeoenvironments in the Lake Barombi Mbo (West Cameroon) deduced from pollen and carbon isotopes of organic matter. Palaeogeogr Palaeoclimatol Palaeoecol 107:65–78

Giresse P, Maley J, Doumenge C, Philippon N, Mahé G, Chepstow-Lusty A, Aleman J, Lokonda M, Elenga H (2018) Paleoclimatic changes are the most probable causes of the rainforest crises 2,600 y ago in Central Africa. Proc Natl Acad Sci 115:E6672–E6673

Gourlet-Fleury S, Mortier F, Fayolle A, Baya F, Ouédraogo D, Bénédet F, Picard N (2013) Tropical forest recovery from logging: a 24 year silvicultural experiment from Central Africa. Philos Trans R Soc B 368:20120302

Griffith DM, Lehmann CE, Strömberg CA, Parr CL, Pennington RT, Sankaran M, Ratnam J, Still CJ, Powell RL, Hanan NP (2017) Comment on the extent of forest in dryland biomes. Science 358, eaao1309

Grollemund R, Branford S, Bostoen K, Meade A, Venditti C, Pagel M (2015) Bantu expansion shows that habitat alters the route and pace of human dispersals. Proc Natl Acad Sci 112:13296–13301

Guillet B, Achoundong G, Happi J, Beyala V, Bonvallot J, Riera B, Mariotti A, Schwartz D (2001) Agreement between floristic and soil organic carbon isotope (13C/12C, 14C) indicators of forest invasion of savannas during the last century in Cameroon. J Trop Ecol 17:809–832

Hanan NP, Sea WB, Dangelmayr G, Govender N (2008) Do fires in savannas consume woody biomass? A comment on approaches to modeling savanna dynamics. Am Nat 171:851–856

Hansen M, Potapov P, Moore R, Hancher M, Turubanova S, Tyukavina A, Thau D, Stehman S, Goetz S, Loveland T (2013) High-resolution global maps of 21st-century forest cover change. Science 342:850–853

Hijmans R, Cameron S, Parra J (2009) WorldClim—global climate data. http://www.worldclim.org

Hirota M, Holmgren M, Van Nes EH, Scheffer M (2011) Global resilience of tropical forest and savanna to critical transitions. Science 334:232–235

Hubau W, Van den Bulcke J, Kitin P, Mees F, Baert G, Verschuren D, Nsenga L, Van Acker J, Beeckman H (2013) Ancient charcoal as a natural archive for paleofire regime and vegetation change in the Mayumbe, Democratic Republic of the Congo. Quat Res 80:326–340

Hubau W, Van den Bulcke J, Van Acker J, Beeckman H (2015) Charcoal-inferred Holocene fire and vegetation history linked to drought periods in the Democratic Republic of Congo. Glob Chang Biol 21:2296–2308

Hurtt G, Chini LP, Frolking S, Betts R, Feddema J, Fischer G, Fisk J, Hibbard K, Houghton R, Janetos A (2011) Harmonization of land-use scenarios for the period 1500–2100: 600 years of global gridded annual land-use transitions, wood harvest, and resulting secondary lands. Clim Chang 109:117–161

IPBES (2018) The IPBES regional assessment report on biodiversity and ecosystem services for Africa. In: Archer E, Dziba L, Mulongoy KJ, Maoela MA, Walters M (eds) Secretariat of the Intergovernmental Science-Policy Platform on Biodiversity and Ecosystem Services, Bonn, Germany

IPCC (2014) Climate change 2014: impacts, adaptation, and vulnerability. Cambridge University Press, Cambridge

Kehoe L, Romero-Muñoz A, Polaina E, Estes L, Kreft H, Kuemmerle T (2017) Biodiversity at risk under future cropland expansion and intensification. Nat Ecol Evol 1:1129–1135

Kleinschroth F, Gourlet-Fleury S, Sist P, Mortier F, Healey JR (2015) Legacy of logging roads in the Congo Basin: how persistent are the scars in forest cover? Ecosphere 6:1–17

Laestadius L, Maginnis S, Minnemeyer S, Potapov P, Saint-Laurent C, Sizer N (2012) Mapping opportunities for forest landscape restoration. Unasylva (FAO) 238:62

Lamb P, Erskine PD, Parrotta JA (2005) Restoration of degraded tropical forest landscapes. Science 310:1628–1632

Laporte NT, Stabach JA, Grosch R, Lin TS, Goetz SJ (2007) Expansion of industrial logging in Central Africa. Science 316:1451–1451

Laurance WF, Arrea IB (2017) Roads to riches or ruin? Science 358:442–444

Laurance WF, Useche DC (2009) Environmental synergisms and extinctions of tropical species. Conserv Biol 23:1427–1437

Laurance WF, Useche DC, Rendeiro J, Kalka M, Bradshaw CJ, Sloan SP, Laurance SG, Campbell M, Abernethy K, Alvarez P (2012) Averting biodiversity collapse in tropical forest protected areas. Nature 489:290–294

Laurance WF, Sayer J, Cassman KG (2014) Agricultural expansion and its impacts on tropical nature. Trends Ecol Evol 29:107–116

Laurance WF, Campbell MJ, Alamgir M, Mahmoud MI (2017) Road expansion and the fate of Africa's tropical forests. Front Ecol Evol 5:75

Lawton R (1982) Natural resources of miombo woodland and recent changes in agricultural and land-use practices. For Ecol Manag 4:287–297
Lebamba J, Vincens A, Lézine A-M, Marchant R, Buchet G (2016) Forest-savannah dynamics on the Adamawa plateau (Central Cameroon) during the "African Humid Period" termination: a new high-resolution pollen record from Lake Tizong. Rev Palaeobot Palynol 235:129–139
Lehmann CE, Anderson TM, Sankaran M, Higgins SI, Archibald S, Hoffmann WA, Hanan NP, Williams RJ, Fensham RJ, Felfili J (2014) Savanna vegetation-fire-climate relationships differ among continents. Science 343:548–552
Letouzey R, Aubréville A (1968) Etude phytogéographique du Cameroun. Lechevalier, Paris
Lewis SL, Lopez-Gonzalez G, Sonké B, Affum-Baffoe K, Baker TR, Ojo LO, Phillips OL, Reitsma JM, White L, Comiskey JA (2009) Increasing carbon storage in intact African tropical forests. Nature 457:1003–1006
Lézine A-M, Holl AF-C, Lebamba J, Vincens A, Assi-Khaudjis C, Février L, Sultan E (2013) Temporal relationship between Holocene human occupation and vegetation change along the northwestern margin of the Central African rainforest. Compt Rendus Geosci 345:327–335
Linder HP (2014) The evolution of African plant diversity. Front Ecol Evol 2:38
Linder HP, de Klerk HM, Born J, Burgess ND, Fjeldså J, Rahbek C (2012) The partitioning of Africa: statistically defined biogeographical regions in sub-Saharan Africa. J Biogeogr 39:1189–1205
Lloyd J, Domingues TF, Schrodt F, Ishida FY, Feldpausch TR, Saiz G, Quesada CA, Schwarz M, Torello-Raventos M, Gilpin M (2015) Edaphic, structural and physiological contrasts across Amazon Basin forest-savanna ecotones suggest a role for potassium as a key modulator of tropical woody vegetation structure and function. Biogeosciences 12:6529–6571
Loarie SR, Duffy PB, Hamilton H, Asner GP, Field CB, Ackerly DD (2009) The velocity of climate change. Nature 462:1052–1055
Malaisse F (2010) How to live and survive in Zambezian open forest (Miombo ecoregion). Presses agronomiques de Gembloux, Gembloux
Maley J (1989) Late quaternary climatic changes in the African rain forest: forest refugia and the major role of sea surface temperature variations. In: Paleoclimatology and paleometeorology: modern and past patterns of global atmospheric transport. Springer, Dordrecht, pp 585–616
Maley J (1996) Fluctuations majeures de la forêt dense humide africaine. L'alimentation en forêt tropicale. Interactions bioculturelles et perspectives de développement. UNESCO-CNRS-ORSTOM, Paris, pp 31–52
Maley J (2002) A catastrophic destruction of African forests about 2,500 years ago still exerts a major influence on present vegetation formations. IDS Bull 33:13–30
Maley J, Willis K (2010) Did a savanna corridor open up across the central African forests 2500 years ago? CoForChange Newsletters (2). http://www.coforchange.eu/products/newsletters
Maley J, Giresse P, Doumenge C, Favier C (2012) Comment on "Intensifying weathering and land use in Iron Age Central Africa". Science 337:1040–1040
Maley J, Doumenge C, Giresse P, Mahé G, Philippon N, Hubau W, Lokonda MO, Tshibamba JM, Chepstow-Lusty A (2018) Late holocene forest contraction and fragmentation in central Africa. Quat Res 89:43–59
Malhi Y, Aragão LE, Galbraith D, Huntingford C, Fisher R, Zelazowski P, Sitch S, McSweeney C, Meir P (2009) Exploring the likelihood and mechanism of a climate-change-induced dieback of the Amazon rainforest. Proc Natl Acad Sci 106:20610–20615
Malhi Y, Gardner TA, Goldsmith GR, Silman MR, Zelazowski P (2014) Tropical forests in the anthropocene. Annu Rev Environ Resour 39:125–159
Mayaux P, Bartholomé E, Fritz S, Belward A (2004) A new land cover map of Africa for the year 2000. J Biogeogr 31:861–877
Mitchard E, Saatchi S, Gerard F, Lewis S, Meir P (2009) Measuring woody encroachment along a forest-savanna boundary in Central Africa. Earth Interact 13:1–29

Mitchard E, Saatchi S, Lewis S, Feldpausch T, Woodhouse I, Sonké B, Rowland C, Meir P (2011) Measuring biomass changes due to woody encroachment and deforestation/degradation in a forest–savanna boundary region of central Africa using multi-temporal L-band radar backscatter. Remote Sens Environ 115:2861–2873

Morin-Rivat J, Fayolle A, Gillet J-F, Bourland N, Gourlet-Fleury S, Oslisly R, Bremond L, Bentaleb I, Beeckman H, Doucet J-L (2014) New evidence of human activities during the holocene in the lowland forests of the Northern Congo Basin. Radiocarbon 56:209–220

Morin-Rivat J, Biwolé A, Gorel A-P, Vleminckx J, Gillet J-F, Bourland N, Hardy OJ, Smith AL, Daïnou K, Dedry L (2016) High spatial resolution of late-Holocene human activities in the moist forests of central Africa using soil charcoal and charred botanical remains. The Holocene 26:1954–1967

Morin-Rivat J, Fayolle A, Favier C, Bremond L, Gourlet-Fleury S, Bayol N, Lejeune P, Beeckman H, Doucet J-L (2017) Present-day central African forest is a legacy of the 19th century human history. eLife 6:e20343

Nacoulma BMI, Schumann K, Traoré S, Bernhardt-Römermann M, Hahn K, Wittig R, Thiombiano A (2011) Impacts of land-use on West African savanna vegetation: a comparison between protected and communal area in Burkina Faso. Biodivers Conserv 20:3341–3362

Nasi R, Taber A, van Vliet N (2011) Empty forests, empty stomachs? Bushmeat and livelihoods in the Congo and Amazon Basins. Int For Rev 13:355–368

Nasi R, Billand A, van Vliet N (2012) Managing for timber and biodiversity in the Congo Basin. For Ecol Manag 268:103–111

Neumann K, Bostoen K, Höhn A, Kahlheber S, Ngomanda A, Tchiengué B (2012a) First farmers in the Central African rainforest: a view from southern Cameroon. Quat Int 249:53–62

Neumann K, Eggert M, Oslisly R, Clist B, Denham T, de Maret P, Ozainne S, Hildebrand E, Bostoen K, Salzmann U (2012b) Comment on "Intensifying weathering and land use in Iron Age Central Africa". Science 337:1040

Ngomanda A, Jolly D, Bentaleb I, Chepstow-Lusty A, Makaya MV, Maley J, Fontugne M, Oslisly R, Rabenkogo N (2007) Lowland rainforest response to hydrological changes during the last 1500 years in Gabon, Western Equatorial Africa. Quat Res 67:411–425

Ngomanda A, Chepstow-Lusty A, Makaya M, Favier C, Schevin P, Maley J, Fontugne M, Oslisly R, Jolly D, Langebroek P (2009) Western equatorial African forest-savanna mosaics: a legacy of late Holocene climatic change? Clim Past 5:647–659

Nicholson SE, Some B, McCollum J, Nelkin E, Klotter D, Berte Y, Diallo B, Gaye I, Kpabeba G, Ndiaye O (2003) Validation of TRMM and other rainfall estimates with a high-density gauge dataset for West Africa. Part I: validation of GPCC rainfall product and pre-TRMM satellite and blended products. J Appl Meteorol 42:1337–1354

Olson DM, Dinerstein E, Wikramanayake ED, Burgess ND, Powell GV, Underwood EC, D'amico JA, Itoua I, Strand HE, Morrison JC (2001) Terrestrial ecoregions of the world: a new map of life on earth: a new global map of terrestrial ecoregions provides an innovative tool for conserving biodiversity. Bioscience 51:933–938

Osborne CP, Charles-Dominique T, Stevens N, Bond WJ, Midgley G, Lehmann CE (2018) Human impacts in African savannas are mediated by plant functional traits. New Phytol 220:10–24

Oslisly R, White L, Bentaleb I, Favier C, Fontugne M, Gillet J-F, Sebag D (2013) Climatic and cultural changes in the west Congo Basin forests over the past 5000 years. Philos Trans R Soc B Biol Sci 368:20120304

Ouikon H (2003) Problématique de développement agricole en République centrafricaine et perspectives de recherche. In: Savanes africaines: des espaces en mutation, des acteurs face à de nouveaux défis. Actes du colloque, Garoua, Cameroun. Cirad-Prasac, Montpellier, 6 p

Pan Y, Birdsey RA, Fang J, Houghton R, Kauppi PE, Kurz WA, Phillips OL, Shvidenko A, Lewis SL, Canadell JG (2011) A large and persistent carbon sink in the world's forests. Science 333:988–993

Pan Y, Birdsey RA, Phillips OL, Jackson RB (2013) The structure, distribution, and biomass of the world's forests. Annu Rev Ecol Evol Syst 44:593–622

Parr CL, Lehmann CE, Bond WJ, Hoffmann WA, Andersen AN (2014) Tropical grassy biomes: misunderstood, neglected, and under threat. Trends Ecol Evol 29:205–213
Pennington RT, Lehmann CE, Rowland LM (2018) Tropical savannas and dry forests. Curr Biol 28:541–545
Phillips OL, Aragão LE, Lewis SL, Fisher JB, Lloyd J, López-González G, Malhi Y, Monteagudo A, Peacock J, Quesada CA (2009) Drought sensitivity of the Amazon rainforest. Science 323:1344–1347
Poorter L, Bongers F, Aide TM, Zambrano AMA, Balvanera P, Becknell JM, Boukili V, Brancalion PH, Broadbent EN, Chazdon RL (2016) Biomass resilience of neotropical secondary forests. Nature 530:211
Poulsen JR, Clark C, Bolker B (2011) Decoupling the effects of logging and hunting on an Afrotropical animal community. Ecol Appl 21:1819–1836
Poulsen JR, Clark CJ, Palmer TM (2013) Ecological erosion of an Afrotropical forest and potential consequences for tree recruitment and forest biomass. Biol Conserv 163:122–130
Proces P, Dubiez E, Bisiaux F, Péroches A, Fayolle A (2017) Production d'Acacia auriculiformis dans le système agroforestier de Mampu, plateau Batéké. République démocratique du Congo Bois et forêts des tropiques 334:23–36
Putz FE, Zuidema PA, Synnott T, Peña-Claros M, Pinard MA, Sheil D, Vanclay JK, Sist P, Gourlet-Fleury S, Griscom B (2012) Sustaining conservation values in selectively logged tropical forests: the attained and the attainable. Conserv Lett 5:296–303
R Core Team (2015) R: a language and environment for statistical computing. R Foundation for Statistical Computing, Vienna
Ratnam J, Bond WJ, Fensham RJ, Hoffmann WA, Archibald S, Lehmann CER, Anderson MT, Higgins SI, Sankaran M (2011) When is a 'forest'a savanna, and why does it matter? Glob Ecol Biogeogr 20:653–660
Reynaud-Farrera I, Maley J, Wirrmann D (1996) Végétation et climat dans les forêts du sud-ouest Cameroun depuis 4770 ans BP: analyse pollinique des sédiments du lac Ossa. C R Acad Sci 322:749–755
Roberts P, Hunt C, Arroyo-Kalin M, Evans D, Boivin N (2017) The deep human prehistory of global tropical forests and its relevance for modern conservation. Nat Plants 3:17093
Ruiz Pérez M, de Blas DE, Nasi R, Sayer JA, Sassen M, Angoué C, Gami N, Ndoye O, Ngono G, Nguinguiri J-C (2005) Logging in the Congo Basin: a multi-country characterization of timber companies. For Ecol Manag 214:221–236
Runge J (2002) Holocene landscape history and palaeohydrology evidenced by stable carbon isotope (δ 13 C) analysis of alluvial sediments in the Mbari valley (5° N/23° E), Central African Republic. Catena 48:67–87
Sala O, Chapin F III, Armesto J, Berlow E, Bloomfield J, Dirzo R, Huber-Sanwald E, Huenneke L, Jackson R, Kinzig A (2000) Global biodiversity scenarios for the year 2100. Science 287:1770
Salzmann U (2000) Are modern savannas degraded forests?—a Holocene pollen record from the Sudanian vagetation zone of NE Nigeria. Veg Hist Archaeobotany 9:1–15
Salzmann U, Hoelzmann P (2005) The Dahomey Gap: an abrupt climatically induced rain forest fragmentation in West Africa during the late Holocene. The Holocene 15:190–199
Sankaran M, Hanan NP, Scholes RJ, Ratnam J, Augustine DJ, Cade BS, Gignoux J, Higgins SI, Le Roux X, Ludwig F, Ardo J, Banyikwa F, Bronn A, Bucini G, Caylor KK, Coughenour MB, Diouf A, Ekaya W, Feral CJ, February EC, Frost PGH, Hiernaux P, Hrabar H, Metzger KL, Prins HHT, Ringrose S, Sea W, Tews J, Worden J, Zambatis N (2005) Determinants of woody cover in African savannas. Nature 438:846–849
Sankaran M, Ratnam J, Hanan N (2008) Woody cover in African savannas: the role of resources, fire and herbivory. Glob Ecol Biogeogr 17:236–245
Scholes R, Archer S (1997) Tree-grass interactions in savannas. Annu Rev Ecol Syst 28:517–544
Schure J, Levang P, Wiersum KF (2014) Producing woodfuel for urban centers in the democratic republic of Congo: a path out of poverty for rural households? World development, forests, livelihoods, and conservation 64, Suppl 1, S80–S90. https://doi.org/10.1016/j.worlddev.2014.03.013

Searchinger TD, Estes L, Thornton PK, Beringer T, Notenbaert A, Rubenstein D, Heimlich R, Licker R, Herrero M (2015) High carbon and biodiversity costs from converting Africa's wet savannahs to cropland. Nat Clim Chang 5:481–486

Shanahan TM, McKay NP, Hughen KA, Overpeck JT, Otto-Bliesner B, Heil CW, King J, Scholz CA, Peck J (2015) The time-transgressive termination of the African Humid Period. Nat Geosci 8:140–144

Sodhi NS, Posa MRC, Lee TM, Bickford D, Koh LP, Brook BW (2010) The state and conservation of Southeast Asian biodiversity. Biodivers Conserv 19:317–328

Sonwa DJ, Walker S, Nasi R, Kanninen M (2011) Potential synergies of the main current forestry efforts and climate change mitigation in Central Africa. Sustain Sci 6:59–67

Spracklen D, Baker J, Garcia-Carreras L, Marsham J (2018) The effects of tropical vegetation on rainfall. Annu Rev Environ Resour 43:193–218

Staver AC (2018) Prediction and scale in savanna ecosystems. New Phytol 219:52–57

Staver AC, Bond WJ (2014) Is there a 'browse trap'? Dynamics of herbivore impacts on trees and grasses in an African savanna. J Ecol 102:595–602

Staver AC, Levin SA (2012) Integrating theoretical climate and fire effects on savanna and forest systems. Am Nat 180:211

Staver AC, Archibald S, Levin S (2011a) Tree cover in sub-Saharan Africa: rainfall and fire constrain forest and savanna as alternative stable states. Ecology 92:1063–1072

Staver AC, Archibald S, Levin SA (2011b) The global extent and determinants of savanna and forest as alternative biome states. Science 334:230–232

Stevens N, Lehmann CE, Murphy BP, Durigan G (2017) Savanna woody encroachment is widespread across three continents. Glob Chang Biol 23:235–244

Swaine M, Hall J (1988) The mosaic theory of forest regeneration and the determination of forest composition in Ghana. J Trop Ecol 4:253–269

ter Steege H, Pitman NC, Killeen TJ, Laurance WF, Peres CA, Guevara JE, Salomão RP, Castilho CV, Amaral IL, de Almeida Matos FD (2015) Estimating the global conservation status of more than 15,000 Amazonian tree species. Sci Adv 1:e1500936

Terborgh J, Nuñez-Iturri G, Pitman NC, Valverde FHC, Alvarez P, Swamy V, Pringle EG, Paine C (2008) Tree recruitment in an empty forest. Ecology 89:1757–1768

Tovar C, Breman E, Brncic T, Harris DJ, Bailey R, Willis KJ (2014) Influence of 1100 years of burning on the central African rainforest. Ecography 37:1139–1148

Tranquilli S, Abedi-Lartey M, Abernethy K, Amsini F, Asamoah A, Balangtaa C, Blake S, Bouanga E, Breuer T, Brncic TM (2014) Protected areas in tropical Africa: assessing threats and conservation activities. PLoS One 9:e114154

Van Gemerden BS, Olff H, Parren MP, Bongers F (2003) The pristine rain forest? Remnants of historical human impacts on current tree species composition and diversity. J Biogeogr 30:1381–1390

Van Reybrouck D (2012) Congo, une histoire. Éditions Actes Sud

Van Vuuren DP, Edmonds J, Kainuma M, Riahi K, Thomson A, Hibbard K, Hurtt GC, Kram T, Krey V, Lamarque J-F (2011) The representative concentration pathways: an overview. Clim Chang 109:5–31

Veldman JW (2016) Clarifying the confusion: old-growth savannahs and tropical ecosystem degradation. Philos Trans R Soc B 371:20150306

Veldman JW, Buisson E, Durigan G, Fernandes GW, Le Stradic S, Mahy G, Negreiros D, Overbeck GE, Veldman RG, Zaloumis NP (2015a) Toward an old-growth concept for grasslands, savannas, and woodlands. Front Ecol Environ 13:154–162

Veldman JW, Overbeck GE, Negreiros D, Mahy G, Le Stradic S, Fernandes GW, Durigan G, Buisson E, Putz FE, Bond WJ (2015b) Tyranny of trees in grassy biomes. Science (New York, NY) 347:484

Veldman JW, Overbeck GE, Negreiros D, Mahy G, Le Stradic S, Fernandes GW, Durigan G, Buisson E, Putz FE, Bond WJ (2015c) Where tree planting and forest expansion are bad for biodiversity and ecosystem services. Bioscience 65:1011–1018

Veldman JW, Silveira FA, Fleischman FD, Ascarrunz NL, Durigan G (2017) Grassy biomes: an inconvenient reality for large-scale forest restoration? A comment on the essay by Chazdon and Laestadius. Am J Bot 104:649–651

Verhegghen A, Mayaux P, De Wasseige C, Defourny P (2012) Mapping Congo Basin vegetation types from 300 m and 1 km multi-sensor time series for carbon stocks and forest areas estimation. Biogeosciences 9:5061

Vincens A, Buchet G, Elenga H, Fournier M, Martin L, de Namur C, Schwartz D, Servant M, Wirrmann D (1994) Changement majeur de la végétation du lac Sinnda (vallée du Niari, Sud-Congo) consécutif à l'assèchement climatique holocène supérieur: apport de la palynologie. Comptes Rendus de l'Académie des Sciences, Paris 318:1521–1526

Vincens A, Schwartz D, Bertaux J, Elenga H, de Namur C (1998) Late Holocene climatic changes in western equatorial Africa inferred from pollen from Lake Sinnda, southern Congo. Quat Res 50:34–45

Vincens A, Schwartz D, Elenga H, Reynaud-Farrera I, Alexandre A, Bertaux J, Mariotti A, Martin L, Meunier J, Nguetsop F (1999) Forest response to climate changes in Atlantic Equatorial Africa during the last 4000 years BP and inheritance on the modern landscapes. J Biogeogr 26:879–885

Vincens A, Buchet G, Servant M, ECOFIT Mbalang Collaborators (2010) Vegetation response to the "African Humid Period" termination in Central Cameroon (7 N)—new pollen insight from Lake Mbalang. Clim Past 6:281–294

Vleminckx J, Morin-Rivat J, Biwolé AB, Daïnou K, Gillet J-F, Doucet J-L, Drouet T, Hardy OJ (2014) Soil charcoal to assess the impacts of past human disturbances on tropical forests. PLoS One 9:e108121

Vodouhê FG, Coulibaly O, Adégbidi A, Sinsin B (2010) Community perception of biodiversity conservation within protected areas in Benin. Forest Policy Econ 12:505–512

Weldeab S, Schneider RR, Müller P (2007) Comparison of Mg/Ca- and alkenone-based sea surface temperature estimates in the fresh water—influenced Gulf of Guinea, eastern equatorial Atlantic. Geochem Geophys Geosyst 8:1–10

White F (1983) The vegetation of Africa, a descriptive memoir to accompany the UNESCO/AETFAT/UNSO vegetation map of Africa. UNESCO, Paris

Willis K, Gillson L, Brncic T (2004) How "virgin" is virgin rainforest? Science (Washington) 304:402–403

Willis K, Bennett K, Burrough S, Macias-Fauria M, Tovar C (2013) Determining the response of African biota to climate change: using the past to model the future. Philos Trans R Soc B Biol Sci 368:20120491

Wittemyer G, Elsen P, Bean WT, Burton ACO, Brashares JS (2008) Accelerated human population growth at protected area edges. Science 321:123–126

Wotzka H-P (2006) Records of activity: radiocarbon and the structure of Iron Age settlement in Central Africa. In: Grundlegungen. Beiträge zur europäischen und afrikanischen Archäologie für Manfred KH Eggert. Francke, Tübingen, pp 271–289

WRI, World Resources Institute (2014) Atlas of forest and landscape restoration opportunities. World Resources Institute, Washington, DC. www.wri.org/resources/maps/Atlas-forest-and-landscape-restoration-opportunities

Youta Happi J (1998) Arbres contre graminées: la lente invasion de la forêt par la savane au Centre-Cameroun. Thèse de Doctorat, Paris IV

Ziegler S, Fa JE, Wohlfart C, Streit B, Jacob S, Wegmann M (2016) Mapping bushmeat hunting pressure in Central Africa. Biotropica 48:405–412

Chapter 10
Forest–Agriculture in the Centre–South Region of Cameroon: How Does Traditional Knowledge Inform-Integrated Management Approaches?

William Armand Mala, Coert J. Geldenhuys, Ravi Prabhu, and François Manga Essouma

10.1 Introduction

The discourse over forest agriculture sustainability in the tropics has been dominated by conventional thinking and western science, during most of the past century (GEF 1993; ASB 1995, 2000; Garrity and Bandy 1996; Van Noordwijk et al. 2001; Palm et al. 2005; COMIFAC 2014). On one hand, it has been defined by ecological thinking based on the theory of equilibrium, and on the other hand by discourses about the effects of forest agriculture on deforestation and biodiversity loss (Toledo et al. 2003; Instone 2003b; Mala et al. 2019; Diaw et al. 2016). However, this 'conventional' thinking has been challenged by new theories that recognize the high degree of uncertainty governing (a) tropical ecosystem functioning (Scoones 1995; Geldenhuys 2000; Holling 2001; Wallington et al. 2005; Mayaux et al. 2013), (b) the human dimensions of tropical forest social–ecological systems (Diaw and Oyono 1998; Oyono 2002; Mala and Oyono 2004; Apusiga 2011) and (c) the linkages between forest resources and agriculture to establish forest–agriculture mosaics (Vosti 1995; Colfer and Byron 2001; Levang et al. 2001; O'brien 2002; Angelsen and Wunder 2003; Armitage 2003; McNeel 2004; Vernooy and Song 2004; Ash et al. 2010; Beddington et al. 2012; Van Vliet et al. 2012) (see also Chap. 9, Vol. 1).

W. A. Mala (✉) · F. Manga Essouma
University of Yaoundé I, Yaoundé, Cameroon

C. J. Geldenhuys
University of Stellenbosch, Stellenbosch, South Africa
e-mail: cgelden@mweb.co.za

R. Prabhu
World Agroforestry Centre, Nairobi, Kenya
e-mail: r.prabhu@cgiar.org

A. Gasparatos et al. (eds.), *Sustainability Challenges in Sub-Saharan Africa I*, Science for Sustainable Societies, https://doi.org/10.1007/978-981-15-4458-3_10

At the same time, there have been efforts to develop practical alternatives to prevailing forest–agriculture practices (usually slash-and-burn) in the humid tropical forests of Sub-Saharan Africa (SSA) (ASB 1995, 2000; Instone 2003a; Gockowski et al. 2005; Palm et al. 2005; IFAD 2012). So far, the conventional thinking and practical solutions have tended to separate forests and agriculture spatially, administratively and conceptually into two separate units (Diaw et al. 1999; Oyono et al. 2003a; Mala 2009; Kaufmann 2014). This has been counter-productive in the small-scale, agricultural settings of the humid tropics, where agricultural systems are embedded within a cropping fallow forest conversion cycle (Diaw 1997; Carrière 1999; Oyono et al. 2003a; Mala et al. 2010, Mala 2016, 2017). As a result, various forest–agriculture technologies seeking to improve soil fertility, crop varieties and the consumption of plant-based protein have been abandoned (Nolte et al. 1997; Mala 2009; Diaw et al. 2018).

In a way, the limitations of the current approaches that have been segregated by agriculture and forest conservation in the humid forest areas of SSA reveal the gaps between (a) the concepts of forest–agriculture or swidden agriculture (Diaw 1997; O'brien 2002; Colfer 2005; Mala 2009), (b) the theories and school of thoughts in ecology and agricultural sustainability (Pretty 2002; Altieri 2002; Altieri et al. 2015) and (c) the forest–agriculture practices and local natural resource management options (Diaw and Oyono 1998; Carrière 1999; Chirwa and Mala 2016; Mbolo Abada et al. 2016).

There are three specific reasons that raise doubts about the ability of conventional forest–agriculture approaches to meet human well-being and biodiversity conservation objectives in the region, namely: (a) the disconnection between rural development and forest management policies, (b) inconsistent scientific thinking and research/management practices and (c) inappropriate innovations at the interface of agriculture and forest conservation (Danielsen et al. 2014; Namirembe et al. 2014) (see also Chap. 3, Vol. 2).

However, there is a gradual shift towards development, transfer and implementation of alternative forest–agriculture policies, practices and technologies (Willems et al. 2015). This often happens at the local level, and sometimes creates enabling conditions for local community empowerment and livelihood improvement (Diaw et al. 1999; Prabhu 2003; Colfer 2005; Berkes 2009; Biggs et al. 2015). Some examples include local and community-based natural resource management options (e.g. community forests) that allow local communities define the use of common pool resources and negotiate appropriate management options (Oyono et al. 2007).

Yet, several contradictions and paradoxes remain at the interface of agriculture and forest conservation in humid tropical forests of SSA, including (a) the rhetoric of biodiversity conservation (Van Vliet et al. 2012; World Bank 2012), (b) the lack of farmer capacity (and of capacity-building efforts) to adapt to new challenges such as climate change (FAO 2010, 2011a; Munang et al. 2013, 2015; Altieri et al. 2015) and (c) the clash between participation and institutional change, with strong social demand for short-term innovations (e.g. agricultural inputs, market opportunities, improved crop varieties) (Cormier-Salem 1999; Conley and Udry 2001; Van

Noordwijk et al. 2001; Ruitenbeck and Cartier 2001; Carlsson et al. 2002; Mala et al. 2004) (see also Chap. 9, Vol. 1; Chap. 3, Vol. 2).

The failure of initially promising forest–agriculture policies, practices and innovations has often been linked to the (a) limited understanding of the possible contribution of traditional knowledge and (b) the false assumptions that management options based on local knowledge are not aligned with conventional/prevailing indicators of agro-ecological sustainability (Smith and McSorely 2000; Palm et al. 2005; Oyono et al. 2007; Malleson et al. 2008; Bellamy and Hill 2010; Zahm et al. 2019). Efforts to design, promote and adopt new agricultural innovations in the region, often clash with such local knowledge and management systems (Bawden 1991; Scoones 1995; Abate et al. 2000; Altieri 2002; Mala 2016) (see Chap. 3, Vol. 2). The fact that local communities and external scientists utilize different knowledge systems and perspectives about the role and functions of biodiversity in agro-ecosystems sometimes contributes to such failures (Haila 1999; Sinclair and Joshi 2001; Altieri 2002; Instone 2003b; Van Mele and Van Chien 2004) (see Chap. 3, Vol. 2).

The above suggest that the complexity of forest-agriculture systems, and the piecemeal research and management approaches combine to currently hamper the development of sustainable solutions for forest–agriculture (Sayer and Campbell 2003; Sanchez et al. 2005; AfDB 2015; Boon 2015). This is a key unresolved sustainability challenge in many areas of tropical SSA (FAO 2011b; AU 2015; FAO/INRA 2016; Diaw et al. 2016, 2018), which has important ramifications for biodiversity conservation (Mala 2017), food security (Lin 2011; IFAD 2012) and poverty alleviation (FAO/INRA 2016). The adoption of the sustainable development goals (SDGs) and their operationalization within the strategic planning of multilateral development banks, bilateral institutions and countries increases the need to deal such challenges in SSA that span multiple SDGs such as SDG1 (no poverty), SDG2 (zero hunger), SDG8 (decent work and economic growth) and SDG15 (life on land), among others (FAO 2010, 2011a, b; Munang et al. 2013; Kaphengst 2014; AfDB 2015; World Bank 2016) (see Chap. 5, Vol. 1).

For example in many parts of the Central Africa region, current agricultural and natural resource management practices and technological innovations related to forest–agriculture[1] fail to achieve improved rural livelihoods, income generation and sustainable land/forest management outcomes (Mala et al. 2004, 2008b; Mala and Oyono 2004; Diaw et al. 2018). At the same time, they do not seem to be successful in conserving biodiversity and ecosystem services in the rapidly changing forest ecosystems of the region (Diaw et al. 2016) (see Chap. 9, Vol. 1). It is thus important to identify promising innovations, policies and practical solutions, and to ensure their wide uptake and sustained use, to tackle effectively the interlinked challenges at the interface of agriculture and conservation in the region.

[1]Such examples include alley cropping, monocultures and shortened fallow agricultural systems using bio-fertilizers and annual and scrubs species such as *Mucuna* sp., *Cajanus* sp. and *Calliandra* sp.

The humid forest zone of southern Cameroon is a particularly relevant area to explore these phenomena. The area contains some of the last undisturbed tropical forests in SSA (Van Germeden et al. 2003; Ernst et al. 2013; Mayaux et al. 2013; Diaw et al. 2018). These forests face increasing pressure from infrastructure development, illegal forest logging, industrial agriculture, intensive farming and mining (Etat des forêts 2015) (see Chaps. 2, 3, 9, Vol. 1). There are concerns over the possibility of substantial loss of forest and agro-diversity (and to an extent of local livelihoods and human well-being), especially considering the ever-increasing population that still relies on forest-based subsistence and cash crop agriculture for its livelihoods and the large-scale adoption of agricultural interventions from other areas (UNDP 2012; Mala 2016). This can have important ramifications for local livelihoods and human well-being (Mala 2016; Mala et al. 2019).

Forest landscape mosaics in this region have been traditionally managed through a cropping fallow forest conversion cycle, which is embedded in social institutions and customary tenure systems (Dounias 1996; Diaw 1997; Carrière 1999; Mala 2009). Agro-diversity has been central for the functioning and organization of those agro-ecosystems, supporting local livelihoods and agro-ecological sustainability at different spatio-temporal scales (Robiglio et al. 2002; Robiglio and Mala 2005; Dounias 1996; Dounias and Hladik 1996; Carrière 1999; Mala et al. 2008a). This contrasts mono-cropping approaches that have recently dominated agricultural policy and technology options in the area (Sanchez et al. 2005; Krause et al. 2013) (see Chap. 3, Vol. 1).

Most of the currently introduced agro-diversity management approaches in the region rely on innovations and practices derived from knowledge systems that are external to the region (e.g. developed countries of the global North, technologies of Sahelian countries tested in forest margins) (see Chap. 3, Vol. 2). Such management approaches usually entail mono-cropping (e.g. for maize) and clash with the local agro-diversity (Mala 2009, 2016). In fact, more often than not, such innovations and practices are applied uncritically in a region characterized by (a) high ecological complexity embedded in dynamic socio-ecological systems; (b) institutional and economic transformations following the devaluation of the local currency (Central African Franc, CFA) and (c) macro-economic reforms in the forestry and cocoa sector following the crisis in the international cocoa market (Ndoye 1997; Diaw et al. 1999; Oyono et al. 2003b) (see also Chap. 9, Vol. 1).

Indeed, there are major knowledge gaps in the region that have often resulted in poor responses and adaptive feedback mechanisms, prohibiting the effective introduction and implementation of agricultural innovations/practices in a way that secures the local agro-diversity and meets the needs of the local communities. In particular, there is a crucial need to understand better the biophysical and socio-economic context of agro-diversity management, as well as how local farmers mobilize local ecological knowledge to develop management practices for land conversion, soil fertility and cultivation of desirable crops (see Chap. 3, Vol. 2). Central to the above is the documentation of conflicts between local and external knowledge systems focusing on agricultural technologies, and how their interplay affects the generation, dissemination and utilization of forest–agriculture innovations (see Chap. 3, Vol. 2).

Considering the above, this study aims at understanding what traditional knowledge is mobilized (and how) in the highly biodiverse forest–agriculture systems of Centre–South Cameroon, and how it can contribute to the development of integrated policy and technology options, and innovations to manage effectively complex forest–agricultural systems in a way that meets local needs and reflects local views. The study is anchored on global policy commitments such as the SDGs (and particularly SDGs 1, 2, 8 and 15 as discussed above) and the current calls for trans-disciplinarity and knowledge integration as advocated by the Intergovernmental Platform on Biodiversity and Ecosystem Services (IPBES), including its African Assessment (Diaz et al. 2015; Diaw et al. 2018) (Chap. 9, Vol. 2).

Section 10.2 outlines the study site, and data collection and analysis approaches. Section 10.3 highlights the main results in terms of (a) human–nature relationships (Sect. 10.3.1); (b) traditional knowledge about forest use, soil fertility and agro-diversity management (Sect. 10.3.2); (c) desirable plant characteristics for household consumption and selling (Sect. 10.3.3); (d) pest and disease management approaches for major crops (Sect. 10.3.4) and (e) social demand and supply of forest–agriculture innovations (Sect. 10.3.5). Section 10.4 puts the main findings into perspective and offers policy and practice recommendations in the context of the SDGs.

10.2 Methodology

10.2.1 Study Site

The study is conducted in three blocks (Yaoundé, Mbalmayo and Ebolowa) within the humid forest zone of southern Cameroon (Fig. 10.1). The blocks follow a gradient of natural resource use management intensification and population density. The intensity of resource use is inversely defined by the length of the fallow period, and it increases from Yaoundé block (3.9 years), to Mbalmayo block (5.4 years) to Ebolowa block (7.5 years). The population density decreases from Yaoundé (30–90 people/km^2), to the Mbamayo block (10–30 people/km^2) and the Ebolowa block (<10 people/km^2) (Gockowski et al. 2005). There is much variation between blocks for key socio-economic variables (Mala 2009), with market access being average in the Ebolowa and Yaounde blocks, and average-to-good in the Mbalmayo block.

The study area is characterized by equatorial climate, with a bimodal rainfall pattern (1350–1900 mm per annum) (see Chap. 9, Vol. 1). Rainfall increases from the northwest to the southeast. Climate-related hazards such as droughts and heavy rains, affect water availability and humidity extremes, which in turn affects the outbreak of fungal/bacterial diseases and insect pests (see below), and essentially shapes local farming strategies and land management (Manyong et al. 1996; Gockowski et al. 2004).

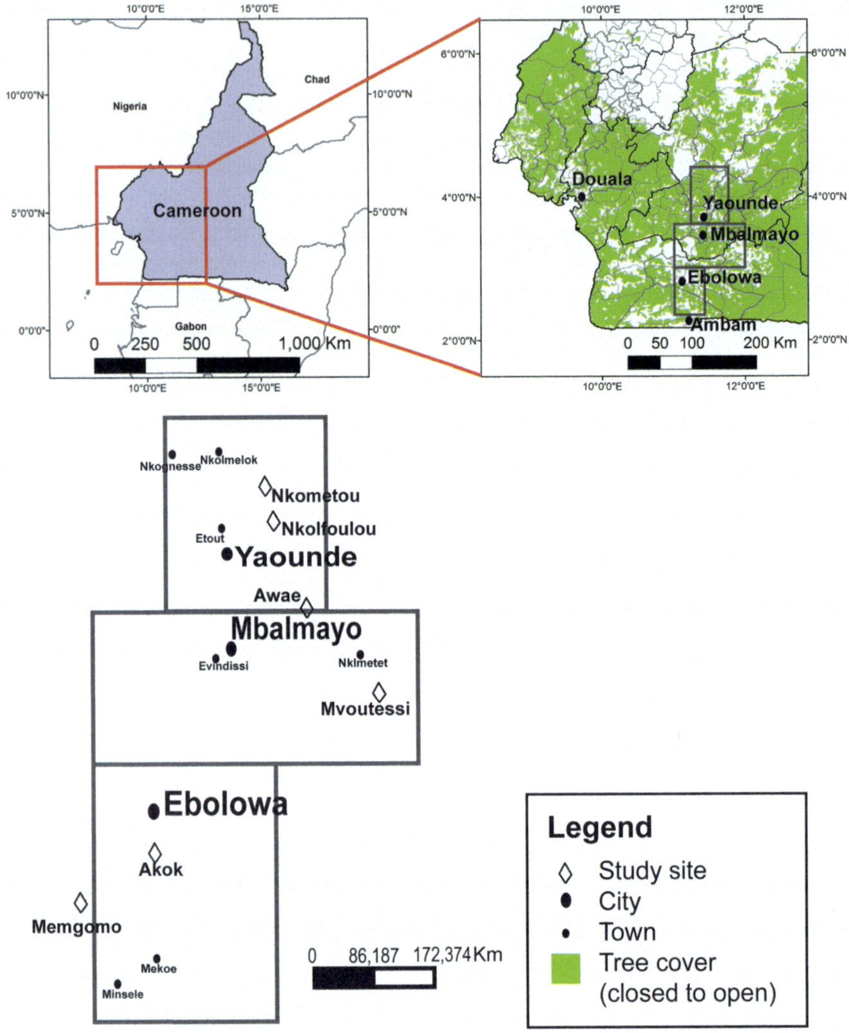

Fig. 10.1 Study areas and block delimitation

The most important agricultural systems across all blocks are food crop systems (e.g. groundnut/cassava-based mixed crop farms), cocoa plantations and plantain/melon farms (see also Chap. 3, Vol. 1). These crops are susceptible to various diseases and pest such as (a) fruit rot (cocoa); (b) nematodes, sigatoka disease and cigar-end diseases (plantains); (c) leaf-spot disease, root rot, rust, aflaroot and rosette-virus diseases (groundnuts); (d) anthracnose, leaf spot and tuber rots (cassava); (e) blast, crown disease, vascular-wilt disease, trunk and bud rots (oil palm) (Westphal et al. 1985). Input intensive, horticultural mono-cropping and maize systems are frequently encountered in the Yaoundé block, which has the best access to urban markets (see below) (see also Chap. 2, Vol. 2).

The intensity of agricultural production and management follows the resource use intensification gradient within the forest margins of southern Cameroon. These differences are very significant when it comes to the use of improved varieties, fertilizers, fungicides and insecticides (Gockowski et al. 2004). In particular, the greatest intensification in the Yaoundé block, with 56%, 39% and 51% of households using improved crop varieties, fertilizers and pesticides, respectively (Gockowski et al. 2004). Purchased agricultural inputs are not common in the Mbalmayo block, while fertilizer/pesticide use is practically non-existent in the Ebolowa block (Gockowski et al. 2004). Practices such as soil tillage before planting, use of ridges and row planting is also more frequent in the Yaoundé block (Gockowski et al. 2004). The number of different farms types is a final aspect related to resource use intensification and farming system diversification (Gockowski et al. 2004). Previous studies have revealed that 62% of the households in the Yaoundé block have five to eight distinct field types, compared to only 28% in the Mbalmayo block and 44% in the Ebolowa block (Gockowski et al. 2004).

10.2.2 Data Collection and Analysis

In order to understand the type, role and outcome of forest–agriculture innovations in the study area, we selected two villages in each block (Sect. 10.2.1). These villages were selected based on the intensity of relevant innovations, and especially villages with a relatively high number of such innovations. In each village we aimed to understand:

- Human–nature relationships (Sect. 10.3.1)
- Traditional knowledge about forest use, soil fertility and agro-diversity management (Sect. 10.3.2)
- Plant characteristics for household consumption and selling (Sect. 10.3.3)
- Pests and disease management approaches for major crops (Sect. 10.3.4)
- Social demand and supply of forest–agriculture innovations (Sect. 10.3.5)

Data were collected through a combination of:

- Household surveys
- Focus group discussions (FGDs)
- Agro-diversity surveys

In each selected village (six villages in the three blocks), we survey five households ($N = 30$ households). In each village, these five households were selected based on their participation in the development and utilization of forest–agriculture innovations. Based on this criterion, the selected households could be divided in three categories: (a) farmers involved in on-farm research, participatory farm research and innovation testing; (b) farmers not directly involved in on-farm research, but that have received benefits from on-farm research and participated in tests; (c) farmers not involved in any research/testing activity (control group). For each village, we compiled a list of households falling in each category, and selected respondents based on the estimated proportion of each group in each village.

Similarly, we conducted 21 focus group discussions (FGDs) involving 15–20 participants in each village. In particular, the household survey respondents formed part of the FGD, and were joined by other local community members that could not participate in the interviews but were designated by the villages informants. These FGDs were used to develop participatory agro-ecological map of each village, and for the general discussion on the demand and supply for each type of innovation.

To establish agro-diversity levels (Sects. 10.3.2–10.3.4), we selected the study systems based on spatio-temporal processes that take place within the cropping-fallow-forest conversion cycle. The main criteria include the:

- Density of crop species
- Number of economic plant species
- Economic importance and density of agricultural and non-agricultural plant species
- Spatial deployment of the cropping fallow forest conversion cycle
- Length of the fallow period corresponding to another rotation

Based on these criteria, we selected five land use systems that were informed by the nine systems identified in the study area through observation, FGDs and surveys. The five land use types followed the cropping fallow forest conversion cycle namely: (a) Cucumeropsis agroforestry systems (i.e. melon seed); (b) mixed food crop agroforestry systems (or mixed groundnut-based crop farms); (c) cocoa agroforestry systems, (d) young pre-forest fallow systems (5 years), and (e) young secondary forest systems (15 years). We collected land use and biophysical data from different plots in these land uses, adapting the multidisciplinary landscape assessment approach of Sheil et al. (2003). In total we sampled 184 plots, distributed as follows:

- Cocoa agroforestry systems: 41 plots with dimensions of 20 m × 20 m (0.04 ha) (total area = 1.64 ha)
- Cucumeropsis agroforestry systems: 36 plots with dimensions of 20 m × 20 m (total area = 1.44 ha);
- Young pre-forest fallows and young secondary forest systems: 30 plots with dimensions of 20 m × 20 m (total area = 2.40 ha both two land uses);
- Mixed food-crop agroforestry systems: 47 plots with dimensions of 40 m × 5 m (total area = 0.94 ha).

In each farm, one to four plots were sampled in cocoa and Cucumeropsis agroforestry systems, depending on total farm size. In each farm, one plot was systematically sampled for the regrowth fallow and regrowth forests. This was to avoid problems related to the boundaries and size of the former farm. Each of the 20 m × 20 m plot was subdivided into four sub-units of 20 m × 5 m (0.01 ha), and each of the 40 m × 5 m plots was subdivided into 10 consecutive 4 m × 5 m subunits (0.002 ha).

In each plot, we collected the following data: (a) local name of the area; (b) local name and scientific name of each plant (the scientific name included the family, genus, species and author); (c) basal diameter of the plant (cm); (d) height of the plant (m); (e) status of the stem after cutting by farmers as follow: plant with entire stem (i.e. which was not cut), plant cut with sprouts developing, plant cut without sprouts developing; (f) characteristics of the woody biomass (i.e. soft, semi-hard, hard).

10.3 Results

10.3.1 Human–Nature Relationships

Based on the in depth household survey, we elicit the social representation and perceptions of human–nature relationships in the study areas. Nature, in the form of forests, animals, rivers and land, is perceived as having many strong relationships with the local communities, essentially providing a vital space for them (Table 10.1). The different components of Nature have specific functions that support important socio-economic activities and environmental processes related to livelihoods (e.g. agriculture, fishing), family life, spiritual/religious values and habitat for different species. Table 10.1 provides an aggregated overview of the main relationships between nature and local communities across all sites, as largely the same relationships emerge in all study villages.

Many of the activities undertaken by local communities are closely linked to natural cycles. In particular we found three broad sub-divisions of time associated with natural resource management:

- Day cycles: morning (kikidigi); midday (zang amos); evening (ngegolè) and midnight (zang alu)
- Moon cycles: full (ngôn nguma) or medium full (ngôn efas nguma)
- Year (mbu) cycles: main dry season (esep), short dry season (oyôn), main rainy season (sugu-oyôn), short rainy season (sugu-esep)

Overall these annual sub-divisions determine agricultural practices, forests product gathering, hunting, fishing, and other livelihood activities (Table 10.2). These times are identified through changes in the local environment (e.g. species prevalence, flowering) and indicate periods when local communities engage in different activities that can often be gender-differentiated (Table 10.2).

It is worth mentioning that there are two dimensions in the social representation of the links between different land uses and natural resources. The first divides space into land use patterns with fixed features such as abandoned villages (bilik), forest (afan), hills (nkol) and rivers (osoë, otõn, ototõn). These land use types can be linked to major local practices such as farming, NFTP harvesting, hunting/trapping and

Table 10.1 Nature components as a vital space in local life

Components	Messages provided to human systems	Local agro-ecological knowledge	Benefits to humans and natural systems
Forests and agricultural landscapes	– Behaviour, physical qualities – Ecological/biophysical responses to climate stress – Attacks by animals and humans – Forest/agriculture uses – Agronomic performance and qualities – Yield and resistance to pests-diseases – Food taste and other qualities	– Regulation in use and maintenance of use – Domestication and plantation – Space markers – Cropping and cultivation	– Landscape characteristics used for human names (e.g. trees (*Bile*), forest (*Afana*), *Gnetum spp.* (*Okoak*), stone (*Akok*)) – Symbolic names to crop varieties and connection to 'spirits' that increase yield and crop quality (e.g. names of wild fruit trees such as *Dracryodes edulis* (*Sa'a*) and *Persea americana* (*Fa*) reflect the abundance and the quality of their fruits)
Animals	– Behaviour, ecology, biology of reproduction and representativeness vis-à- vis other animals – Insect movement	– Domestication, hunting, trapping, consumption for good health and protection against certain diseases	– Animal names used a names for humans (e.g. panthers (Ze), elephant (*Zok*)) – Names of natural phenomena (e.g. transition between seasons such as dry season (*Esep*)) are linked to the movements of animals/insects and the spread of human diseases
Rivers	– Behaviour of rivers – Abundance of specific fishery products	– Fishing – Space marker	– River names are linked to local knowledge of bio-indicators (e.g. of dominant fish species, niche of animals, other social events) – Fish names used as human names
Land and hills	– Complexity of nature and ecological niches	– Land uses – Space markers	– Lands, hills, mountains are personalized to maintain communication channels

fishing. The second dimension relates to the local knowledge of ecological dynamics and/or vegetation regeneration. This dimension is essential for deciding where human activities will take place within the landscape.

Table 10.2 Seasonality of natural resource management activities

Season	Sub-season	Local name	Indicator	Activities	
				Men	Women
Dry season	Short dry season	*Oyôn*	– Presence of caterpillars – Abundance of wild fruits	– Forest clearing – Tree felling	– Food crop harvesting – Forest patch clearing – Food crop cultivation – Forest product harvesting and commercialization
	Long dry season	*Esep*	– Migration of insects, birds and wildlife – Falling of tree leaves	– Forest patch clearing – Hunting of small rodents	– Traditional fishing
Rainy season	Short rainy season	*Sugu esep*	– Flowering of cocoa plantations	– Cocoa plantation maintenance	– Intensive food crop farming and harvesting – Forest product harvesting
	Long rainy season	*Sugu oyôn*		– Cocoa harvesting and commercialization – Intensive hunting/trapping	– Cocoa harvesting support – Food crop harvesting

10.3.2 Traditional Knowledge About Forest Use and Forest Agriculture

10.3.2.1 Forest Use

As identified in Sect. 10.3.1, the forest plays a very important role in the life of local communities. However, based on interviews, we identify that forests are not homogeneous, but have different characteristics and cater for different human needs and uses (Table 10.3). Overall, forests can be roughly divided into virgin forest (afan adam) and different types of secondary forest (Table 10.3). Virgin forest has never experienced human activity in the collective memory, and there is no history of human disturbance (although it is associated with some uses). Different types of secondary forest include old secondary forest (Ilion nyok afan/mbiam), intermediate secondary forest (nnom ekorogo), pre-secondary forest fallow (ekorogo) and young fallow (npwaman ekorogo) (Table 10.3).

Table 10.3 Local classification of forest types, defining characteristics and forest uses

Forest types	Local name	Characteristics	Forest uses
Virgin forest	*Afan adam*	– Total absence of human disturbance – Contains large animals – High abundance and diversity of animals	– Commercial hunting/trapping – Collection of commercial NTFPs
Old secondary forest	*Ilion nyog afan/ Mbiam*	– Contains large animals – High abundance and diversity of animals – Located faraway from villages – Few indications of human activity (e.g. 'huts' for seasonal migration) – Presence of isolated old oil palm trees (*nfon alen*) – Presence of *Irvingia* spp. and *Cola acuminata*	– Commercial food-crop agriculture (mainly with *Cucumeropsis manii* or *Musa spp.*) – Commercial hunting/trapping – Collection of NTFPs
Intermediate secondary forest	*Nnom ekorogo*	– Abundance of mature oil palm trees – Abundance of *Musa* spp. and *Macaranga* spp. – Abundance of rodents and other small mammals attracted by farming	– Commercial food crop agriculture (mainly with *C. manii* or *Musa* spp.) – Domestic hunting/trapping – Intensive collection of Raphia and Palm tree wines
Pre-secondary forest fallow and/or young fallow	*Ekorogo-mpwaman ekorogo*	– Abundance of oil palm seedlings and young trees (*Eleais guineensis*) – Abundance of *Chromoleana odorata* in agricultural fields close to villages – Abundance of Maranthaceae in agricultural fields far from villages – Dispersed presence of food crops such as *Musa* spp. and *Cassava* spp. – Abundance of rodents and other small mammals attracted farming	– Food crop agriculture where indicators of fertility are abundant – Intensive collection of Raphia and Palm tree wines – Intensive collection of NTFPs – Domestic trapping

10.3.2.2 Soil Classification and Fertility

Based on the different interviews, we identify two main soil classification systems (Table 10.4). The first system relates to the type of forest cover and contains four

Table 10.4 Soil classification system and perceptions of soil fertility

Category	Classification	Local name	Perceived soil fertility	Perceived crop suitability	Respondents (%)
Land use	Forest soil	*Si mefane*	High	*Cucumeropsis*, plantain and cocoa agroforesty systems	100
	Fallow soil	*Si bikorogo*	Low to high	Mixed food crop systems	85
	Swamp soil	*Si elobi*	High	Food crop systems	85
	Hill soil	*Si minkol*	High	Several farming systems	75
Color	Black soil	*Evindi si*	High	Several farming systems	85
	Grey soil	*Nselek si*	Low	Less used for cropping	80
	Brown soil	*Avie si*	High	Mixed food crop and agro-forestry systems	75
	Mixed brown–black soil	*Evindi-avie si*	Low	Used by community members that do not have available land	65

Table 10.5 Appropriate agricultural land uses in relation to local soil classes

Agricultural land use	Local soil classes (% responses)			
	Brown soil *(avie si)*	Black soil *(evindi si)*	Black–brown soils *(evindi-avie si)*	Sandy soils *(nselek si)*
Cocoa-plantain systems	0	100	0	0
Plantain, cassava, cocoyam and maize systems	39	61	0	0
C. mannii (melon seed), plantain and cocoyam systems	19	78	3	0
Tomato farms	0	100	0	0
Mixed food crop systems	22	70	5	3

soil classes: forest soil (si mefane), fallow or pre-forestry soil (si bikorogo), marshy soil (si elobi) and hill soil (si minkol). The second system is based on four different soil colours: brown soil (avié si), black soil (evindi si), mixed black–brown soil (evindi-avié si) and grey soil (nselek si). All of these forest soil types are perceived to be highly fertile. Based on the perceptions of soil fertility, each soil class is related to a particular agricultural cropping system (Table 10.5).

Household surveys (30 respondents) also suggest that the local soil classification influences the selection of forestland for cropping (Table 10.5). For example, respondents considered black soil (evindi si) good for all cropping systems, with all (100%) indicating them as suitable for cocoa-plantain cropping systems and tomato farms. Other farmers consider black soil suitable for plantain, cassava, cocoyam and maize (61%) or *Cucumercopsis mannii* (ngon), plantain and cocoyam systems (78%) and mixed crop systems (70%). On the other hand, brown soil (*avie si)* is less frequently associated with agricultural systems such as plantain, cassava,

Table 10.6 Soil fertility and suitability for selected land uses (in %)

	Cocoa agroforestry systems									
Blocks	Pa	Mu	Ts	Co	Ma	Sc	Pe	Fi	Tc	Afv
Ebolowa	68.2	0	20.6	0	11.2	0	0	0	0	0
Mbalmayo	21.0	31.0	19.7	0	3.5	5.2	5.5	0	7.2	6.90
Yaoundé	5.6	13.1	0	33.6	0	17.8	9.4	20.6	0	0
Humid forest zone	34.9	17.0	16.5	5.9	5.6	5.6	4.3	3.6	3.4	3.3
	Cucumeropsis agroforestry systems									
Blocks	Mu	Fi	Pa	Ma	Tc	Sc	Pr	Afv	Cp	Hd
Ebolowa	20.2	0	10.6	20.1	18.5	15.4	15.0	0	0	0
Mbalmayo	19.2	28.4	25.3	0	0	0	0	12.2	9.8	5.1
Yaoundé	0	73.5	8.9	0	0	0	0	0	0	17.7
Humid forest zone	18.4	17.5	17.0	9.8	9.1	7.6	7.3	5.4	4.4	3.5
	Pre-forest young fallow systems									
Blocks	Co	Hd	Ma	Ar	Mu	Fi	Mr	Ts	St	Sc
Ebolowa	36.4	19.2	20.2	8.9	5.7	0	0	9.4	0	0
Mbalmayo	24.1	12.4	2.2	16.1	14.6	0	10.8	0	10.2	9.7
Yaoundé	37.1	18.0	0	0	9.4	31.6	3.9	0	0	0
Humid forest zone	32.3	16.6	10.0	9.7	9.5	6.2	4.5	4.3	3.5	3.4
	Young secondary forest systems									
Blocks	Mu	Ma	Fi	Hd	Fu	Dh	Pa	Af	Afv	Pe
Ebolowa	71.2	21.7	0	0	0	7.1	0	0	0	0
Mbalmayo	39.4	6.1	10.0	11.8	15.0	5.6	0	7.2	0	4.9
Yaoundé	17.1	0	22.8	12.4	0	0	27.9	0	16.4	3.4
Humid forest zone	50.0	12.1	7.6	6.6	5.9	5.3	4.5	2.8	2.7	2.5

Note: *Af* Alchornea floribunda, *Afv* age of fallow/vegetation, *Ar Aframomum* spp., *Co Chromolaena odorata*, *Cp Ceiba petandra*, *Dh* Depth of humus, *Fi* Ficus spp., *Fu* Funtumia spp., *Hd* Haumania danckelmanniana, *Ma* Macaranga spp., *Mr Myrianthus aboreus*, *Mu Musanga cecropioides*, *Pa Pycnanthus angolensis*, *Pe* Presence of earthworms, *Pr Parkia* spp., *Sc* Soil color, *St Solanum torvum*, *Tc Triplochyton scleroxylon*, *Ts Terminalia superba*

cocoyam and maize (39%), mixed food crops (22%) or *C. mannii* (ngon), plantain and cocoyam systems. Black–brown (evindi-avie si) and sandy soils (neslek si) are generally less associated with these agricultural systems (Table 10.5).

In essence, agricultural activities are managed locally through a multi-criteria approach based on the presence of a series of local bio-indicators defined as markers used to trace land suitability of lands for agriculture and associated practices. These markers include the presence of specific plants species, earthworm activity, vegetation age, forest cover and soil colour and quality (Table 10.6). This approach relies heavily on local knowledge about land use history, use of wild plants and crops, interactions between crops (and between crops and other wild plant species), tree size and future uses of existing land use types. Table 10.6 outlines the presence of these bio-indicators in five key land uses, namely (a) cocoa agroforestry systems; (b) *Cucumeropsis* agroforestry systems; (c) pre-forest young fallow systems; (d) young secondary forest systems.

10.3.3 Agro-diversity Management

Twenty-six crop species were recorded with a total of 55 cultivars i.e. separate genetic entities (Table 10.7). There is a relatively high mean number of cultivars per crop species within the study area, with six cultivars for cassava and five cultivars for plantain. About 12% of crop species have four or more cultivars, 8% have three, 58% have two cultivars and 23% have just one cultivar.

Traditional forest–agriculture practices follow the forest-cropping fallow-forest conversion cycle, which begins with forest clearing and burning for developing Cucumeropsis agroforestry systems (Sect. 10.1). Traditional practices influence decisions over the conservation of various forest tree species within farming plots during this cycle. Through ecological surveys, we assess below the size of the remaining trees and agro-diversity in terms of tree species.

Generally speaking, smaller plants are cleared, and larger trees are left relatively intact within cleared plots. In particular, Fig. 10.2 implies that plant size influences

Table 10.7 Crop species and cultivars per species within the mixed food-crop agroforestry system

Crop species	Common name	Local name	Number of cultivars
Manihot esculenta	Cassava	Mbon	6
Musa paradisiaca	Plantain	Ekon	5
Arachis hypogea	Groundnut	Owondo	4
Zea mays	Maize	Fon	3
Solanum nigrum	Eggplant	Zom	3
Vernonia amygdalina	Ndole	Metet	2
Musa sapientum	Banana	Odjoé	2
Ipomea batatas	Sweet potatoe	Meboura	2
Xanthosoma sagittifolium	Macabo	Akaba	2
Allium spp	Onion	Ayan	2
Amaranthus spp	Amarantha	Folong	2
Carica papaya	Papaya	Fofo	2
Corchorus olitorius		Tegue	2
C. mannii	Melon seed	Ngôn	2
Cucurbita spp	Squash	Ndzeng	2
Hibiscus esculentus	Gumbo	Etetam	2
Solanum incanum	Eggplant	Zong	2
Capsicum spp.	Pepper	Ondondo	2
Lycopersicon esculentum	Tomatoe	Ngoro	2
Dioscorea sp	Yam	Ekoara	2
Colocasia esculenta	Cocoyam	Atu	1
Talinum triangulare	Epinard	Elók soup	1
Solanum aethiopicum	Eggplant	Zom nnam	1
Nicotiana tabacum	Tabaco	Ta'a	1
Cucumis sativa	Cucumber	Ombalak	1
Solanum tuberosum	Potatoe	Atora	1

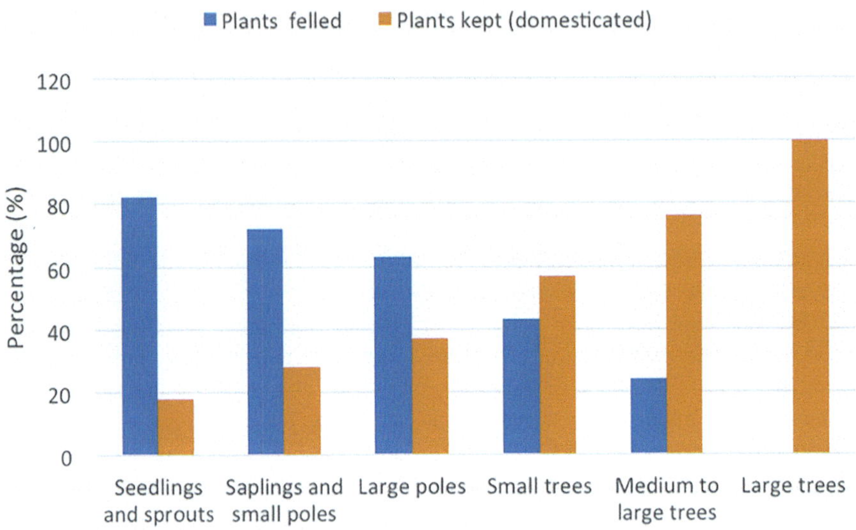

Fig. 10.2 Size of cleared and retained plants within Cucumeropsis farms. Note: Results refer only to Cucumeropsis agroforestry systems, as this is the first land use in the forest-cropping fallow-forest conversion cycle

the decisions of whether to fall or keep plants in Cucumeropsis agroforestry systems. Large trees are almost always kept, while seedlings, sprouts, saplings and small poles are extensively cleared.

However, it is not only the plant size that determines the decision of whether it will be cleared or not, but also its species and the underlying land use. Our inventory shows that the total stem density of the top ten retained species varies substantially between land uses (Table 10.8). Assuming that particularly important plant species have at least ≥60 stems/ha, then the most important plant species in each land use are:

- Cocoa agroforestry systems: only two species with densities ≥60 stems/ha; *P. americana* (67 stems/ha) and *E. guineensis* (64 stems/ha)
- *Cucumeropsis* agroforestry systems: at least 10 species with densities ≥60 stems/ha ranging between 165 stems/ha (*Musanga cecropiodes*) and 64 stems/ha (*Myrianthus arboreus*)
- Mixed food crop agroforestry systems: eight species with densities ≥60 stems/ha ranging from 295 stems/ha (*Tabernaemontana* sp.) to 72 stems/ha (*Celtis* sp.)
- Young pre-forest fallow systems: five species with densities ≥60 stems/ha ranging from 100 stems/ha (*Macaranga* sp) to 66 stems/ha (*Didelotia letouzeyi*)
- Young secondary forests: six species with densities ≥60 stems/ha ranging from 151 stems/ha (*Funtumia* sp.) to 63 stems/ha (*Tabernaemontana* sp., *M. arboreus)*

Most of these species are fast-growing, while four as common (i.e. encountered amongst top ten species in all five land uses), namely *Albizia* sp., *Tabernaemontana* sp., *E. guineensis* and *Macaranga* sp.) (Table 10.8). In terms of end use these plant

Table 10.8 Total stand density for all plant size categories of the top ten tree species across land uses

Cocoa agroforestry systems		*Cucumeropsis* agroforestry systems		Mixed food crop agroforestry systems		Young pre-forest fallows		Young secondary forests	
Tree species	Density (stems/ha)	Tree species	Density (stems/ha)	Tree species	Density (stems/ha)	Tree species	Density (stems/ha)	Tree species	Density (stems/ha)
P. americana [C1]	67.2	*Musanga cecropioides* [C2]	165.0	*Tabernaemontana* sp[C5]	295.3	*Macaranga* sp. [C5]	100.0	*Funtumia* sp. [C4]	151.0
E. guineensis [C5]	64.1	*Macaranga* sp. [C5]	159.0	*D. letouzeyi* [C3]	134.0	*E. guineensis* [C5]	87.0	*Macaranga* sp. [C5]	103.0
Margaritaria discoides [C2]	55.8	*Albizia* sp. [C5]	159.0	*Albizia* sp. [C5]	116.4	*M. arboreus* [C3]	86.0	*E. guineensis* [C5]	87.0
D. edulis [C1]	55.5	*Funtumia* sp. [C4]	130.2	*Pseudospondias* sp. [C1]	110.0	*Albizia* sp. [C5]	76.0	*Albizia* sp. [C5]	79.0
Funtumia sp. [C4]	51.3	*Tabernaemontana* sp. [C5]	103.6	*Alchornea floribunda* [C2]	77.8	*D. letouzeyi* [C3]	66.0	*Tabernaemontana* sp. [C5]	63.0
Macaranga sp. [C5]	41.8	*Voacanga africana* [C2]	102.3	*Rauwolfia vomitoria* [C1]	77.6	*Antiaris africana* [C2]	56.0	*M. arboreus* [C3]	63.0
D. letouzeyi [C3]	41.4	*E. guineensis* [C5]	92.7	*Macaranga* sp. [C5]	75.8	*Tabernaemontana* sp. [C5]	44.0	*Voacanga africana* [C2]	59.0
Albizia sp. [C5]	40.7	*Pycnanthus angolensis* [C1]	83.8	*Celtis* sp. [C2]	72.0	*Celtis* sp. [C2]	44.0	*Oncoba welwitschii* [C1]	40.0
Musa sp.[C1]	33.3	*T. superba* [C1]	79.5	*E. guineensis* [C5]	48.1	*Alchornea floribunda* [C2]	40.0	*Musanga cecropioides* [C2]	40.0
Tabernaemontana sp. [C5]	27.7	*M. arboreus* [C3]	63.6	*Antiaris africana* [C2]	45.9	*Funtumia* sp. [C4]	39.0	*Margaritaria discoides* [C2]	38.0
Stems/ha for top 10 species (% all stems)	478.8 (30.4)		1145.4 (36.3)		1052.8 (66.2)		1432.5 (38.9)		1750.0 (35.4)
Total stems/ha for all species	1572.8		3152.2		1589.8		3685.0		4947.5

Note: *C5* common across five land uses, *C4* common across four land uses, *C3* common across three land uses, *C2* common across two land uses, *C1* common in one land use

Table 10.9 Consumption of crops and forest products across households (% of households)

Agricultural and forest products	Ebolowa	Mbalmayo	Yaoundé	Humid forest zone
Cassava (and derived products)	73.3 (33)*	50.0 (21)	94.3 (33)	71.3 (87)
Groundnuts	40.0 (18)	31.0 (13)	31.4 (11)	34.4 (42)
Plantain	33.3 (15)	33.3 (14)	31.4 (11)	32.8 (40)
Cocoyam (and derived products)	28.9 (13)	33.3 (14)	31.4 (11)	31.4(38)
NTFPs	6.7 (3)	64.2 (27)	8.6 (3)	26.2 (32)
Palm trees (and derived products)	6.7 (3)	16.7 (7)	40.0 (14)	19.7 (24)
Horticultural crops	6.7 (3)	19.0 (8)	31.4 (11)	18.0 (22)
Maize (and derived products)	8.9 (4)	14.3 (6)	22.9 (8)	14.8 (18)
Melon seed	28.9 (13)	7.1 (3)	0.0 (0)	13.1 (16)
Sweet potatoes	6.7 (3)	7.1 (3)	20.0 (7)	10.7 (13)
Yam	0.0 (0)	9.5 (4)	8.6 (3)	7.4 (9)
Mushrooms	0.0 (0)	7.1 (3)	0.0 (3)	4.9 (6)

Note: Numbers in brackets indicate the actual number of respondents

species include non-timber fruit species (e.g. *P. americana, Mangifera indica, E. guineensis, D. edulis, Trichoscypha acuminate, Irvingia gabonensis*), as well as valuable timber species (e.g. *Disthemonathus benthamianus, Lophira alata, Pterocarpus soyauxii, Milicia excelsa, T. superba, T. scleroxylon*).[2]

10.3.4 Desirable Plant Characteristics for Household Consumption and Selling

We identify 12 main groups of agricultural and forest products that contribute to household consumption needs (Table 10.9). Cassava (71.3% of households), groundnuts (34.4%), plantain (32.8%) and cocoyam (and associated products) (31.4%) are popular across all blocks. Other plants such as NTFPs, oil palms (and associated products), horticultural crops, maize (and associated products), sweet potatoes and melon seeds are quite important only in some blocks (Table 10.9).

In particular, household survey respondents indicated that cassava (and derived products) contribute substantially to household consumption in the Yaoundé block (94.3% of households), with decreasing importance in the Ebolowa and Mbalmayo blocks. Groundnuts, plantain and cocoyam (and derived products) have similar importance across the three blocks. NTFPs importance is relatively higher in the Mbalmayo block, the importance of palm trees (and derived products), horticultural crops, maize (and derived products) and sweet potatoes is relatively higher in the Yaoundé block, while the importance of melon seed is relatively higher in the

[2]Even though several timber species have been domesticated over time, they cannot be commercially used unless a logging title is issued following the Cameroon Forest Law of 1994.

Table 10.10 Desirable crop characteristics for household use and market selling

Crop qualities	Responses
Household use	
High yield for crop processing	88.5 (108)
High yield for production	79.5 (97)
Good taste	59.8 (73)
Good crop development	23.8 (29)
Resistance to pests/diseases	23.0 (28)
Selling	
Taste	70.5 (86)
Good price	51.6 (63)
Weight	48.4 (59)
Health and appearance	42.6 (52)
Derived products	42.6 (52)
Easy to sell	9.8 (12)

Note: Numbers in brackets indicate the actual number of respondents

Ebolowa block (Table 10.9). It is worth noting that crops with higher number of cultivars or varieties such as cassava, groundnuts and plantain tend to be more important for household consumption.

In terms of the preferred crop qualities, we identify a divergence of qualities for crops destined for household consumption and market sales (Table 10.10). The most important crop qualities for household consumption include high yield for crop processing (88.5%); high yield for crop production (79.5%) and good taste (59.8%). Others include good crop development (23.8%) and resistance to pests/diseases (23%). On the other hand, some of the main desirable attributes for marketed crops include taste (70.5%), ability to fetch high prices (51.6%), weight (48.4%), healthy appearance (42.6%) and derived products (42.6%) (Table 10.10). It is worth noting that there is some variation in these crop qualities between the three blocks (see Table 10.11).

10.3.5 Pests and Disease Management Approaches for Major Crops

Pests and diseases affect most of the main food and cash crops identified in Sect. 10.3.3. This can have important implications for household livelihoods, whether these crops are used for household consumption or selling. Our household survey reveals that the perceptions about the effect of pests/diseases on the yields of (and income from) the main food and cash crops varied over time in the study areas.

According to Fig. 10.3, there is strong consensus among respondents that pests/diseases have a high effect on cocoa yield and income, both during the time of survey (87%) and 10 years prior to it (75%). Respondents suggest that the overall

Table 10.11 Desirable crop characteristics for household use and market sales (% respondents)

Crop qualities	Ebolowa	Mbalmayo	Yaoundé	Humid forest zone
Household use				
High yield for crop processing	84.4 (38)	88.1 (37)	94.3 (33)	88.5 (108)
High yield for production	82.2 (37)	69.0 (29)	88.6 (31)	79.5 (97)
Good taste	57.8 (26)	54.7 (23)	68.6 (24)	59.8 (73)
Good crop development	26.7 (12)	16.7 (07)	28.6 (10)	23.8 (29)
Resistance to pests/diseases	22.2 (10)	28.6 (12)	17.1 (06)	23.0 (28)
Sales				
Taste	53.3 (24)	73.8 (31)	88.6 (31)	70.5 (86)
Good price	57.8 (26)	76.2 (32)	14.3 (05)	51.6 (63)
Weight	31.1 (14)	57.1 (24)	60.0 (21)	48.4 (59)
Health and appearance	37.8 (17)	40.5 (17)	51.4 (18)	42.6 (52)
Derived products	53.3 (24)	42.9 (18)	28.6 (10)	42.6 (52)
Easy to sell	11.1 (05)	9.5 (04)	8.6 (03)	9.8 (12)

Note: Numbers in brackets indicate actual number of respondents

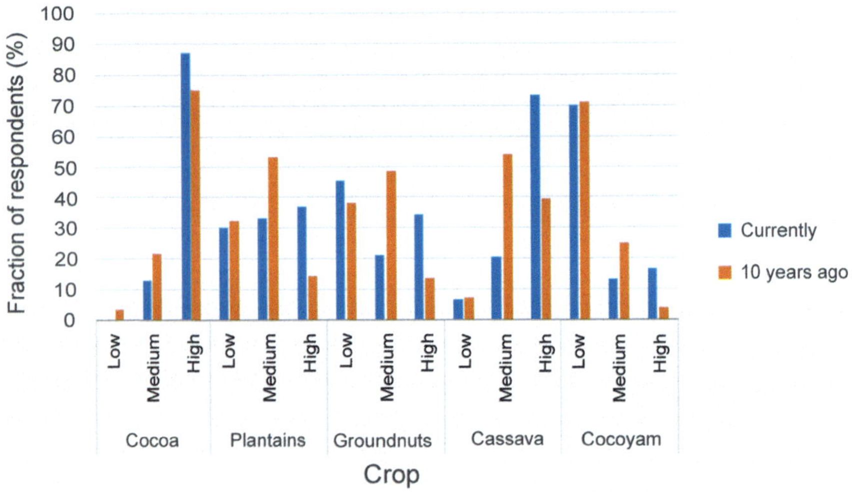

Fig. 10.3 Incidence of crop pests and diseases

effect has increased slightly in the 10 years prior to the interviews. There is also strong consensus that that pests and diseases had a high effect on cassava yield/income at the time of the survey, and that this effect had increased in the 10 years prior to the survey. On the contrary, there is a strong consensus that pests and diseases do not have a high effect on cocoyam yield/income over time. Responses for plantains and groundnuts are more mixed, but suggest the increasing effects of pests and diseases in the 10 years prior to the interviews.

Depending on the crop type, there are different pest and disease management approaches, both on whether to take action and the action itself. The main actions adopted in the study areas include (a) use of modern pesticides, (b) abandonment of

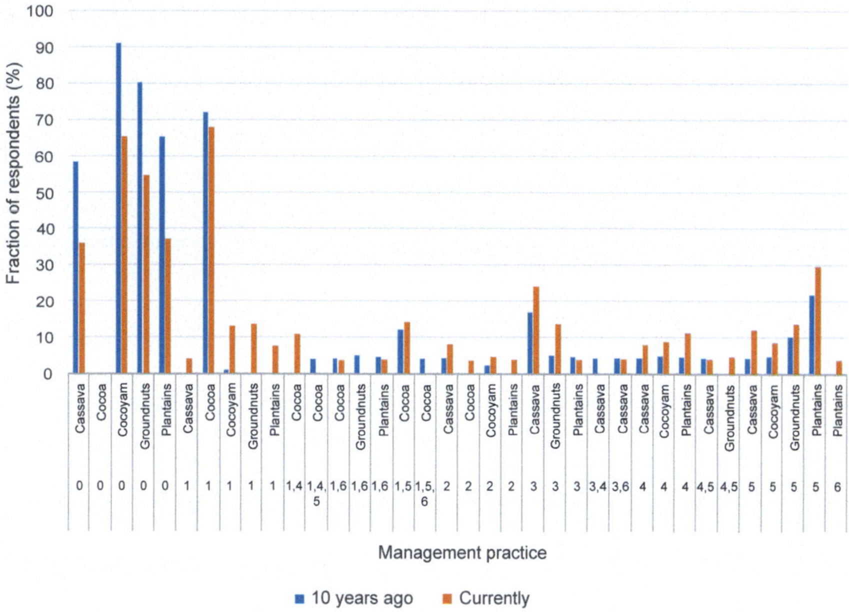

Fig. 10.4 Prevalence of different pest and disease management options for the main crops. Note: Management actions: 0 = Nothing/no action; 1 = Modern pesticide use; 2 = Abandonment of crop variety; 3 = Introduction of new crop variety; 4 = Introduction of improved variety; 5 = Introduction of improved farming method; 6 = Use of local pesticides

crop varieties, (c) introduction of new crop varieties and/or tree species, (d) introduction of improved crop varieties, (e) introduction of improved farming methods and (f) use of local pesticides.

Overall, the general trend among local farmers is to take management actions only for cocoa, and no action for cocoyam, groundnuts, plantain and cassava (Fig. 10.4). However, for some crops there are multiple actions taken, while there is a change over time towards the most preferred form of action. It is worth noting that although eight management actions have been used for cocoa, the use of modern pesticides is by far the most prevalent.

Other prevalent pest and disease management approaches at the time of interviews included (a) the introduction of new crop varieties for cassava and groundnuts, (b) the use of modern pesticides for cocoyam, plantain and groundnuts and (c) the introduction of improved farming methods for cassava, cocoyam, plantain and groundnuts.

10.3.6 Social Demand and Supply of Forest Agriculture Innovations

Based on household surveys in six villages (30 respondents in total) we identify 12 forest–agriculture innovations that were introduced in the study sites in the

10 years before the interviews, with nine (75%) having a technical orientation (crop varieties) (Table 10.12). Introduction of cassava varieties (31.4%) was the most prevalent throughout the period, followed by maize varieties (18.1%), improved farming techniques (13.3%) and seed treatment techniques (10.3%) (Table 10.12). Four categories of forest–agriculture innovations have been abandoned, all of which have a technical orientation. In particular, most farmers mentioned the abandonment of exotic agroforestry trees (38.3%) as the only important abandoned innovation. Five categories of forest–agriculture innovations have been adopted and/or have affected household income, with the most adopted innovations being oil palm varieties (50.5%) and cassava varieties (22.1%) (Table 10.12).

The FGDs suggest that communities in each block articulate different social demands for improvements (Table 10.13), with priorities depending on the respondents' perception and knowledge of market access, and technical/market/socio-

Table 10.12 Deployment of forest–agriculture innovations (% of households)

Type	Innovations	Period			
		Y-0	Y-5	Y-10	Overall
Innovations introduced					
T	Varieties of cassava	25.9	33.3	34.8	31.4
T + C	Varieties of maize	13.0	15.4	26.1	18.1
T	Improved farming techniques	18.5	12.8	8.7	13.3
T	Techniques of seed treatment	13.0	18.0	0.0	10.3
T	Varieties of sweet and Irish potatoes	7.4	5.1	8.7	7.1
C	Crop processing and post-harvest technologies	5.6	5.1	8.7	6.5
T	Agroforestry trees (*Calliandra* and *Inga spp.*)	1.9	5.1	4.4	3.8
T	Varieties of plantain	3.7	2.6	0.0	2.1
T + C	Varieties of groundnuts and soybean	3.7	2.6	0.0	2.1
T + SO	Management of crop nurseries	1.9	0.0	4.4	2.1
T	Oil palm varieties	0.0	0.0	4.4	1.5
T + C	Horticultural crops	3.7	0.0	0.0	1.2
Innovations abandoned					
T	Agroforestry trees (*Calliandra* and *Inga spp.*)	25.0	40.0	50.0	38.3
T	Techniques for treatment of seeds	12.5	20.0	0.0	10.8
T	Improved farming techniques	12.5	20.0	0.0	10.8
T	Varieties of cassava	25.0	0.0	0.0	8.3
Innovations adopted					
T + C + SO	Improved oil palm varieties	36.4	40.0	75.0	50.5
T	Cassava varieties	36.4	30.0	0.0	22.1
T	Plantain varieties	0.0	0.0	25.0	8.3
T	Techniques–treatment of seeds	9.1	10.0	0.0	6.4
T + C	Soybean varieties	0.0	10.0	0.0	3.3

Note: *T* technical, *C* commercial, *SO* socio-organizational, *Y-0* current, *Y-5* 5 years ago, *Y-10* 10 years ago

Table 10.13 Social demand for improvements

Ebolowa	Mbalmayo	Yaoundé
Food crops[C2]	Cocoa[C3]	Cocoa[C3]
Cocoa[C3]	Oil palm plantations[C3]	Micro-credit[C1]
Poultry breeding[C3]	Food crops[C2]	Oil palm[C3]
Sustainable forest management options (NTFPs and community forests)[C2]	Pork breeding[C3]	Maize[C1]
Cucumeropsis farms[C2]	Poultry breeding[C3]	Tomatoes[C1]
Oil palm plantations[C3]	Sustainable forest management options (NTFPs and community forests)[C2]	Aquaculture/fish ponds[C3]
Fruit trees[C1]	*Cucumeropsis* farms[C2]	Plantain[C1]
Aquaculture/fish ponds[C3]	Fruit trees[C1]	Poultry breeding[C3]
Pork breeding[C3]	Aquaculture/fish ponds[C3]	Pork breeding[C3]
Honey production[C3]	Honey production[C3]	Honey production[C3]

Notes: *C3* common for the three blocks, *C2* common for two blocks, *C1* specific for one block

organizational constraints. Of the ten priority innovations cited in each block, six are common to all blocks, including cocoa production, oil palm production, poultry and pork breeding, honey production and aquaculture/fish ponds (Table 10.13). Three improvements are common in two sites, i.e. food crops, and sustainable forest management (incl. NTFPs, community forests, *Cucumeropsis* farms). Only a few improvements are specific to a single block, e.g. micro-credit, maize production, plantains production and tomato production (for the Yaoundé block) and fruit trees (for the Ebolowa block) (Table 10.13).

The local surveys elicited the different types of policy and technology options both from the social demand side and the technology/policy supply side. Overall, 13 research and development (R&D) themes were identified across the study area, related to different technical (T), commercial (C) and socio-organizational (SO) themes (Table 10.14). Policy and technology options related to cocoa were ranked first both for the social demand side and the technology/policy supply side (Table 10.14). For the rest of the options, there seems to be a disjoint between the social demand for improvements (i.e. specific innovations and technology options), and their R&D supply in the study area. While innovations in the demand side tend to have a good representation of technical, commercial and socio-organizational innovations, innovations in the supply side focused more on technical issues such as crop variety development and farming system improvement (Table 10.14). This implies the strong focus of existing innovations on agriculture rather than the integration of forest and agriculture issues.

Table 10.14 Social demand and supply for innovations and technology options

Social demand	Supply
1. Cocoa (T + C + SO)	1. Cocoa (T + C + SO)
2. Aquaculture/fish ponds (T)	2. Plantain (T)
3. Oil palm plantations (T)	3. Maize (T)
4. Pork rearing (T)	4. Groundnuts (T)
5. Poultry (T)	5. Cassava (T)
6. Honey production (T)	6. Cocoyam (T)
7. Food crops (T + C + SO)	7. Soil fertility and improved fallow systems (T)
8. Sustainable forest management[C2] (T + C + SO)	8. Tomatoes (T)
9. Maize[C1] (T + C + SO)	9. Soybean (T)
10. Micro-credit[C1] (T + C + SO)	10. Sustainable forest management (T)
11. Plantain[C1] (T + C + SO)	11. Oil palm plantations (T)
12. Tomatoes[C1] (T + C + SO)	12. Small-scale livestock (T)
13. Fruit trees[C1] (T + C + SO)	13. Honey production (T)

Note: *T* technical, *C* commercial, *SO* socio-organizational, *C2* common for two blocks, *C1* specific for one block

10.4 Discussion

10.4.1 Understanding Human–Nature Relationships in Humid Tropical Forests

Our study finds that the communities in the study area have strong nature-human relationships in that forest knowledge and the local perceptions of nature are not isolated from their conception of the world (Sect. 10.3.1). Instead, they are linked to both the human and spiritual worlds through multiple avenues (see Table 10.1, Sect. 10.3.1). Enhancing human well-being and ensuring livelihoods is central to these relationships, as these conceptions of human well-being and livelihoods are not just something abstract, but are shared life objectives within groups, or between community members. This effort to achieve livelihoods and well-being is strongly linked to the effective mobilization of local knowledge systems for the identification, classification and utilization of natural resources (Chazdon et al. 2009; Mala et al. 2008a).

Moreover, the social definition of forest is based on the uses and practices (often associated with the existence of forests and natural resources), and not on fixed parameters prevalent in technical definitions (see Table 10.2). This finding reflects studies in other similar forest socio-ecological systems in West Kalimantan (Colfer and Byron 2001) and southern Cameroon (Dounias and Hladik 1996; Oyono 2002; Mala and Oyono 2004). All of the above suggest the existence of a social definition of forests in many of the areas where local communities depend substantially on them for their livelihoods.

These local forest knowledge systems can interpret the responses of the natural environment, and thus guide resource management practices for farming, hunting and the collection of forest products (Dounias and Hladik 1996; Carrière 1999; Kanmegne 2004). The ability of farmers to interpret environmental responses also affects the integration of forest conservation and agriculture aspects, based on the coexistence of trees, crops and other species of flora and fauna. In this sense, the management of biodiversity and natural resources is based on certain ecological beliefs, ideological values and perceptions, which collectively guide human activities to enable the effective coexistence of plants and wildlife, in order to maintain land/forest productivity and facilitate vegetation recovery (Mala 2016; 2017).

This evolving knowledge has gradually led to the domestication of many local crop species and varieties, affecting substantially the structure, composition and diversity of forest landscape mosaics (see also Dounias 1996; Dounias and Hladik 1996; Carrière 1999). In this respect, the forest landscape mosaics in the humid forest zone of southern Cameroon are not only the outcomes of biophysical processes, but are also influenced, to a large extent, by human activity (Van Germeden et al. 2003) (see also Chap. 9, Vol. 1). Furthermore, this local knowledge has shaped in many ways agricultural practices, for example by informing local land use decisions, forest uses, pest/disease management and understanding of soil fertility (Sects. 10.3.2–10.3.4). This highlights the coherence and ecological rationality of many traditional practices related to agriculture and natural resource management, which comes in contrast to external expert approaches and conceptions towards agro-ecological sustainability that have often stigmatized the functionality of these relationships (GEF 1993; ASB 1995; Nolte et al. 1997) (Sect. 10.1).

Our findings also suggest that ecological transformation is predetermined by only one of the previous transformation (e.g. forest farm, fallow, cocoa farms), or it is based on a series of transformed ecological units. Many of the ecological dynamics of the forest landscape mosaics are effectively due to the succession of human transformation, rather than just the result of natural processes (see also Chap. 9, Vol. 1). For example, when the local farmers clear the forest to make a farm, this marks the beginning of a new process of ecological transformation (Sect. 10.3.3). This eventually affects agro-ecological resilience through multiple factors ranging from changes of vegetation structure and floristic composition and diversity, to the structuring of human–nature relationships as exemplified in other parts of Cameroon (Dounias 1996; Carrière 1999; Mala and Oyono 2004; Mala 2009, 2016, 2017) (see also Chap. 6, Vol. 2).

The above challenge reflects to some degree the conceptions of biodiversity management evolving around the theory of equilibrium in natural processes (Olsson et al. 2004; Wallington et al. 2005; Roux et al. 2006) (Sect. 10.1). In fact these tight nature–human relationships and extensive agro-diversity management approaches are rooted in a rich body of local knowledge, which suggests the strong interdependency and inter-connectivity between land use components (e.g. farms, fallows, forest) (Léonard and Oswald 1996; Mala et al. 2010; Kaufmann 2014). This inter-connectivity makes a strong case against the segregation of forest conservation and agricultural issues within the forest landscape mosaics of humid tropical SSA.

It is also worth mentioning that the local agro-diversity (and related knowledge/management) covers a large spectrum of plant species (both crop and non-crop) (Sects. 10.3.1–10.3.4). However, only few of the many different crop species are used during the development of forest–agriculture policy, innovations and technological options in the region (Sect. 10.3.5). Many local crop species and varieties are still poorly recognized by the conventional agricultural R&D approaches, which instead promote more popular crops and cropping systems such as maize monocultures (Sects. 10.1 and 10.3.5). Most of these approaches are often ill adapted to local conditions, the socio-economic context of the humid tropics and farmers' needs (Campbell et al. 2010; Willems et al. 2015).

It is important to understand better the interactions discussed above, if we are to develop sustainable forest–agriculture innovations in Cameroon (and more broadly tropic Central Africa). It would require moving beyond the conventional conceptualization of anthropogenic activity, as having only negative effects on agro-ecosystem dynamics (GEF 1993; ASB 1995, 2000). This understanding has, so far, been based on the artificial separation of 'agricultural' and 'forest' issues, and has seriously fragmented policy formulation, and the development of agricultural innovations and technologies (Sect. 10.1). This has contributed substantially to the current failure to link forestry and agricultural management in the complex social–ecological systems of Cameroon (Colfer 2005; Mala et al. 2008a, b; Mala 2009).

Overall, we argue that it is not enough to use just a single approach for tackling sustainability challenges in these agro-ecosystems and developing feasible innovations (Prabhu 2003; Robiglio et al. 2002; Colfer 2005; Robiglio and Mala 2005). Instead, what is clearly required at the forest margins of Central Africa are approaches that can integrate and use different flexible management processes/approaches during the design and analysis of interventions. This would be necessary for tackling effectively coupled sustainability issues (see Sect. 10.4.2).

10.4.2 Reconciling Trade-offs Between Forest–Agriculture and Biodiversity Conservation

Inherently some agricultural practices and forest uses are mutually exclusive. In particular, there is no single approach for the joint harvest of timber and non-timber products, and satisfy at the same time household livelihoods related to agriculture (e.g. crops for household use and selling) and industrial development (e.g. small-scale agroforestry, forest enterprises). More importantly many of these practices have important trade-offs with biodiversity conservation (see also Chaps. 1, 3, Vol. 1; Chaps. 3, 6, Vol. 2).

Forest landscape mosaics encompass a complex combination of agricultural and non-agricultural land uses, with the latter often containing different food/cash crops, trees, poles, saplings and seedlings, which fulfil a range of biophysical and socio-economic functions (Gockowski and Dury 1999; Mala et al. 2008b) (Sects. 10.3.2–

10.3.3). High-value timber species are often maintained within farms (Sect. 10.3.2) to meet household needs, and not necessarily to generate external income through selling. These NTFPs play an instrumental role for forest livelihoods, when compared to other forest products such as timber and fuelwood. In fact their contribution to household income can be as high as 45% in some localities of South Cameroon, where indigenous communities live a nomadic lifestyle (Van Dijk 1999; Mala 2009; Diaw et al. 2016).

As already discussed throughout this chapter, local agro-diversity knowledge is critical for sustainable land management approaches that reconcile livelihood needs and biodiversity conservation (Lefroy et al. 1999; Michon and De Laforesta 1999; Wiersum 2004; Mala et al. 2008a, b) (see Chap. 3, Vol. 2). However, it is not straightforward to avoid trade-offs in such complex forest mosaics with multifunctional crops and trees. There is a need to develop management approaches that are on the one hand able to reconcile local needs with biodiversity conservation.

Overall, our results suggest that the concept of forest–agriculture could deliver broader benefits to local communities, compared to the prevailing slash-and-burn agriculture practices. This could be achieved by reconciling forest livelihoods and biodiversity conservation, and would require reversing the current bias in how agricultural impacts on livelihoods, biodiversity and agro-ecosystem dynamics are understood in humid tropical forests. Forest–agriculture innovations related to cocoa, plantains, maize, peanuts and cassava are particularly relevant, considering their high socio-economic importance in the study area (Gockowski et al. 2004, 2005) (Sect. 10.3.5). Many of these innovations are generally developed or tested by disqualifying local ecological knowledge on agro-diversity and land use management, something that does not provide credit to the farmers to adopt them.

All of the above suggest that despite the significant effort to provide forest–agriculture innovations in the study area, there is a disjoint with what the local farmers actually need (Sect. 10.3.5). This is highlighted in the relatively high rates of abandonment of certain innovations related to soil fertility and agroforestry. This was possibly due to the fact that some of the innovations have sought to improve soil fertility through the selection of plant species (e.g. *Calliandra* spp., *Mucuna* spp. *Inga edulis*, see Nolte et al. 1997; FAO 1999; Kanmegne 2004; Mala 2009), often overlapping with the range of domesticated crop species (Dounias and Hladik 1996; Carrière 1999; Mala 2016, 2017).

The lack of property rights is one of the factors that might curtail the effectiveness of forest–agriculture. Although local farmers in the study areas have access and user rights, they do not have property rights. Essentially, the forests (and timber trees wherein) are the property of the State, as stipulated in the Cameroon Forest Law of 1994, even if the farmers have domesticated the plant species and managed the land over long periods of time. This affects considerably the number and types of plant species encountered in farming plots. In fact, the lack of stable land tenure reduces the interest and willingness of farmers to maintain timber species within their plots, as these species can only be commercialized if farmers hold appropriate permits. Such insecure rights over land and trees may create pre-conditions for the illegal exploitation of timber, because according to the Cameroon Forest Law, even the use of timber for domestic purposes and housing construction requires the authorization of a forest officer. As a result farmers' decision-making is constrained between conserving the trees that are the property of the state, cutting selectively trees, selling the standing trees to artisanal sawyer obtaining some minor income or retaining the valued trees species for their progeny. Most of the above options may have negative biodiversity conservation outcomes. More importantly, some of the recent developments based on market-based conservation instruments such as Payments for Ecosystem Services (PES) and Reducing Emissions from Deforestation and Forest Degradation (REDD+) schemes may also be developed at smaller scales. A concrete example of such Access and Benefits Sharing schemes in Cameroon have those related to *Echinops* and *Monidia*, which have covered a large number of farmers and surface area as a means of reinforcing forest-agriculture and biodiversity conservation (Bellamy and Hill 2010).

To achieve the effective uptake of forest–agriculture (and related innovations), there is a need for flexible and participatory approaches that engage meaningfully the knowledge and perspectives of local communities. Such approaches should start from identifying potential conflicts, and then evaluating and monitoring how innovative agricultural and forestry practices are implemented. Different organizations such as multilateral funders (e.g. African Development Bank, Work Bank) and science–policy interfaces (e.g. IPBES) have proposed relevant approaches (Munang et al. 2013, 2015; World Bank 2012, 2016; Bechini et al. 2017).

10.4.3 Linking External and Local Perspectives and Knowledge Systems

When promoting the conservation and sustainable use of biodiversity in the humid tropical forests of SSA, agro-diversity needs to be considered alongside forest tree species and animals (Sect. 10.3.3). To achieve this, it is imperative to move beyond the current disjointed understanding and management of agricultural and forest landscape elements (Sect. 10.4.1). The current lack of integrated approaches at the interface of forest, agriculture and biodiversity conservation (compounded by weak tenure systems and little consideration of local knowledge/practices and decision-making processes), has curtailed the possibility of achieving some short of sustainable slash-and-burn agriculture in the region (Sect. 10.4.2) (Diaw 1997; Diaw and Oyono 1998; Oyono et al. 2003a; Diaw et al. 2005).

However, by viewing forest–agriculture practices through the lens a forest-cropping fallow-forest conversion cycle, can partially help establish a better understanding of practices that have had positive biodiversity conservation outcomes (including agro-diversity) (Mbolo Abada et al. 2016; Chirwa and Mala 2016). Such a broader understanding of biodiversity conservation can create the conditions for forest regeneration (Carrière 1999; Mala 2009; Van Vliet et al. 2012) and also offer opportunities to develop broader conservation policies and approaches (e.g. market-based conservation instruments such as REDD+) (FAO 2010, 2011a; Ernst et al. 2013; AfDB 2015; Amin 2016; Mala et al. 2019).

A key aspect of our study was the elucidation of how 'external' and 'local' knowledge systems are currently used for forest–agriculture. Often the prevailing management approaches and agricultural innovations are dominated by knowledge generated outside the region (Sect. 10.3.5), which to some extent gives credence to the notion that power differentials dictate which knowledge perspectives are to be used, as external knowledge tends to be viewed more 'authoritative' and hence 'powerful' (Mala 2009; Diaw et al. 2005; Diaw and Oyono 1998).

Based on the above we argue that in order to effectively promote forest-agriculture it is important to create productive linkages and positive reinforcements between the two knowledge systems. This aligns well with current calls about the need to link better such different knowledge systems to achieve interlinked sustainability challenges such as food security, livelihoods resilience and biodiversity conservation. The IPBES conceptual framework (Diaz et al. 2015; Diaw et al. 2018) could offer one feasible option to achieve the effective dialogue between these different knowledge systems (see also Chap. 9, Vol. 2).

10.4.4 Policy and Practice Recommendations

Our findings indicate that in Centre–South Cameroon there are both strong human–nature relationships and diverse agricultural practices based on local and external knowledge systems. Our study is thematically situated at the interface of deforestation and forest degradation on the one hand, and rural livelihoods and poverty alleviation on the other hand. Thus it is strongly related to multiple SDGs, and particularly SDG1 (no poverty), SDG2 (zero hunger), SDG8 (descent work and economic growth) and SDG15 (life on land). Even though the interface of these themes is complicated in the forest areas of Central Africa as discussed throughout this chapter, there is an urgent need to develop solutions to ensure progress for international commitments such as the SDGs, Convention on Biological Diversity (CBD), Paris Agreement, and Africa Vision 2063. Below we outline some policy and practice recommendations based on our study.

First, the strong human–nature relationships highlight the need to design policy instruments that consider and promote the use of forest–agriculture approaches anchored on Indigenous and Local Knowledge (ILK). This can be a good first step to develop innovations that reflect the local contexts, needs and priorities. This can also go a long way towards improving the science–policy interface when developing forest–agriculture policies and technology options in the humid forest areas of Central Africa.

Second, forest–agriculture practices and interventions are embedded in complex socio-ecological systems that contain multiple interlinked components, and cater for multiple human needs. When developing policy and technology options it is important to retain this coupled understanding of traditional forest–agriculture landscape mosaics. Failure to employ such a coupled understanding increases the risk of perpetuating the sharp distinction between forests and agricultural systems, which has largely contributed to the current failure to seriously tackle deforestation, food insecurity and poverty alleviation in the region.

Third, in order to achieve effective biodiversity conservation, especially within forest–agriculture mosaics outside of protected areas, would most likely need the reconciliation of multiple competing objectives. Deploying integrated approaches that engage local communities would be necessary to achieve this reconciliation. For example, ILK practices anchored on the coexistence and utilization of different plant species within farms could promote some of the multiple sustainability objectives discussed above such as biodiversity conservation, food security and sustained livelihoods. The development of novel policies and innovations would require the better articulation, understanding and integration of these trade-offs in decision-making processes at different scales ranging from the farm to the international level.

Fourth, there is a need for adaptive management practices considering the rapid demographic, socio-economic and environmental change (including climate change) in the humid forest areas of Central Africa. Access to climate information and capacity-building efforts that would allow learning through experimentation will be necessary to assist local communities deal with the high degree of uncertainty that characterizes agro-ecosystems in the region.

10.5 Conclusions

This study sought to understand how traditional knowledge is mobilized in the highly biodiverse forest–agriculture systems of Centre–South Cameroon, and how it can contribute to the development of integrated policy and technology options. Through household surveys and FGDs we track the strong and well-established human–nature relationships that guide how local knowledge systems inform livelihoods strategies and natural resources management. Two good indications of these strong relationships are (a) the social definition of forest based on forest uses rather than simply biophysical characteristics and (b) the ILK-based multi-criteria approaches used for the management of forest–agriculture systems.

Overall, agro-diversity management is a crucial activity throughout the study area. Agro-diversity management is informed by a deep knowledge of the biophysical characteristics and socio-economic functions of tree species and crops, including their qualities for subsistence and selling. This attest to the capacity of local communities to conserve biodiversity outside of protected areas, as a means of sustaining livelihoods. However, many of the agricultural innovations disseminated in the recent past have neither met social demand nor reflected local knowledge and worldviews. Most of these forest–agriculture innovations are entirely dominated by technical approaches that are at odds with local sensibilities. This has possibly contributed to the high abandonment rates of some of these innovations.

We argue that it is important to engage more meaningfully ILK for the development and management of forest–agriculture systems, and the development of relevant innovations. This is very consistent both with both the current global policy agendas (e.g. SDGs, CBD, Africa Union Vision 2063, Paris Agreement) and the stronger push for science–policy interfaces that draw from different knowledge systems.

References

Abate T, Van Huis A, Ampofo JKO (2000) Pest management strategies in traditional agriculture: an African perspective. Ann Rev Entomol 45:631–659

AfDB (2015) African ecological futures. AfDB, Abidjan. https://www.afdb.org/fileadmin/uploads/afdb/Documents/GenericDocuments/xwwf_african_futures_report_english-lo-rez.pdf

Altieri AM (2002) 2002. Agro-ecology: the science of natural resource management for poor farmers in marginal environments. Agric Ecosyst Environ 93:1–24

Altieri MA, Nicholls CI, Henao A et al (2015) Agroecology and the design of climate change-resilient farming systems. Agron Sustain Dev 35:869–890. https://doi.org/10.1007/s13593-015-0285-2

Amin A (2016) Exploring the role of economic incentives and spillover effects in biodiversity conservation policies in sub-Saharan Africa. Ecol Econ 127:185–191. https://doi.org/10.1016/j.ecolecon.2016.03.018

Angelsen A, Wunder S (2003) Exploring the forest-poverty link: key concepts, issues and research implications, CIFOR occasional paper# 40, CIFOR, Bogor

Apusiga A (2011) Indigenous knowledge, cultural values and sustainable development in Africa. Paper presented at 2nd annual ibadan sustainable development summit, Nigeria, August 2011

Armitage DR (2003) Traditional agro-ecological knowledge, adaptive management and the socio-politics of conservation in Central Sulawesi, Indonesia. Environ Conserv 30(1):70–90

ASB (1995) Alternatives to slash and burn phase I report: forest margins benchmark of Cameroon, alternatives to slash and burn programme. IITA Humid Forest Ecoregional Centre, Yaoundé

ASB (2000) Alternatives to slash and burn phase II report: forest margins benchmark of Cameroon, alternatives to slash and burn programme. IITA Humid Forest Ecoregional Centre, Yaoundé

Ash N, Blanco B, Brown C, Garcia K, Henrichs T, Lucas N, Raudsepp-Hearne C, Simpson RD, Scholes R, Tomich TD, Vira B, Zurek M (2010) Ecosystems and human wellbeing: a manual for assessment practitioners. Island Press, Washington DC. http://www.ecosystemassessments.net/resources/ecosystems-and-human-well-being-amanual-for-assessment-practitioners.pdf

AU (2015) Agenda 2063: the Africa we want. African Union Commission, Addis Ababa. http://archive.au.int/assets/images/agenda2063.pdf

Bawden RJ (1991) Systems thinking and practice in agriculture. J Dairy Sci 1991(74):2362–2373

Bechini et al. (2017) Improving access to research outcomes for innovation in agriculture and forestry: the VALERIE project. Italian J Agron 12:756 85–95

Beddington J, Asaduzzaman M, Clark M, Fernández A, Guillou M, Jahn M, Erda L, Mamo T, Van Bo N, Nobre CA, Scholes R, Sharma R, Wakhungu J (2012) Achieving food security in the face of climate change: final report from the commission on sustainable agriculture and climate change. In: CGIAR research program on climate change, agriculture and food security (CCAFS), Copenhagen, Denmark. www.ccafs.cgiar.org/commission

Bellamy JJ, Hill K (2010) National capacity self-assessments: results and lessons learned for global environmental sustainability. Global Support Programme, Bureau for Development Policy, United Nations Development Programme, New York. https://www.thegef.org/sites/default/files/publications/NCSA-SR-web-100913_2.pdf

Berkes F (2009) Evolution of co-management: role of knowledge generation, bridging organizations and social learning. J Environ Manag 90:1692–1702. https://doi.org/10.1016/j.jenvman.2008.12.001

Biggs R, Schlüter M, Schoon ML (Eds.). (2015) Principles for building resilience: sustaining ecosystem services in social-ecological systems. Cambridge University Press, Cambridge. https://doi.org/10.1017/CBO9781316014240

Boon EK (2015) Integrating environment into decision making in Africa. In: Boon EK (ed) Area studies (regional sustainable development review): Africa—volume II. EOLSS Publications, Accra

Campbell BM, Sayer JA, Walker B (2010) Navigating trade-offs: working for conservation and development outcomes. Ecol Soc 15(2)

Carlsson B, Jacobson S, Holmén M, Rickne A (2002) Innovations systems: analytical and methodological issues. Res Policy 31:233–245

Carrière S (1999) Les orphelins de la forêt: influence de l'agriculture itinérante sur brûlis de Ntumu et des pratiques agricoles associées sur les dynamiques forestières du sud Cameroun. Thèse de Doctorat, Université de Montpellier, Montpellier, 448p

Chazdon RL, Harvey CA, Komar O, Griffith DM, Ferguson BG, Ramos MM, Morales H, Nigh R, Soto-Pinto L, Van Breugel M, Philpott SM (2009) Beyond reserves: a research agenda for conserving biodiversity in human-modified tropical landscape. Biotropica 41(2):142–153

Chirwa P, Mala WA (2016) On farm trees: towards the promotion and development of a traditional or local silviculture in tropical and subtropical forest regions. Special issue. Agrofor Syst 90 (4):555–561

Colfer CJP, Byron Y (eds) (2001) People managing forests: the links between human well-being and sustainability. RFF/CIFOR, Washington DC/Bogor

Colfer CJP (2005) The complex forest: communities, uncertainties and adaptive collaborative management. RFF/CIFOR, Washington DC, 370p

COMIFAC (2014) Plan de convergence pour la gestion durable des écosystèmes forestiers d'Afrique centrale 2015–2025. http://pfbccbfp.org/docs/key_docs/Plan_de_Convergence_COMIFAC_2015-2025_02072014_fr.pdf

Conley T, Udry C (2001) Social learning through networks: the adoption of new agricultural technologies in Ghana. Am J Agric Econ 83(3):668–673

Cormier-Salem MC (1999) Innovations et relations société-environnement. In: Chauveau JP, Cormier-Salem MC, Mollard E (eds) Innovations en agriculture: questions de méthodes et de terrains d'observation. IRD Editions, Paris, pp 127–140

Danielsen F, Jensen PM, Burgess ND, Altamirano R, Alviola PA, Andrianandrasana H, Brashares JS, Burton AC, Coronado I, Corpuz N, Enghoff M, Fjeldså J, Funder M, Holt H, Hübertz H, Jensen AE, Lewis R, Massao J, Mendoza MM, Ngaga Y, Pipper CB, Poulsen MK, Rueda RM, Sam MK, Skielboe T, Sørensen M, Young R, Enghoff M (2014) A multicountry assessment of tropical resource monitoring by local communities. Bioscience 64(3):236–251. https://doi.org/10.1093/biosci/biu001

Diaw MC, Colfer CJP, Prabhu R (1999) Pour le suivi participatif et la cogestion adaptive des forêts tropicales africaines. CIFOR, Bogor, 14p

Diaw MC, Nguiebouri J, Gagoe Tchoko J, Ngo Baneg M-F, Vambi B, Nlend E, Keugni M-C (2016) Pour une inscription du paradigme participatif dans l'économie sociale et la transformation structurelle: le cas des forets modèles. in G. Buttoud et J.C. Nguinguiri, Editeurs, La gestion inclusive des forêts d'Afrique centrale. Passer de la participation au partage des pouvoirs, FAO/CIFOR, Libreville/Bogor. pp. 187–201

Diaw MC, Oyono PR (1998) Dynamiques et représentations des espaces forestiers au Sud Cameroun: pour une relecture sociale du paysages. Arbres Forêts et Communautés Rurales 15 (16):36–43

Diaw MC, Tiani AM, Nguiebouri J, Jum C, Akwah G, Mala WA, Robiglio V (2005) Adaptive collaborative management in diverse contexts, case 5 Cameroon-Central and South. In: Colfer CJ (ed) The complex forest, communities, uncertainty and adaptive collaborative management. RFF/CIFOR, Washington DC/Bogor, pp 238–251

Diaw MC, de Morais Luis T, Harhash KA, Andriamaro L, Archer E, Batisani N, Bornman T, Fuashi NA, Golden C, Hamed Y, Ivey P, Lindley S, Mulongoy KJ, Onyige CD, Matlhola DM, Belhabib D (2018) Setting the scene. In: Archer E, Dziba LE, Mulongoy KJ, Maoela MA, Walters M (eds) The regional assessment report on biodiversity and ecosystem services for Africa. IPBES, Bonn, pp 1–76

Diaw MC (1997) Si, Nda bot and Ayong: shifting cultivation, land use and property rights in southern Cameroon. Rural development forestry network paper 21e

Diaz et al (2015) The IPBES conceptual framework—connecting nature and people. Curr Opin Environ Sustain 14:1–16

Dounias E, Hladik CM (1996) Les agroforêts Mvae et Yassa au Cameroun Littoral: functions socio-culturelle, structure and composition floristique. In: Hladik CM, Hladik A, Pagezy H, Linares OF, Koppert GJA, Alain F (eds) L'alimentation en forêt tropicale: interactions bioculturelles et perspectives de développement: 2. Bases culturelles des choix alimentaires et stratégies de développement Source. UNESCO, Paris, pp 1103–1126. (L'Homme et la Biosphère). ISBN 92-3-203381-X

Dounias E (1996) Recrus forestiers post-agricoles: perceptions et usages chez les Mvae du Sud-Cameroun. Journal d'Agriculture Tropicale et de Botanique Appliquée 38:153–178

Ernst C, Mayaux P, Verhegghen A, Bodart C, Christophe M, Defourny P (2013) National forest cover change in Congo Basin: deforestation, reforestation, degradation and regeneration for the years 1990, 2000 and 2005. Glob Chang Biol 19(4):1173–1187. https://doi.org/10.1111/gcb.12092

Etat des forêts (2015) Les forêts du Bassin du Congo - Forêts et changements climatiques. In: de Wasseige C, Tadoum M, Eba'a Atyi R, et Doumenge C (eds). – 2015. Weyrich. Belgique. 128p

FAO (1999) Sustaining agricultural biodiversity and agro-ecosystem functions: farming systems approaches for the sustainable use and conservation of agricultural biodiversity and agro-ecosystems. In: Report international technical workshop organized jointly by the food and agriculture organization of the United Nations (FAO) and the secretariat of the convention on biological diversity, FAO, Rome, 2–4 December 1998

FAO (2010) "Climate-smart" agriculture: policies, practices and financing for food security, adaptation and mitigation. FAO, Rome. http://www.fao.org/docrep/013/i1881e/i1881e00.htm

FAO (2011a) FAO-adapt: framework programme on climate change adaptation. http://www.fao.org/climatechange/27594-03ecd7bd225b93086e7dca3944de64307.pdf

FAO (2011b) Save and grow: a policymaker's guide to the sustainable intensification of smallholder crop production. http://www.fao.org/ag/save-and-grow/

FAO/INRA (2016) Innovative markets for sustainable agriculture—how innovations in market institutions encourage sustainable agriculture in developing countries. Loconto A, Poisot AS, Santacoloma P. (eds.) Rome, Italy

Garrity D, Bandy D (1996) A global focus on alternatives to slash and burn. In: LiYu X, YuangZhang Y, JinHui Q (eds.), Alternatives to slash-and-burn agriculture. Proceedings of an International symposium held in Kunming, China, 6–16 March 1995, pp 52–58

GEF (1993) Global alternatives to slash and burn. Global Environment Facility (GEF)/United Nations Development Program, New York

Geldenhuys CJ (2000) Ecological processes the basis for sustainable nature conservation. Plant Theory 23:2–4

Gockowski J, Dury S (1999) The economics of cocoa-fruit agroforests in southern Cameroon. In: IUFRO/ICRAF/CATIE international conference on multi-strata agroforests. CATIE, Turrialba

Gockowski J, Tonye J, Baker D, Legg C, Weise S, Ndoumbé M, Tiki-Manga T Fouaguégué A (2004) Characterization and diagnosis of farming systems in the ASB forest margins benchmark of southern Cameroon. IITA social sciences working paper no 1. IITA, Ibadan, 67p

Gockowski J, Tonye J, Diaw MC, Hauser S, Kotto-Same J, Njomgang R, Moukam A, Nwaga A, Tiki-Manga T, Tondoh J, Tchoundjeu Z, Weise S, Zapfack L (2005) The forest margins of Cameroon. In: Palm CA, Vosti AS, Sanchez AP, Polly JE (eds) Slash-and-Burn agriculture: the search for alternatives. Columbia University Press, New York, pp 305–331

Haila Y (1999) Biodiversity and the divide between culture and nature. Biodivers Conserv 8:165–181

Holling C (2001) Understanding the complexity of economic, ecological, and social systems. Ecosystems 11:41–47

IFAD (2012) Adaptation for smallholder agriculture programme (ASAP). IFAD, Rome

Instone L (2003a) Shaking the ground of shifting cultivation: or why (do) we need alternatives to slash-and-burn? In: Resource management in Asia-Pacific program working paper no. 43. Research School of Pacific and Asian Studies/The Australian National University, Camberra

Instone L (2003b) T(r)opical translations: reterritorialising the space of biodiversity conservation. In: Resource management in Asia-Pacific program working paper no.46. Research School of Pacific and Asian Studies/The Australian National University, Camberra

Kanmegne J (2004) Slash and burn agriculture in the humid forest zone of southern Cameroon: soil quality dynamics, improved fallow management and Farmers' perception. PhD Dissertation, Wageningen University, Wageningen, 184p

Kaphengst T (2014) Towards a definition of global sustainable land-use? A discussion on theory, concepts and implications for governance. EcoLogic, Berlin

Kaufmann JC (2014) Contrasting visions of nature and landscapes. In: Scales IR (ed) Conservation and environmental management in Madagascar. Routledge, London and New York, pp 320–341. ISBN 9781136309083

Krause M, Lotze-Campen H, Popp A, Dietrich JP, Bonsch M (2013) Conservation of undisturbed natural forests and economic impacts on agriculture. Land Use Policy 30(1):344–354

Lefroy EC, Hobbs RJ, O'Connor MH, Pate JS (eds.). (1999) Agriculture as a mimic of natural ecosystems. Agrofor Sys 45:275–289

Léonard É, Oswald M (1996) 1996. Une agriculture forestière sans forêt. Changements agro-écologiques et innovations paysannes en côte d'ivoire. Nat Sci Soc 4(3):202–215

Levang P, Michon G, De Forestra H (2001) De la jachère arborée aux agro-forêts, des stratégies paysannes adaptées à des milieux de fertilité médiocre. In: CIRAD and Ministère de la coopération (dir.), Actes du Séminaire sur la Fertilité du milieu et Stratégies paysannes sous les tropiques humides du 13–17 Novembre 1995. CIRAD, Montpellier, pp 228–235

Lin BB (2011) Resilience in agriculture through crop diversification: adaptive management for environmental change. BioSci 61(3):183–193

Mala WA (2017) Chapter 11 coexistence of forests and agriculture: repercussions for the availability of NWFPs. In: Living in and from the forests of Central Africa. FAO, Rome, pp 81–87. ISBN 978-92-5-109489-1

Mala AW (2016) Savoirs et pratiques traditionnels et locaux en agroécologie et conservation de la biodiversité: zones forestières du Centre et Sud Cameroun. In: Roué M, Césard N, Adou Yao YC, Oteng-Yeboah A (eds) Indigenous and local knowledge of biodiversity and ecosystem services in Africa. Knowledges of nature, vol 8. UNESCO, Paris, pp 55–85

Mala WA, Geldenhuys CJ, Prabhu R (2008a) A predictive model of local agricultural biodiversity knowledge management in southern Cameroon. Biodiversity 9(1 & 2):96–101

Mala WA, Geldenhuys CJ, Prabhu R (2010) Local conceptualisation of nature, forest knowledge systems and adaptive management in southern Cameroon. Indilinga Afr J Indig Know Syst 9 (2):172–184

Mala WA, Oyono PR (2004) Social dimensions of local practices of natural resource management in the Central Africa region. Res Rev 20(2):23–31

Mala WA, Oyono PR, Ebene Y, Diaw MC (2004) Assessment of the context of slash-and-burn agriculture and the dynamics of landscape changes in southern Cameroon. Unpublished report. CIFOR, Yaoundé

Mala WA, Prabhu R, Geldenhuys CJ, Diaw MC (2008b) Agricultural biodiversity knowledge systems and its adaptive capacity to respond to household consumption and market preferences. In: Parrotta JA, Jinlong L, Heok-Choh S (eds) Forest management and poverty alleviation: roles of traditional forest-related knowledge. IUFRO world series no. 21. IUFRO, Vienna, pp 114–115

Mala AW (2009) Knowledge systems and adaptive collaborative management of natural resources in southern Cameroon: decision analysis of agrobiodiversity for forest-agriculture innvations. PhD dissertation. Stellenbosch University, Stellenbosch. 253p

Mala AW, Oumarou hindou I. et Diaw MC (2019) Les savoirs sur la nature pour inspirer la transformation sociale et économique. Revue Liaison Énergie-Francophonie. Numéro 111, 1er trimestre 2019, pp 17–20

Malleson R, Asaha S, Sunderland T, Burnham P, Egot M, Obeng-Okrah K, Ukpe I, Miles W (2008) A methodology for assessing rural livelihood strategies in West/Central Africa: lessons from the field. Article. Ecol Environ Anthropol 4(1)

Manyong MV, Smith J, Weber GK, Jagtap SS, Oyewole B (1996) Macrocharacterization of agricultural systems in Central Africa: an overview. Resource and crop management research monograph no. 22. IITA, Ibadan. 58p

Mayaux P, Pekel JF, Desclee B, Donnay F, Lupi A, Achard F, Clerici M, Bodart C, Brink A, Nasi R, Belward A (2013) State and evolution of the African rainforests between 1990 and 2010. Philos Trans R Soc B 368(1625). https://doi.org/10.1098/rstb.2012.0300

Mbolo Abada MM, Zekeng JC, Mala AW, Fobane JL, Djomo Chimi C, Ngavounsia T, Tamandjong V (2016) The role of cocoa agroforestry systems in conserving forest tree diversity in the central region of Cameroon. Agrofor Syst 90(4):577–590

McNeel JA (2004) Nature vs. nurture: managing relationships between forests, agroforestry and wild biodiversity. Agrofor Syst 61:155–165

Michon G, De Laforesta G (1999) Agroforests: incorporating a forest vision in agroforestry. In: Buck LE, Lassoie JP, Fernandes ECM (eds) Agroforestry systems. CRC Press, Boca Raton, pp 381–406

Munang R et al (2013) Climate change and ecosystem-based adaptation: a new pragmatic approach to buffering climate change impacts. Curr Opin Environ Sustain 5(1):67–71. https://doi.org/10.1016/j.cosust.2012.12.001

Munang R, Mgendi R, Alverson K, O'Brien-Onyeka M, Ochieng C, Molua E, Ottichilo W, Mugerwa K, Bunch R, Bogale A, Bekele W, Sudarkasa M, Badiane O, Ayuk E, Amis M, Tabot-Tabot E, Kinney K, Kwangwari C, Luganda P, Aidoo K, Mulangu F, Kariuki J (2015) Ecosystem-based adaptation (EbA) for food security in Africa: towards a comprehensive strategic framework to upscale and out-scale EbA-driven agriculture in Africa. United Nations Environment Programme (UNEP), Nairobi

Namirembe S, Leimona B, van Noordwijk M, Bernard F, Bacwayo KE (2014) Co-investment paradigms as alternatives to payments for tree-based ecosystem services in Africa. Curr Opin Environ Sustain 6:89–97

Ndoye O (1997) The impacts of macroeconomic and agricultural policies on forest conditions in Cameroon. CIFOR, Yaoundé

Nolte C, Duguma B, Wendt J (1997) Research and development of fallows in the humid forest zone of Cameroon. A working paper prepared for the launching of EPHTA Humid Forest Margins Benchmark, Yaoundé, Hilton, 26–28 May 1997

O'brien W (2002) The nature of shifting cultivation: stories of harmony, degradation, and redemption. Hum Ecol 30(4):483–502

Olsson P, Folke C, Berkes F (2004) Adaptive co-management for building resilience in socio-ecological systems. Environ Manag 34:75–90

Oyono PR, Akwah G, Diaw MC, Jum C, Tiani A-M, Mala WA, Assembe S, Nguiébouri J, Ebene Y, Efoua S, Kouna C (2003b) Developing collaborative monitoring for adaptive co management of tropical forests (ACM). Final report—Cameroon. Center for International Forestry Research (CIFOR), Yaoundé

Oyono PR, Diaw MC, Efoua S, Mala WA, Assembe S (2007) Issue-appropriate collaborative monitoring: examples from forest management in Cameroon. In: Gujit I (ed) Learning from collaborative monitoring: triggering adaptation in adaptive and collaborative management (ACM). CIFOR, Bogor

Oyono PR, Mala WA, Tonye J (2003a) Rigidity versus adaptation: contribution to the debate on agricultural viability and forest sustainability in southern Cameroon. Cult Agric 25(2):32–40

Oyono PR (2002) Usages Culturels de la Forêt au Sud-Cameroun: Rudiments d'Ecologie Sociale et Matériau pour la Gestion du Pluralisme. Africa 57(3):334–355

Palm AC, Vosti AS, Sanchez AP, Ericksen JP (eds.) (2005) Slash and burn agriculture: search for alternatives. Columbia University Press, New York, 461p

Prabhu R (2003) Developing collaborative monitoring for adaptive collaborative management of African forests. Final report, 1 January 2000–31 December 2002. CIFOR, Harare

Pretty J (2002) Agri-culture: reconnecting people, land and nature. Earthscan, London

Robiglio V, Mala WA, Diaw MC (2002) Mapping landscapes: integrating GIS and social science methods to model human-nature relationships in southern Cameroon. Small Scale For Econ Manag Policy J 2(2):171–184

Robiglio V, Mala WA (2005) Between local and expert knowledge: integrating participatory mapping and GIS for the implementation of integrated forest management options. Case study of Akok—Cameroon. For Chron 81(3):392–397

Roux DJ, Rogers KH, Biggs HC, Ashton PJ, Sergeant A (2006) Bridging the science-managing divide: moving from unidirectional knowledge transfer to knowledge interfacing and sharing. Ecol Soc 11(1):4

Ruitenbeck J, Cartier C (2001) The invisible wand: adaptive co-management as an emergent strategy in complex bio-economic systems. Occassional paper no. 34. Centre for International Forestry Research, Bogor

Sanchez AP, Palm AC, Vosti AS, Tomich PT, Kayoski J (2005) Alternatives to slash and burn: challenge and approaches of an international consortium. In: Palm AC, Vosti AS, Sanchez AP, Ericksen JP (eds) Slash and burn agriculture: search for alternatives. Columbia University Press, New York, pp 3–37

Sayer JA, Campbell BS (2003) The science of sustainable development: local livelihoods and global environment. Cambridge University Press, Cambridge

Scoones I (1995) Living with uncertainty: new directions in pastoral development in Africa. Intermediate Technology Publications, London

Sheil D et al (2003) A new type of sample unit for the efficient assessment of diverse tree communities in complex forest landscapes. J Trop For Sci 15(1):117–135

Sinclair FL, Joshi L (2001) Taking local knowledge about trees seriously. In: Forestry, forest users and research: new ways of learning. ETFRN, Wageningen, pp 45–61

Smith HA, McSorely R (2000) Intercropping and pest management: a review of major concepts. Am Entomol 46:154–161

Toledo VM, Ortiz-Espejel B, Cortès L, Moguel P, Ordonez M (2003) The multiple uses of tropical forests by indigenous peoples in Mexico: a case of adaptive management. Conserv Ecol 7(3)

UNDP (2012) Africa human development report 2012 towards a food secure future. United Nations Development Programme Regional Bureau for Africa (RBA) 1 UN Plaza, New York, NY 10017, USA. 190p. Web: https://unp.un.org, www.undp.org/africa and www.afhdr.org

Van Dijk JFW (1999) An assessment of non-wood forest products resources for the development of sustainable commercial extraction. In: TCH S, Clark LE, Vantomme P (eds) The NWFP of Central Africa: current research issues and prospects for conservation and development. FAO, Rome

Van Germeden BS, Olff H, Parren PEE, Brongers F (2003) The pristine rainforest? Remnants of historical impacts of current trees species composition and diversity. J Biogeogr 30:1381–1390

Van Mele P, Van Chien H (2004) Farmers, biodiversity and plant protection: developing a learning environment for sustainable tree cropping systems. Int J Agric Sustain 2(1):67–75

Van Noordwijk M, Kurniatun H, Weise S (2001) Sustainability of tropical land use systems following forest conversion. ASB lecture note 3. ICRAF, Bogor

Van Vliet N et al (2012) Trends, drivers and impacts of changes in swidden cultivation in tropical forest-agriculture frontiers: a global assessment. Glob Environ Chang 22(2):418–429

Vernooy R, Song Y (2004) New approaches to supporting the agricultural biodiversity important for sustainable rural livelihoods. Int J Agric Sustain 2(1):55–66

Vosti S (1995) The Role of agriculture in saving the rain forest. 2020 Brief 9, February 1995. International Food Policy Research Institute, Washington

Wallington TJ, Hobbs RJ, Moore SA (2005) Implication of current ecological thinking for biodiversity conservation: a review of the salient issues. Ecol Soc 10(1):15

Westphal E, Embrechts J, Ferwerda JD, Van Gils-Meeus EAH, Mutsaers WJH, Westphal-Stevels CMJ (1985) Cultures vivrières tropicales avec référence spéciale au Cameroun. Pudoc Wageningen/CAP Documentation, Wageningen. 514p

Wiersum KF (2004) Forest gardens as an 'intermediate' land uses system in the nature-culture continuum: characteristics and future potential. Agrofor Syst 61:123–134

Willems DJM, Koenderink NJJP, Top JL (2015) From science to practice: bringing innovations to agronomy and forestry. J Agric Inf 6:85–95

World Bank (2012) World Bank VP eyes opportunities for sustainable growth in Africa. The World Bank, Washington, DC

World Bank (2016) World development indicators 2016: featuring the sustainable development goals. World Bank, Washington, DC

Zahm F, Alonso Ugaglia A, Barbier J-M, Boureau H, Del'homme B, Gafsi M, Gasselin P, Girard S, Guichard L, Loyce C, Manneville V, Menet A, Redlingshöfer B (2019) Évaluer la durabilité des exploitations agricoles. La méthode IDEAv4, un cadre conceptuel combinant dimensions et propriétés de la durabilité. Cah Agric 28:5

Index

A. Gasparatos et al. (eds.), *Sustainability Challenges in Sub-Saharan Africa I*, Science for Sustainable Societies, https://doi.org/10.1007/978-981-15-4458-3

The manufacturer's authorised representative in the EU is Springer Nature Customer Service Centre GmbH, Europaplatz 3, 69115 Heidelberg, Germany. If you have any concerns regarding our products, please contact ProductSafety@springernature.com

Printed and bound by CPI Group (UK) Ltd, Croydon, CR0 4YY

30/09/2026

02225299-0006